Contributions to Statistics

http://www.springer.de/egi-bin/search_book.pl?series=2912

Anthony C. Atkinson
Peter Hackl
Werner G. Müller

Editors

mODa 6 – Advances in Model-Oriented Design and Analysis

Proceedings of the 6th International Workshop
on Model-Oriented Design and Analysis held
in Puchberg/Schneeberg, Austria,
June 25–29, 2001

**With 25 Figures
and 22 Tables**

Physica-Verlag

A Springer-Verlag Company

Series Editors
Werner A. Müller
Martina Bihn

Editors
Prof. Anthony C. Atkinson
Department of Statistics
The London School of Economics and Political Science
Houghton Street
London WC2A 2AE, United Kingdom
a.c.atkinson@lse.ac.uk

Prof. Peter Hackl
Dr. Werner G. Müller
Vienna University of Economics
and Business Administration
Institute of Statistics
Augasse 2-6
1090 Vienna, Austria
peter.hackl@wu-wien.ac.at
werner.mueller@wu-wien.ac.at

ISSN 1431-1968
ISBN 3-7908-1400-8 Physica-Verlag Heidelberg New York

Cataloging-in-Publication Data applied for
Die Deutsche Bibliothek – CIP-Einheitsaufnahme
Advances in model oriented design and analysis: proceedings of the 6th International Work-
shop on Model Oriented Design and Analysis held in Puchberg/Schneeberg, Austria, June
25–29, 2001; with 22 tables / mODa 6. Anthony C. Atkinson ... (ed.). – Heidelberg; New
York: Physica-Verl., 2001
   (Contributions to statistics)
   ISBN 3-7908-1400-8

Physica-Verlag Heidelberg New York
a member of BertelsmannSpringer Science+Business Media GmbH

© Physica-Verlag Heidelberg 2001
Printed in Germany

Softcover Design: Erich Kirchner, Heidelberg

SPIN 10834948   88/2202-5 4 3 2 1 0 – Printed on acid-free paper

Preface

This book includes many of the papers presented at the 6th International workshop on Model Oriented Data Analysis held in June 2001. This series began in March 1987 with a meeting on the Wartburg near Eisenach (at that time in the GDR). The next four meetings were in 1990 (St Kyrik monastery, Bulgaria), 1992 (Petrodvorets, St Petersburg, Russia), 1995 (Spetses, Greece) and 1998 (Marseilles, France). Initially the main purpose of these workshops was to bring together leading scientists from 'Eastern' and 'Western' Europe for the exchange of ideas in theoretical and applied statistics, with special emphasis on experimental design. Now that the separation between East and West is much less rigid, this exchange has, in principle, become much easier. However, it is still important to provide opportunities for this interaction.

MODA meetings are celebrated for their friendly atmosphere. Indeed, discussions between young and senior scientists at these meetings have resulted in several fruitful long-term collaborations. This intellectually stimulating atmosphere is achieved by limiting the number of participants to around eighty, by the choice of a location in which communal living is encouraged and, of course, through the careful scientific direction provided by the Programme Committee.

It is a tradition of these meetings to provide low cost accommodation, low fees and financial support for the travel of young and Eastern participants. This is only possible through the help of sponsors and outside financial support was again important for the success of the meeting. The organisers are grateful to The European Union (contract # HPCF-CT 2000 00045) and to GlaxoSmithKline for their generous support.

As the papers collected here show, the present meeting at Puchberg, Austria from the 25th to the 29th of June 2001, once more brings together statisticians from around the world. The contributions to this volume were carefully selected by the editors from those submitted. All have been refereed. They have been arranged in alphabetical order of author, but some patterns of topic are clear.

As is usual in MODA conferences, many of the papers are on aspects of optimum design. Designs for various aspects of problems involving nonlinear models are considered by Fan and Chaloner; Haines, Clarke, Gows and Rosenberger; Martínez, Ortiz and Rodríguez; Rosenberger, Haines and Perevozskaya; Melas and by Torsney and López-Fidalgo. Optimum designs are also described for a variety of linear situations, some nonstandard, by Biedermann and Dette – nonparametric regression; Ceranka and Katluska – optimum weighing designs; Downing, Fedorov and Leonov – parameterised variance functions; Großmann, Graßhoff, Holling and Schwabe – paired comparisons and by Tack and Vandebroek – trend robust designs, a paper which can be compared with the more classical characterizations by Afsarinejad. Pronzato studies optimum designs when the experimental conditions cannot be exactly controlled; Mizera and Müller explore the breakdown of designs when l_1 estimators are used – D-optimum designs behave well; Hilgers shows that some mixture designs are not D-optimum and Boer, Hendrix and Rasch find designs for spatial monitoring. Others papers with a spatial component include Felsenstein; Hoffman, Lee and Williams; and Uciński, who considers the determination of optimum sensor motion.

Papers less directly focused on design include those of Cohen, Di Bucchianico and Riccomagno and of Pistone, Riccomagno and Wynn, which use Gröbner basis methods to determine the family of models that can be fitted to a given design. Another group is concerned with inference, in some cases in quite a general framework. These include Galtchouk; Malyutov and Tsitovitch; and Yin, Kelly and Dowell, whose application is image processing. Läuter investigates the distribution of residuals and Vuchkov and Boyadjieva discuss methods of quality improvement. Finally there is a group of papers, in addition to some of those on optimum design for nonlinear models, which addresses problems in medical statistics. These are Dragolin and Fedorov; D'yachkov, Macula, Torney and Vilenkin on screening trials; and Mazzaro, Pesarin and Salmaso on permutation tests. Finally, Hardwick, Oehmke and Stout consider the medically important problem of the design of adaptive experiments when the response is delayed.

The increase in medical applications, compared with previous conferences, is an indication of the active involvement of our sponsor GlaxoSmithKline in the conference. We have already mentioned the importance of their sponsorship and of that of the European Union. We are particularly grateful to the latter for their support at a time when relationships between Austria and some member countries of the EU were under appreciable strain. Several regular attendees at these conferences indicated they would have reservations in attending the conference under the prevailing conditions. The Programme Committee therefore decided to issue the statement that is included on the next page of these Proceedings.

The editors are most grateful to the contributors for their high quality papers and to the referees for their committed and timely work. A list of both can be found at the end of this volume. We also acknowledge the secretarial and organisational aid of S. Gmeiner-Wallner, P. Murphy, S. Steffek and G. Sedlacek.

A.C. Atkinson, P. Hackl, W.G. Müller,   Vienna, February 2001

From **25th to 29th of June 2001** it was planned to hold the sixth workshop on Model Oriented Data Analysis (mOda6), a gathering of around 75 specialised statisticians, in **Puchberg/ Schneeberg, Austria**. In February 2000 a new Austrian government was formed: a coalition between the conservative People's Party and the so-called Freedom Party, the latter in the past heavily criticised for anti-European, xenophobic and racist declarations of its leading representatives. The inclusion of this party in the government was internationally condemned as represented by statements and action of the European Union's institutions, individual member countries, other states, groups and individuals.

Due to this development the steering committee of this workshop discussed at its meeting of 13th of April 2000, whether the event should be moved to another country. After carefully balancing the issues the committee feels that it can make its views felt more effectively by the sentiments expressed in this declaration, which will be published, nationally and internationally, to draw attention to its stand.

MODA statement concerning the political situation in Austria

An important MODA motive has always been to foster the multilateral aspect of our science and international friendship among scientists. The committee and organisers of MODA6 do not wish the meeting to be associated with a movement opposing such ideas. They will ensure this by maximum possible separation of science and politics, specifically by the following measures

- funding from official governmental and other political sources will neither be pursued nor accepted;

- officials of local or national government will be neither invited nor accepted to participate either formally or via social functions, excursions, etc;

- no political activity (leaflets, posters, etc.) will be allowed at the conference site.

We, the undersigned members of this committee whole-heartedly condemn any political forces, which incite and exploit hostile emotions towards foreigners or minorities. To actively prevent the workshop being publicly exploited by such a group, any information transfer about the conference will refer to this statement.

In making this stand we support those in Austria opposed to the unfortunate aspects of the new political climate.

> *A.C. Atkinson, V.V. Fedorov, C.P. Kitsos, H. Läuter,*
> *W.G. Müller, A. Pàzman, L. Pronzato, R. Schwabe,*
> *B. Torsney, I.N. Vuchkov, H.P. Wynn, A. Zhigljavsky*

Contents

Trend-free Repeated Measurement Designs

K. Afsarinejad

ABSTRACT: The existence and non-existence of trend-free repeated measurement designs are investigated. Two families of efficient / optimal repeated measurement designs which are very popular among experimenters are shown to be trend-free or trend-free with respect to treatments.

KEYWORDS: changeover; crossover; optimal design; orthogonal polynomial; carry over effect; trend-free.

1 Introduction

In many industrial, medical and agricultural experiments, treatments are applied to each experimental unit sequentially in time or space. There is a possibility that a systematic effect, or trend, influences the observations in addition to the experimental unit, treatment and carry over effects. This type of effect should be taken into account both when the experiment is planned and when the results are analysed. Another problem, which may occur in clinical trials, is that periods may not be quite well defined since patients are not recruited simultaneously. A period is not usually a single given calendar date and for some patients in a repeated measurement experiment, period one can be later than period two was for the others. In such cases, a model with a fixed effect for periods is not suitable. An alternative would be to replace the period effects by an orthogonal polynomial for each patient.

One way to account for the presence of trends is to use analysis of covariance, treating trend values as covariates. However, one may use suitable designs in the presence of trends to avoid the complications of analysis of covariance and increase design efficiencies.

Some studies in this direction have been made by Atkinson and Donev (1996), Box (1952), Box and Hay (1953), Bailey, Cheng and Kipnis (1992),

mODa6, A.C.Atkinson, P.Hackl and W.G.Müller, eds., Physica, Heidelberg, 2001.

Bradley and Yeh (1985), Bradley and Odeh (1988), Chai and Majumdar (1993), Cheng (1985), Cheng and Jacroux (1988), Coster (1993), Coster and Cheng (1988), Cox (1951,1952,1958), Daniel (1976,chapter 15), Daniel and Wilcoxon (1966), Githinji and Jacroux (1998), Hill (1960), Jacroux (1993, 1994), Jacroux and Ray (1990), Joiner and Campbell (1976), Lin and Dean (1991), Mukerjee and Sengupta (1994), Ogilvie(1963), Phillips (1964,1968a,1968b), Prescott (1981), Stufken (1988), Taylor (1967), Yeh and Bradley (1983), Yeh, Bradley and Notz (1985), Whittinghill (1989) and Wilkie (1987).

In the present paper, we consider two families of repeated measurement designs (RMDs) and show that efficient/optimal trend-free RMDs exist. In Section 2 we give the notation and basic definitions for RMDs. The model and conditions for trend-free RMDs are discussed in Section 3. Finally, Section 4 shows that some families of efficient/optimal RMDs which are very popular among experimenters are trend-free or trend-free with respect to treatments.

2   Notation and definitions

In an RMD there are t treatments, n experimental units and p periods. Each experimental unit receives one treatment during each period. Thus a RMD denoted by: RMD (t, n, p) is a $p \times n$ array with treatments as entries, rows representing periods and columns as experimental units. The set of all such arbitrary RMDs is designated by $\Omega_{t,n,p}$. Let us formally define some useful concepts.

Definition 2.1. A design $d \in \Omega_{t,n,p}$ is said to be uniform on periods if each treatment is administered γ times during each period.

Definition 2.2. A design $d \in \Omega_{t,n,p}$ is said to be balanced for carry over effects if in the $p \times n$ array d each treatment is preceded by each other treatment (not including itself) λ times.

Definition 2.3. A design $d \in \Omega_{t,n,p}$ is said to be strongly balanced for carry over effects if each treatment is preceded by all the other treatments (including itself)λ times.

Definition 2.4. A design $d \in \Omega_{t,n,p}$ is called circular balanced if the collection of ordered pairs $(d(i,j), d(i+1,j)), 1 \leq i \leq p, 1 \leq j \leq n$ contains each pair of distinct treatments λ times.

Definition 2.5. A design $d \in \Omega_{t,n,p}$ is called circular strongly balanced if the collection of ordered pairs $(d(i,j), d(i+1,j)), 1 \leq i \leq p, 1 \leq j \leq n$ contains each ordered pair of treatments (distinct or not) λ times.

In the last two definitions, when $i = p, i + 1 = 1$, since we count the pairs $(d(p, j), d(1, j))$ as well. Here we are interested in constructing minimal size designs, i.e., those designs which are balanced and require the minimum possible number of experimental units.

3 Model and conditions for trend-free RMDS

We assume that, within each experimental unit there is a common polynomial trend of order k on the p periods, which can be expressed by the orthogonal polynomials $\phi_\alpha(l), \alpha = 1, \ldots, k$, on $l = 1, \ldots, p$, where ϕ_α is a polynomial of degree α. The polynomials $\phi_1(l), \ldots, \phi_k(l)$, satisfy

$$\sum_{l=1}^{p} \phi_\alpha(l) = 0, \qquad \sum_{l=1}^{p} \phi_\alpha(l)\phi_{\alpha'}(l) = \delta_{\alpha\alpha'},$$

Where $\delta_{\alpha\alpha'}$ denotes the Kronecker delta, $\alpha\alpha' = 1, \ldots, k$.

Let $X_\beta = [I_n \otimes 1_p], X_\theta = [1_n \otimes \Phi_k]$, where 1_n is a $n \times 1$ vector of unit elements, I_n is a $n \times n$ identity matrix, $\otimes$ denotes a Kronecker product and Φ_k is a $p \times k$ matrix with elements $\phi_\alpha(l)$ in row l and column $\alpha, l = 1, \ldots, p, \alpha = 1, \ldots, k$. Then listing the response variables in the vector Y in order of period position within successive experimental units, the standard model for a RMD with trend is:

$$E(Y) = X_\mu\mu + X_\tau\tau + X_\rho\rho + X_\beta\beta + X_\theta\theta. \tag{3.1}$$

Here μ is a constant, $X_\mu = 1_{np}$, is τ is the $t \times 1$ vector of treatment parameters, $X'_\tau = [\Delta_1, \Delta_2, \ldots, \Delta_n]$ where for $i = 1, \ldots, n, \Delta'_n$ is a $p \times t$ matrix whose (k, j)th element is unity if treatment j is applied in period k, and zero otherwise, ρ is the $t \times 1$ vector of carry over effects, $X'_\rho = [\prod_1, \prod_2, \ldots, \prod_n]$ where for $i = 1, \ldots, n, \prod'_i$ is a $p \times t$ matrix whose (k, j) element is unity if treatment k is applied to the period preceding the $j - th$ period, and zero otherwise. Furthermore β is an $n \times 1$ vector of experimental unit parameters and θ is a $p \times 1$ vector of orthogonal polynomial coefficients.

Now we are in a position to define trend-free RMDs.

Definition 3.1. An RMD is trend-free with respect to treatments if $X'_\tau X_\theta = 0$.

Definition 3.2. An RMD is trend-free with respect to carry over effects if $X'_\rho X_\theta = 0$.

Definition 3.3. An RMD is trend-free if it is trend-free with respect to both treatments and carry over effects.

For design d let μ_{dil} and ν_{djl} denote the number of times treatment i appears in period l and the number of times carry over j appears in period l, respectively. A design is linear trendfree with respect to treatments if and only if

$$\sum_{l=1}^{p} \mu_{dil}\phi_l(l) = o, \qquad i = 1,\ldots,t. \tag{3.2}$$

A design is linear trend-free with respect to carry over effects if and only if

$$\sum_{l=1}^{p} \nu_{djl}\phi_l(l) = o, \qquad j = 1,\ldots,t. \tag{3.3}$$

The polynomial, $\phi_1(l)$ satisfies $\phi_1(l) = -\phi_1(p - l + 1)$. For k odd $\phi_1\{(k + 1)/2\} = 0$. Clearly (3.2) and (3.3) are true whenever

$$\mu_{dil} = \mu_{di(p-l+1)} \text{ and } \nu_{djl} = \nu_{dj(p-l+1)} \quad l = 1,\ldots,[(p+1)/2], i, j = 1,\ldots,t, \tag{3.4}$$

where [ ] denotes the largest integer less than or equal to $(p+1)/2$. Conditions (3.2) and (3.3) do not imply condition (3.4). Condition (3.4) is necessary and sufficient for a design to be odd-degree trend-free (Lin and Dean, 1991, corollary 2.1.2). Now suppose we have t treatments, each replicated r times. The following theorem is a modification to Corollary (2.1.1) of Lin and Dean (1991).

Theorem 3.1. A design $d \in \Omega_{t,n,p}$ which is trend-free with respect to treatments (carry overs) for a $(p - 1)$th order polynomial trend exists if and only if it is possible to arrange the treatments (carry overs) so that each treatment (carry overs) occurs $p^{-1}r$ times in each period.

4   Main results

In this section we are going to obtain some classes of trend-free RMDs and RMDs which are trend-free with respect to treatments. First we shall consider RMDs which are trend-free with respect to treatments.

Theorem 4.1. All the balanced or strongly balanced RMDs which are uniform on periods are also trend-free with respect to treatments.

Proof: Each treatment appears the same number of times in each period so by Theorem (3.1) they are trend-free with respect to treatments.

There are two families of balanced RMDs.

Family 1: $n = p = t$. Balanced RMDs exists for all even t. In the literature balanced RMDs are known as Williams designs. The method of construction due to Bradley (1958) is presented below.

Construction 4.1. Construct a $t \times t$ table in which columns represents experimental units and rows represent periods. Number the t experimental units successively from 1 to t. Assign integer 1 to t to the t cells in the first column by entering successive numbers in every other cell from top to bottom beginning with the first. Reverse the direction, going from bottom to top, once the end of a column has been reached, making sure that the return starts from the t-th cell in the column. Finally, complete each row in a cyclic manner.

Example 4.1. Let $t = 4$. Then the above procedure produces the following design.

$$
\begin{array}{cccc}
1 & 2 & 3 & 4 \\
4 & 1 & 2 & 3 \\
2 & 3 & 4 & 1 \\
3 & 4 & 1 & 2.
\end{array}
$$

If t is odd, then a balanced RMD does not exist for $t = 3, 5, 7$. Balanced RMDs exist for $t = 9, 15, 21, 27, 39, 55$ and 57. For other values of t the existence and non-existence of balanced RMDs is not known. One can trivially construct balanced RMDs with twice the number of experimental units for all odd t. An easily remembered method is given next.

Construction 4.2. Construct a design for t experimental units analogous to method 3.1 outlined for t even. Similarly, construct for the remaining t experimental units a design by letting the first column be the reverse of the first column of the design constructed for the initial t units. The following example illustrates the above method.

Example 4.2. Let $t = 5$. Then the above procedure produces the following design.

$$
\begin{array}{cccccccccc}
1 & 2 & 3 & 4 & 5 & 3 & 4 & 5 & 1 & 2 \\
5 & 1 & 2 & 3 & 4 & 4 & 5 & 1 & 2 & 3 \\
2 & 3 & 4 & 5 & 1 & 2 & 3 & 4 & 5 & 1 \\
4 & 5 & 1 & 2 & 3 & 5 & 1 & 2 & 3 & 4 \\
3 & 4 & 5 & 1 & 2 & 1 & 2 & 3 & 4 & 5.
\end{array}
$$

Unfortunately these designs are not trend-free with respect to carry over effects. To get around this problem one can consider the class of circular balanced RMDs. In this class the structure of carry over effects and treatments is the same and therefore designs in this class are trend-free. One can introduce a preperiod (period 0) giving all the period carry over effects. The period 0 would not be used in analysis. Circular RMDs exist whenever $t > 2$. Sonnemann, quoted in Kunert (1985) , gave the following construction method for circular balanced uniform RMDs with a minimum number of experimental units whenever $t > 2$ is an even integer.

Construction 4.3. Let $t = 2k$ and obtain the first set of t columns by developing the column $(2k \quad 1 \quad 2k-1 \quad 2k-2 \quad 3 \ldots k+1 \quad k)'$ and call it D_0. Let $\pi = (1 \quad 2 \ldots k-1 \quad 2k-1 \ldots k)$ and let $D_i = \pi^i D_0, i = 1, 2, \ldots, t-2$. Then $D_0, D_1, \ldots, D_{t-2}$ can be juxtaposed to obtain the required RMD.

Example 4.3. Let $k = 3, t = 6$; then $\pi = (12543)$ and

$$
D_0 = \begin{matrix} 6 & 1 & 2 & 3 & 4 & 5 \\ 1 & 2 & 3 & 4 & 5 & 6 \\ 5 & 6 & 1 & 2 & 3 & 4 \\ 2 & 3 & 4 & 5 & 6 & 1 \\ 4 & 5 & 6 & 1 & 2 & 3 \\ 3 & 4 & 5 & 6 & 1 & 2 \end{matrix}
\qquad
D_1 = \begin{matrix} 6 & 2 & 5 & 1 & 3 & 4 \\ 2 & 5 & 1 & 3 & 4 & 6 \\ 4 & 6 & 2 & 5 & 1 & 3 \\ 5 & 1 & 3 & 4 & 6 & 2 \\ 3 & 4 & 6 & 2 & 5 & 1 \\ 1 & 3 & 4 & 6 & 2 & 5 \end{matrix}
$$

$$
D_2 = \begin{matrix} 6 & 5 & 4 & 2 & 1 & 3 \\ 5 & 4 & 2 & 1 & 3 & 6 \\ 3 & 6 & 5 & 4 & 2 & 1 \\ 4 & 2 & 1 & 3 & 6 & 5 \\ 1 & 3 & 6 & 5 & 4 & 2 \\ 2 & 1 & 3 & 6 & 5 & 4 \end{matrix}
\qquad
D_3 = \begin{matrix} 6 & 4 & 3 & 5 & 2 & 1 \\ 4 & 3 & 5 & 2 & 1 & 6 \\ 1 & 6 & 4 & 3 & 5 & 2 \\ 3 & 5 & 2 & 1 & 6 & 4 \\ 2 & 1 & 6 & 4 & 3 & 5 \\ 5 & 2 & 1 & 6 & 4 & 3 \end{matrix}
$$

$$
D_4 = \begin{matrix} 6 & 3 & 1 & 4 & 5 & 2 \\ 3 & 1 & 4 & 5 & 2 & 6 \\ 2 & 6 & 3 & 1 & 4 & 5 \\ 1 & 4 & 5 & 2 & 6 & 3 \\ 5 & 2 & 6 & 3 & 1 & 4 \\ 4 & 5 & 2 & 6 & 3 & 1. \end{matrix}
$$

Afsarinejad (1990), using disjoint directed Hamiltonian cycles, constructed circular balanced uniform RMDs with a minimum number of experimental units whenever t is an odd number.

Construction 4.4. Let $t = 2k+1$ and label the t treatments by $\#0, 1, 2, \ldots, 2k$. Then construct $2k$ columns as follows:

$$C_i^+ = (0 \quad i \quad i+1 \quad i-1 \quad i+2 \quad i-2 \ldots i+(k-1) \quad i-(k-1) \quad i+k)'$$
$$C_i^- = (0 \quad i+k \quad i-(k-1) \quad i+(k+1) \ldots i-2 \quad i+2 \quad i-1 \quad i+1 \quad i)'$$

for $1 \le i \le k$. All the elements except 0 are taken as the positive integers $1, 2, \ldots, 2k$ (mod $2k$). These $2k$ columns are initial columns of $2k$ squares, which can be developed in turn to obtain $2k$ squares. Juxtaposing these squares gives a circular balanced uniform RMD with a minimum number of experimental units.

Example 4.5. Let $t = 5$. Then the four columns are:

$$C_1^+ = (01243)'$$
$$C_2^+ = (02314)'$$
$$C_1^- = (03421)'$$
$$C_2^- = (04132)'.$$

The required design is:

$$
\begin{array}{l}
0\ 1\ 2\ 4\ 3\ 0\ 2\ 3\ 1\ 4\ 0\ 3\ 4\ 2\ 1\ 0\ 4\ 1\ 3\ 2 \\
1\ 2\ 4\ 3\ 0\ 2\ 3\ 1\ 4\ 0\ 3\ 4\ 2\ 1\ 0\ 4\ 1\ 3\ 2\ 0 \\
2\ 4\ 3\ 0\ 1\ 3\ 1\ 4\ 0\ 2\ 4\ 2\ 1\ 0\ 3\ 1\ 3\ 2\ 0\ 4 \\
4\ 3\ 0\ 1\ 2\ 1\ 4\ 0\ 2\ 3\ 2\ 1\ 0\ 3\ 4\ 3\ 2\ 0\ 4\ 1 \\
3\ 0\ 1\ 2\ 4\ 4\ 0\ 2\ 3\ 1\ 1\ 0\ 3\ 4\ 2\ 2\ 0\ 4\ 1\ 3 \\
\end{array}
$$

A shortcoming with the designs in family one is that each experimental unit is used for t tests. That is each experimental unit must receive all the treatments. This may not be possible in some experiments, such as drug testing or other medical experiments. In these situations, the experimenter has to search for his design among the designs in family two.

Family 2: $p < t < n$. Afsarinejad (1983) constructed all the possible balanced minimal RMDs and strongly balanced minimal RMDs. These designs are all trend-free with respect to treatments.

Construction 4.5. A balanced minimal RMD can be obtained by developing, in turn, mod t, each of the following λ_1 columns.

$$
\begin{array}{ccccc}
c_1 & c_p & \cdots & c_{(i-1)p-(i-2)} & \cdots & c_{(\lambda 1-1)p-(\lambda 1-2)} \\
c_2 & c_{p+1} & \cdots & c_{(i-1)p-(i-3)} & \cdots & c_{(\lambda 1-1)p-(\lambda 1-3)} \\
\vdots & \vdots & & \vdots & & \vdots \\
c_p & c_{2p+1} & \cdots & c_{ip-(i-1)} & \cdots & c_t,
\end{array}
$$

where $(c_1, c_2, \ldots, c_t) =$

$$
\begin{array}{ll}
(1, t, 2, t-1, 3, \ldots, t/2, (t+2)/2) & \text{if } t \text{ is even} \\
(1, t, 3, t-2, 5, \ldots, (t+1)/2, (t+5)/2, \ldots, t, 1) & \text{if } t = 4\gamma + 1 \\
(1, t, 3, t-2, 5, \ldots, (t-1)/2, (t+3)/2, \ldots, t, 1) & \text{if } t = 4\gamma + 3.
\end{array}
$$

The set $U_{i=2}^t (c_i - c_{i-1})$ contains each non-zero number mod t and so each pair of distinct treatments will appear once in the final design.

Example 4.6. Let $t = 10$ and $p = 4$. Then the above procedure produces the following column: $(1, 10, 2, 9, 3, 8, 4, 7, 5, 6)$ and the design is presented below.

```
1 2 3 4 5 6 7 8 9 10 9 10 1 2 3 4 5 6 7 8 4 5 6 7 8 9 10 1 2 3
10 1 2 3 4 5 6 7 8 9 3 4 5 6 7 8 9 10 1 2 7 8 9 10 1 2 3 4 5 6
2 3 4 5 6 7 8 9 10 1 8 9 10 1 2 3 4 5 6 7 5 6 7 8 9 10 1 2 3 4
9 10 1 2 3 4 5 6 7 8 4 5 6 7 8 9 10 1 2 3 6 7 8 9 10 1 2 3 4 5
```

Now, let $t = 9$ and $p = 5$. Then the above procedure produces the following column: $(1, 9, 3, 7, 5, 7, 3, 9, 1)$ and the design is given below.

```
1 2 3 4 5 6 7 8 9 5 6 7 8 9 1 2 3 4
9 1 2 3 4 5 6 7 8 7 8 9 1 2 3 4 5 6
3 4 5 6 7 8 9 1 2 3 4 5 6 7 8 9 1 2
7 8 9 1 2 3 4 5 6 9 1 2 3 4 5 6 7 8
5 6 7 8 9 1 2 3 4 1 2 3 4 5 6 7 8 9.
```

The following construction method due to Afsarinejad (1983) establishes the existence of strongly balanced minimal RMDs whenever $p < t$.

Construction 4.6. The construction method is the same as Construction (3.5) with the exception that the column is:

$(c_1, c_2, \ldots, c_t, c_{t+1}) =$

$$
\begin{array}{ll}
(1, t, 2, t - 1, \ldots, t/2, (t + 2)/2, (t + 2)/2) & \text{if } t \text{ is even} \\
(1, t, 3, t - 2, \ldots, (t + 5)/2, (t + 1)/2, (t + 1)/2, \ldots, t, 1) & \text{if } t = 4\gamma + 1 \\
(1, t, 3, t - 2, \ldots, (t - 1)/2, (t + 3)/2, (t + 3)/2, \ldots, t, 1) & \text{if } t = 4\gamma + 3.
\end{array}
$$

The set $U_2^{t+1}(c_i - c_{i-1})$ contains each number mod t and so each ordered pair of treatments will appear precisely once in the final design.

Example 4.7. Let $t = 10$ and $p = 6$. Then, $(c_1, c_2, \ldots, c_{11}) = (1, 10, 2, 9, 3, 8, 4, 7, 5, 6, 6)$ and the design obtained is presented below.

```
1 2 3 4 5 6 7 8 9 10 8 9 10 1 2 3 4 5 6 7
10 1 2 3 4 5 6 7 8 9 4 5 6 7 8 9 10 1 2 3
2 3 4 5 6 7 8 9 10 1 7 8 9 10 1 2 3 4 5 6
9 10 1 2 3 4 5 6 7 8 5 6 7 8 9 10 1 2 3 4
3 4 5 6 7 8 9 10 1 2 6 7 8 9 10 1 2 3 4 5
8 9 10 1 2 3 4 5 6 7 6 7 8 9 10 1 2 3 4 5.
```

Now, let $t = 9$ and $p = 4$. Then $(c_1, c_2, \ldots, c_{10}) = (1, 9, 3, 7, 5, 5, 7, 3, 9, 1)$ and the design is given below.

1 2 3 4 5 6 7 8 9 7 8 9 1 2 3 4 5 6 7 8 9 1 2 3 4 5 6
9 1 2 3 4 5 6 7 8 5 6 7 8 9 1 2 3 4 3 4 5 6 7 8 9 1 2
3 4 5 6 7 8 9 1 2 5 6 7 8 9 1 2 3 4 9 1 2 3 4 5 6 7 8
7 8 9 1 2 3 4 5 6 7 8 9 1 2 3 4 5 6 1 2 3 4 5 6 7 8 9.

Again, these designs are not trend-free with respect to carry over effects. Designs, which are trend-free with respect to both treatment effects and carry over effects, exist and they are called circular balanced RMDs.

Fortunately, circular balanced RMDs can be constructed using nearest neighbour designs. The construction of these designs was given by Rees (1967) and has been considered by several authors, for example, Hwang (1973) , Hwang and Lin (1974, 1976, 1977, 1978), Lawless (1971), Street (1982) and Street and Street (1987, chapter 14). We shall present a construction method and, for the rest of the methods, we refer interested readers to the above mentioned authors. All that is needed is to construct an initial block to satisfy the following conditions: (I) all the differences, forward and backward, must be distinct; (II) the sum of the forward differences must be zero (mod t). Then use this initial block to develop a design cyclically. Now find the mirror image of this initial block and develop another design cyclically. These two designs together constitute a circular balanced RMD.

Construction 4.7. Given $t = 4\gamma + 3$ and $p = 2\gamma + 1$, circular RMDs exists whenever t is a prime power. Let x be a primitive root. Then the initial block for the first part of design would be : $(x^0, x^2, x^4, \ldots, x^{4\gamma})$ and the initial block for the second part of the design would be $(x^{4\gamma}, \ldots, x^4, x^2, x^0)$.

Example 4.7. Let $t = 7$ and $p = 3$. The initial block of the first part is $(1, 2, 4)$ and the initial block of second part is $(4, 2, 1)$. The design is:

1 2 3 4 5 6 7 4 5 6 7 1 2 3
2 3 4 5 6 7 1 2 3 4 5 6 7 1
4 5 6 7 1 2 3 1 2 3 4 5 6 7.

References

Afsarinejad, K. (1983). Balanced repeated measurements designs. *Biometrika 70, 563-568.*

Afsarinejad, K. (1990). Circular balanced uniform repeated measurements designs, II. *Statistics and Probability Letters 9, 141-143*

Atkinson, A.C. and Donev A.N (1996). Experimental designs optimally balanced for trend. *Technometrics 38, 333-341*

Bailey, R.A., Cheng, C.S. and Kipnis, P. (1992). Construction of trend-resistant factorial designs. *Statistica Sinica 2, 393-411*

Box, G.E.P. (1952). Multi-factor designs of first order. *Biometrika 39, 49-57*

Box, G.E.P. and Hay, W.A. (1953). A statistical design for the removal of trends occurring in a comparative experiment with an application in biological assay *Biometrics 9, 304-319*

Bradley, J.V. (1958). Complete counterbalancing of immediate sequential effects in a Latin square design. *J. Amer. Statist. Assoc. 53, 525-528*

Bradley, R.A., Yeh, C.M. (1980). Trend-free block designs: Theory. *Ann. Statist. 8, 883-893*

Bradley, R.A. and Odeh, R.E. (1988). A generating algorithm for linear trend-free and nearly linear trend-free block designs. *Commun. Statist. -Simula. 17, 1259-1280*

Chai, F.S. and Majumdar, D. (1993). On the Yeh-Bradley conjecture on linear trend-free block designs. *Ann. Statist. 21, 2087-2097*

Cheng, C.S. (1985). Run order of factorial designs. In *Proceedings of the Berkeley conference in honor of Jerzy Neyman and Jack Kiefer* (vol. 2), eds. L.M. Lecam and R.A. Olshen, Belmont, CA: Wadsworth, 619-633

Cheng, C.S. and Jacroux, M. (1988). The construction of trend-free run orders of two-level factorial designs. *J. American Statistical Association 83, 1152-1158*

Coster, D.C. (1993). Trend-free run orders of mixed-level fractional factorial designs. *Ann. Statist. 21, 2072-2086*

Coster, D.C. and Cheng, C.S. (1988). Minimum cost trend-free run orders of fractional factorial designs. *Ann. Statist. 16, 1188-1205*

Cox, D.R. (1951). Some systematic experimental designs. *Biometrika 38, 312-323*

Cox, D.R. (1952). Some recent work on systematic experimental designs. *J. Roy. Statist. Soc. Ser. B 14, 211-219*

Cox, D.R. (1958). The planning of experiments. *Wiley: New York*

Daniel, C. (1976). Applications of statistics to industrial experimentation. *Wiley: New York*

Daniel, C. and Wilcoxon, F. (1966). Factorial 2p-q plans robust against linear and quadratic trends. *Technometrics 8, 259-278*

Githinji, F. and Jacroux, M. (1998). On the determination and construction of optimal designs for comparing a set of test treatments with a set of controls in presence of a linear trend. *J. Statist. Plann. Inference 66, 161-174*

Hill, H.M . (1960). Experimental designs to adjust for time trends. *Technometrics 2, 67-82*

Hwang, F.K. (1973). Constructions for some classes of neighbour designs. *Ann. Statist. 1, 786-790*

Hwang, F.K. and Lin, S. (1974). A direct method to construct triple systems. *J. Combinatorial Theory. A 17, 84-94*

Hwang, F.K. and Lin, S. (1976). Construction of 2-balanced (n,k,l) arrays. *Pacific J. Maths. 64, 437-453*

Hwang, F.K. and Lin, S. (1977). Neighbour designs. *J.Combinatorial Theory A 23, 303-313*

Hwang, F.K. and Lin, S. (1978). Distributions of integers into k-tuples with prescribed conditions. *J. Combinatorial Theory A 25, 105-116*

Jacroux, M. (1993). On the construction of trend-resistant designs for comparing a set of test treatments with a set of controls. *J. American Statistical Association 88, 1398-1403*

Jacroux, M. (1994). On the construction of trend resistant mixed level factorial run orders. *Ann. Statist. 22, 904-916*

Jacroux, M. and Ray, R.S. (1990). On the construction of trend-free run orders of treatments. *Biometrika 77, 187-191*

Joiner, B.L. and Campbell, C. (1976). Designing experiments when run order is important. *Technometrics 18, 249-259*

Kunert, J. (1985). Optimal repeated measurements designs for correlated observations and analysis by weighted least squares. *Biometrika 72, 375-389*

Lawless, J.F . (1971). A note on certain types of BIBD's balanced for carry over effects. *Ann. Math. Statist. 42, 1439-1441*

Lin, M. and Dean, A.M. (1991). Trend-free block designs for varietal and factorial experiments. *Ann. Statist. 19, 1582-1596*

Mukerjee, R. and Sengupta, S. (1994). A-optimal run orders with a linear trend. *Austral. J. Statist. 36, 115-122*

Ogilvie, J.C. (1963). A simple method for the elimination of individual trends in the analysis of balanced sets of Latin squares. *Biometrics 19,264-272*

Phillips, J.P.N. (1964). The use of magic squares for balancing and assessing order effects in some analysis of variance designs. *Appl. Statist. 13,67-73*

Phillips, J.P.N. (1968a). A simple method of constructing certain magic rectangles of even order. Math. *Gazette 52, 9-12*

Phillips, J.P.N. (1968b). Methods of constructing one-way and factorial designs balanced for trend. *Appl. Statist. 17, 162-170*

Prescott, R.J. (1981). The comparison of success rates in cross-over trials in the presence of an order effect. *Appl. Statist. 30, 9-15*

Rees, D.H. (1967). Some designs of use in serology. *Biometrics, 23,779-791*

Street, A.P. (1982). A survey of neighbour designs. *Congressus Numerantium 34,119-155*

Street, A.P. and Street, D.J. (1987). Combinatorics of experimental design. *Clarendon Press, Oxford*

Stufken, J. (1988). On the existence of linear trend-free block designs. *Commun. Statist. Meth. 17, 3857-3863*

Taylor, J. (1967). The value of orthogonal polynomials in the analysis of change-over trials with dairy cows. *Biometrics 23, 297-311*

Yeh, C.M. and Bradley, R.A. (1983). Trend-free block designs: existence and construction results. *Commun. Statist. -Theor. Meth. 12, 1-24*

Yeh, C.M., Bradley, R.A. and Notz, W.I. (1985). Nearly trend-free block designs. *J.American Statistical Association 80, 985-992*

Whittinghill, D.C. (1989). A note on trend-free block designs. *Commun. Statist. -Theory Meth. 18, 277-285*

Wikie, D. (1987). Analysis of factorial experiments and least squares polynomial fitting by the method of orthogonal polynomials for any spacing of the level of the independent variables. *J. Appl. Statist. 14, 83-89*

Williams, E.J. (1949). Experimental designs balanced for the estimation of carry over effects of treatments. *Austral.J.Sci.Res. A2, 149-168*

Minimax Optimal Designs for Nonparametric Regression — A Further Optimality Property of the Uniform Distribution

S. Biedermann
H. Dette

ABSTRACT: In the common nonparametric regression model $y_{i,n} = g(t_{i,n}) + \sigma(t_{i,n})\varepsilon_i$, $i = 1, \ldots, n$, with i.i.d. noise and nonrepeatable design points t_i, we consider the problem of choosing an optimal design for the estimation of the mean function g. A minimax approach is adopted which searches for designs minimizing the maximum of the asymptotic integrated mean squared error, where the maximum is taken over an appropriately bounded class of functions (g, σ). The minimax designs are found explicitly, and for certain special cases the optimality of the uniform distribution can be established.

KEYWORDS: kernel estimation; locally optimal designs; mean squared error; minimax designs; nonparametric regression

1 Introduction

Consider the common nonparametric regression model

$$Y_{i,n} = g(t_{i,n}) + \sigma(t_{i,n})\varepsilon_{i,n} \qquad i = 1, \ldots, n, \qquad (1.1)$$

where the $\varepsilon_{i,n}$ form a triangular array of rowwise independent identically distributed random variables with variance 1 and mean 0, g, σ are unknown smooth functions and $\{t_{i,n} \mid i = 1, \ldots, n\}$ is a fixed design in the interval $[0, 1]$. Much effort has been devoted to the problem of estimating the mean function (see e.g. the recent monographs of Härdle, 1990, Wand and Jones, 1995, Fan and Gijbels, 1996), and many of the developed es-

mODa6, A.C.Atkinson, P.Hackl and W.G.Müller, eds., Physica, Heidelberg, 2001.

timation methods are meanwhile standard methods in applied regression analysis.

Although it is well known (and intuitively clear) that the asymptotic properties of the various nonparametric estimators depend sensitively on the underlying designs, the problem of designing experiments in the nonparametric setup (1.1) has found much less attention in the literature. This is mainly due to the fact that – as in the case of nonlinear regression models – the optimality criteria usually depend on the unknown mean and variance function. Müller (1984) studied optimal designs for estimating derivatives of the mean function by minimizing the asymptotic integrated mean squared error of a kernel estimate. He showed that the optimality criteria are local in the sense of Chernoff (1953) and that the asymptotically optimal designs depend on the unknown mean and variance function. Because of these difficulties Cheng, Hall and Titterington (1998) proposed a sequential approach for defining an optimal design measure in the context of local linear regression.

A different approach was used by Müller (1996) who considered the situation, where the unknown response function has to be predicted at a finite number of specified points. This author proposed to minimize a weighted average of the variances of local linear estimates at these points (see Fan and Gijbels, 1996, or Wand and Jones, 1995).

A further optimality criterion is considered in the present paper, which is different from the methods considered by the aforementioned authors and mainly motivated by two observations. On the one hand, the designs proposed by Müller (1984) are not robust with respect to misspecifications of the mean and variance function. On the other hand, there are many situations where sequential designs cannot be applied (e.g. experiments in agriculture). For these reasons, we propose a minimax approach which seeks for designs minimizing the maximum of the asymptotic integrated mean squared error, where the maximum is taken over certain Sobolev balls for the mean and variance function. Section 2 states some basic terminology from nonparametric kernel estimation, which is necessary for the formulation of the minimax optimality criterion. The main results can be found in Section 3 showing that the minimax optimal designs are equal to the weight function used in the definition of the integrated mean squared error. This highlights the particular role of the uniform distribution, which turns out to be the minimax optimal design with respect to the classical integrated mean squared error criterion.

2 Asymptotic representation of the integrated mean squared error of a nonparametric regression estimator

For the introduction of the optimality criterion, we need an asymptotic representation of the integrated mean squared error of a nonparametric estimate of the mean function g, which is nowadays standard in mathematical statistics (see Härdle, 1990, Wand and Jones, 1995 or Fan and Gijbels, 1996). We assume that the mean function is k-times continuously differentiable, i.e. $g \in C^k[0,1]$, and the variance function is Lipschitz continuous of order γ, i.e. $\sigma \in \mathrm{Lip}_\gamma[0,1]$ for some $\gamma \in (0,1]$. Following Gasser and Müller (1984), we consider

$$g_{n,v}(t) = \frac{1}{h^{v+1}(t)} \sum_{i=1}^{n} \int_{s_{i-1,n}}^{s_{i,n}} K_v\left(\frac{t-x}{h(t)}\right) dx \cdot Y_{i,n} \qquad (2.1)$$

as an estimate of the v-th derivative $g^{(v)}$ of the regression function ($v = 0, 1, \ldots, k-1$). Here K_v is a Lipschitz continuous function with compact support, say $[-1,1]$, such that

$$\frac{(-1)^j}{j!} \int_{-1}^{1} K_v(x)x^j\,dx = \begin{cases} 0 & \text{if } 0 \le j < k; j \ne v \\ 1 & \text{if } j = v \\ B & \text{if } j = k \end{cases} \qquad (2.2)$$

where $B \ne 0$. The triangular array $\{s_{i,n} \mid i = 0, \ldots, n\}$ is defined by $s_{0,n} = 0, s_{n,n} = 1, s_{i,n} = (t_{i,n} + t_{i+1,n})/2$ ($i = 1, \ldots, n-1$), and the design points $t_{i,n}$ are supposed to satisfy a Sacks and Ylvisaker (1970) condition

$$\int_0^{t_{i-1,n}} f(t)\,dt = \frac{i-1}{n-1} \qquad i = 1, \ldots, n \qquad (2.3)$$

with a positive design density f.

Note that a triangular array of random variables is considered in model (1.1) in contrast to the formulation $y_i = g(t_i) + \sigma(t_i)\varepsilon_i$, which is maybe more familiar to most workers in experimental design. The reason is a much simpler formulation of the asymptotic properties of the estimate (2.1). For example, if $\tilde{t}_i = \frac{i-1}{n-1}$ would denote a uniform design, an additional observation at a point t^*, say, would destroy uniformity. On the other hand, the formulation (2.3) using the array $t_{i,n}$ would yield the same design density independent of $n \in \mathbb{N}$.

The quantity $h(t)$ denotes the bandwidth which may depend on a specific point t (local bandwidth). The locally optimal bandwidth minimizes the asymptotic mean squared error at the point t and is given by

$$h^*(t) = \left\{ \frac{(2v+1)V\sigma^2(t)}{2(k-v)nB^2 f(t)(g^{(k)}(t))^2} \right\}^{1/(2k+1)} \qquad (2.4)$$

where $V = \int_{-1}^{1} K_v^2(x)dx$ (see Gasser and Müller, 1984), and it is assumed that $g^{(k)}(t) \neq 0$. Insertion of the optimal bandwidth in the weighted integrated mean squared error

$$E \int_0^1 \{\hat{g}_{n,v}(t) - g^{(v)}(t)\}^2 w(t)dt$$

yields (using an appropriate modification of the kernel at the boundary, see Müller, 1984)

$$n^{2(k-v)/(2k+1)} E \int_0^1 \{\hat{g}_{n,v}(t) - g^{(v)}(t)\}^2 w(t)dt = c \cdot \Psi_{g,\sigma}^{(v)}(f) + o(1), \quad (2.5)$$

where w is a positive continuous weight function on the interval $[0,1]$, c is a constant (independent of f) and

$$\Psi_{g,\sigma}^{(v)}(f) = \int_0^1 \left\{\frac{\sigma^2(t)w(t)}{f(t)}\right\}^{2(k-v)/(2k+1)} \{(g^{(k)}(t))^2 w(t)\}^{(2v+1)/(2k+1)} dt.$$

$$(2.6)$$

3 Optimal designs minimizing the maximum integrated mean squared error

Müller (1984) determined the design density f which minimizes the criterion (2.6) for fixed g, σ^2, i.e.

$$f^*(t) = \frac{\varphi(t)}{\int_0^1 \varphi(x)dx} \qquad (3.1)$$

where $\varphi(t) = [(\sigma^2(t))^{2(k-v)}(g^{(k)}(t))^{2(2v+1)}(w(t))^{2k+1}]^{1/(4k+1-2v)}$. The optimal design is local in the sense of Chernoff (1953) and might be not robust with respect to misspecifications of the variance and mean function (see also Example 3.3). For this reason, we propose a minimax criterion for the determination of optimal designs and call a design density f^* minimax optimal for the estimation of the vth derivative of the regression function if it minimizes

$$\max\{\Psi_{g,\sigma}^{(v)}(f) \mid (g,\sigma) \in \mathcal{F}\}. \qquad (3.2)$$

Here $\mathcal{F}$ is an appropriate class of functions given either by

$$\mathcal{F}_2 = \left\{(g,\sigma) \in C^{(k)}[0,1] \times \text{Lip}_\gamma[0,1] \; \middle| \qquad (3.3) \right.$$

$$\left. \int_0^1 \sigma^2(t)dt \leq \varepsilon, \int_0^1 (g^{(k)}(t))^2 w(t)dt \leq \eta\right\}$$

or by

$$\mathcal{F}_\infty = \left\{ (g,\sigma) \in C^{(k)}[0,1] \times \operatorname{Lip}_\gamma[0,1] \; \Big| \right. \tag{3.4}$$

$$\left. \sup_{t\in[0,1]} |\sigma^2(t)| \le \varepsilon, \; \sup_{t\in[0,1]} |g^{(k)}(t)|^2 w(t) \le \eta \right\}.$$

Theorem 3.1 *Let $\mathcal{F}$ be either defined by (3.3) or (3.4). Then the design with density*

$$f^*(t) = \frac{w(t)}{\int_0^1 w(x)dx}$$

is optimal with respect to the minimax criterion (3.2).

Proof: We consider only the case $\mathcal{F} = \mathcal{F}_2$, the corresponding result for the sup-norm is proved by similar arguments. Without loss of generality it is assumed that $\int_0^1 w(t)dt = 1$, i.e. $f^* = w$. The proof is performed in two steps showing

(1) $\qquad \sup\{\Psi_{g,\sigma}^{(v)}(f^*) \mid (g,\sigma) \in \mathcal{F}_2\} \;\; = [\varepsilon^{2(k-v)}\eta^{2v+1}]^{1/(2k+1)}$

(2) $\qquad \forall f \;\; \exists (g,\sigma) \in \mathcal{F}_2 : \;\; \Psi_{g,\sigma}^{(v)}(f) \ge [\varepsilon^{2(k-v)}\eta^{2v+1}]^{1/(2k+1)}$

(1) The first part follows by a direct application of Hölder's inequality observing that with the notation $p = \frac{2k+1}{2(k-v)}, q = \frac{2k+1}{2v+1}$ the integrated mean squared error is given by

$$\Psi_{g,\sigma}^{(v)}(w) = \int_0^1 |\sigma^2(t)|^{1/p} |(g^{(k)}(t))^2 w(t)|^{1/q} dt \tag{3.5}$$

$$\le \left\{ \int_0^1 \sigma^2(t)dt \right\}^{1/p} \left\{ \int_0^1 (g^{(k)}(t))^2 w(t)dt \right\}^{1/q} \le \varepsilon^{1/p}\eta^{1/q},$$

and that there is equality in (3.6) for the functions

$$\sigma(t) \equiv \sqrt{\varepsilon}; \quad g(t) = \sqrt{\eta} \int_0^t \int_0^{s_1} \cdots \int_0^{s_{k-1}} \frac{ds}{\sqrt{w(s)}} \tag{3.6}$$

(note that $(g,\sigma) \in \mathcal{F}_2$).

(2) Let f denote an arbitrary positive density on the interval $[0,1]$ and let

$$p' = 1 + \frac{1}{p} = \frac{4k+1-2v}{2k+1}; \quad q' = 1 + p = \frac{4k+1-2v}{2(k-v)} \tag{3.7}$$

(note that $1/p' + 1/q' = 1$). For the functions

$$\sigma_*(t) = \sqrt{\varepsilon w(t)}; \quad g_*(t) = \sqrt{\eta}\frac{t^k}{k!} \quad (3.8)$$

it follows (observing $\int_0^1 w(t)dt = 1$) that $(g,\sigma) \in \mathcal{F}_2$ and

$$\Psi_{g_*,\sigma_*}^{(v)}(f) = \varepsilon^{1/p}\eta^{1/q}\int_0^1 |\frac{w(t)}{f(t)}|^{1/p}w(t)dt$$

$$= \varepsilon^{1/p}\eta^{1/q}\int_0^1 (w(t))^{p'}\frac{dt}{|f(t)|^{p'/q'}} = \varepsilon^{1/p}\eta^{1/q}\|\frac{w}{|f|^{1/q'}}\|_{p'}^{p'}$$

$$\geq \varepsilon^{1/p}\eta^{1/q}\left\{\frac{\|w\|_1}{\|f^{1/q'}\|_{q'}}\right\}^{p'} = \varepsilon^{1/p}\eta^{1/q}$$

where $\|s\|_p = (\int_0^1 |s(t)|^p dt)^{1/p}$ denotes the L^p-norm with respect to the Lebesgue measure and the inequality follows again from Hölder's inequality. This proves (2) and completes the proof of Theorem 3.1.

Example 3.2. Note that for a constant weight function the uniform design turns out to be minimax optimal. This result can be explained intuitively, because in this case the optimization problem is invariant with respect to shifts of the design space. Consequently, the optimal design must be the uniform measure.

Example 3.3. Consider the case $k = 2, v = 0$ and $w(x) \equiv 1$. The locally optimal design (with respect to the asymptotic integrated MSE) is given by (3.1) and we are interested how a misspecification of the mean or variance function affects the performance of this design. To this end we consider the following scenario

$$
\begin{array}{lll}
(1) & g(x) = e^x; & \sigma(x) = 1 \\
(2) & g(x) = e^x; & \sigma(x) = e^x \\
(3) & g(x) = e^x; & \sigma(x) = e^{2x} \\
(4) & g(x) = \sin x; & \sigma(x) = 1 \\
(5) & g(x) = \sin x; & \sigma(x) = e^x \\
(6) & g(x) = \sin x; & \sigma(x) = e^{2x}.
\end{array}
$$

For a misspecification we investigate four cases

$$
\begin{array}{lll}
(A) & g \text{ correct}; & \sigma \text{ constant} \\
(B) & g \text{ correct}; & \sigma^{-1} \text{ instead of } \sigma \\
(C) & g^2 \text{ instead of } g; & \sigma \text{ correct} \\
(D) & e^g \text{ instead of } g; & \sigma \text{ correct}
\end{array}
$$

and calculate the locally optimal design from Müller (1984). Note that in cases (A) and (B) the variance function is misspecified, while (C) and (D) correspond to an incorrect assumption for the mean function. The corresponding results are listed in Table 1.1, which also contains a column for the minimax optimal design which turns out to be the uniform design in these cases. The table shows the asymptotic integrated MSE obtained by the particular design f^* in (3.1) for a misspecification relative to the asymptotic integrated MSE obtained by the design f^* for the correct regression and variance function. The last column shows the corresponding ratio for the minimax optimal design.

	A	B	C	D	minimax
1	1	1	0.4650	0.4625	0.9970
2	0.9562	0.8374	0.4519	0.4494	0.9326
3	0.9573	0.8406	0.4404	0.4381	0.9066
4	1	1	0.7193	0.6676	0.9776
5	0.9600	0.8502	0.8289	0.7977	0.8997
6	0.8658	0.5643	0.9379	0.7894	0.7894

TABLE 1. Asymptotic relative efficiency of the optimal design f^* in (3.1) for various misspecifications of the mean and variance function. The last column: minimax design obtained from Theorem 3.1.

We observe that a misspecification of the mean and variance function has a serious impact on the performance of the resulting design. This dependency is even stronger, if the mean function is not adequately specified. The minimax design has a reasonable performance in all considered cases. For these reasons, the locally optimal designs proposed by Müller (1984) should not be used in practice unless information about the variance and mean structure is available. Otherwise the minimax designs proposed in this paper are recommended.

Acknowledgments: The authors would like to thank I. Gottschlich for typing most parts of this paper with considerable technical expertise. The financial support of the Deutsche Forschungsgemeinschaft (SFB 475, Reduction of complexity in multivariate data structures) is gratefully acknowledged.

References

Cheng, M.-H., Hall, P., and Titterington, D.M. (1998). Optimal design for curve estimation by local linear smoothing. *Bernoulli*, **4**, 3-14.

Chernoff, H. (1953). Locally optimal design for estimating parameters. *Annals of Mathematical Statistics*, **24**, 586-602.

Fan, J., Gijbels, I. (1996). *Local Polynomial Modelling and its Applications*. Chapman and Hall, London

Gasser, T. and Müller, H.G. (1984). Estimating regression functions and their derivatives by kernel functions. *Scandinavian Journal of Statistics*, **11**, 171-185.

Härdle, W. (1990). *Applied Nonparametric Regression*. Cambridge University Press, Boston.

Müller, H.G. (1984). Optimal designs for nonparametric kernel regression. *Statistics & Probability Letters*, **2**, 285-290.

Müller, W.G. (1996). Optimal design for local fitting. *Journal of Statistical Planning and Inference*, **55**, 389-397.

Sacks, J. and Ylvisaker, D. (1970). Designs for regression problems for correlated errors. *Annals of Mathematical Statistics*, **41**, 2057-2074.

Wand, M.P. and Jones, M.C. (1995). *Kernel Smoothing*. Chapman & Hall, London.

Optimization of Monitoring Networks for Estimation of the Semivariance Function

E.P.J. Boer
E.M.T. Hendrix
D.A.M.K. Rasch

ABSTRACT: The optimal adjustment of an existing monitoring network for estimation of the semivariance function by means of optimal design of experiments is discussed. The difference between neglecting and including correlation between point pairs, from which the semivariance function is estimated, is visualized for a simple adjustment of a monitoring network. A branch-and-bound algorithm is applied to calculate an exact optimal configuration of monitoring sites (design). For a case study it is shown that the optimal design is robust against misspecified parameter values and model choice.

KEYWORDS: geostatistics; semivariance function; optimal design of experiments; monitoring network

1 Introduction

A problem frequently encountered in practice is the adjustment of an existing monitoring network. Monitoring networks have to be enlarged or reduced, mostly depending on the budgets of the research. One of the biggest problems in designing or adjusting a monitoring network is that the objective of measuring is often ambiguous. A clear example of this can be found in geostatistical analyses. In general, the semivariance function (characterizing spatial correlation) is estimated for kriging (spatial prediction), but the optimal monitoring network for estimation of the semivariance function is not the same as the optimal network for kriging (Zimmerman and Homer, 1991). In this paper the focus is on the objective of designing an optimal monitoring network for estimation of the semivariance function.

mODa6, A.C.Atkinson, P.Hackl and W.G.Müller, eds., Physica, Heidelberg, 2001.

The criterion for optimality used in this paper is based on the classical theory of optimal design of experiments (Fedorov, 1972).

Early discussion of optimal design for estimation of the semivariance function can be found in Russo (1984). He suggests a criterion which attempts to modify the spatial configuration of points in such a way that distances between point pairs (lags) are as much as possible equally distributed among the several distance classes. The results of this approach hardly depend on the model of the semivariance function.

Zimmerman and Homer (1991) applied the classical theory of optimal experimental design for estimation of the semivariance function. The design problem consists of adding q new sites, to an existing monitoring network $B_F = \{s_1, ..., s_n\}$, from a set $B_P = \{s_{n+1}, ..., s_{n+Q}\}$ of potential sites.

In this paper we would like to add three new aspects to the use of optimal designs for estimation of the semivariance function. In the first place, a visualization of a simple adjustment of a monitoring network is given. Secondly, a branch-and-bound algorithm is applied to calculate an exact optimal configuration of monitoring sites, given a certain criterion and a set of possible monitoring sites. Finally, a robustness study of the optimal design against misspecified parameter values and model choice is done. All is elaborated for a case study introduced by Cressie *et al* (1990).

2 Theory

2.1 The semivariance function

A number of papers have been written about design aspects when a certain semivariance function is assumed (e.g. Cressie *et al*, 1990). However, usually the semivariance function needs to be estimated; which can be considered as a first step in geostatistical analyses.

The random variable Z at a site s in a domain $D \subset \mathbb{R}^2$ is described as (Cressie, 1991)

$$Z(s) = \mu(s) + \delta(s) \tag{2.1}$$

where $\mu(s)$ is the deterministic part of $Z(s)$ and $\delta(s)$ is a spatially dependent zero-mean stochastic process. This paper focuses on the optimal estimation of the unknown spatial correlation of $\delta(s)$. The spatial correlation between points can be quantified by means of the semivariance function:

$$\gamma(h) = \frac{1}{2}\text{var}[\delta(s_1) - \delta(s_2)] \tag{2.2}$$

where it is assumed that the variance of differences depends only on the (Euclidean) distance $h = \| s_1 - s_2 \|$ between sites s_1 and s_2. Three parametric semivariance functions $\gamma(h, \theta)$ are used in this study: the spherical, exponential and Gaussian semivariance functions (Isaaks and Srivastava,

1989). The semivariance can be estimated from pairs $\{h_k, \hat{\gamma}_k\}$, derived by measuring for every point pair (s_i, s_j)

$$h_k = \| s_i - s_j \|, \quad k = 1, ..., \frac{1}{2}n(n-1) = N.$$

and

$$\hat{\gamma}_k = \frac{1}{2}[\delta(s_i) - \delta(s_j)]^2, \quad i < j; \quad i, j = 1, ..., n. \tag{2.3}$$

$\delta(s)$ can be estimated by detrending the data. All the pairs $\{h_k, \hat{\gamma}_k\}$ can be displayed as a scatter plot (variogram cloud). Müller (1999) show that a direct fit of a parametric semivariance function is often preferable. The observations of the semivariance function (Equation 2.3) are often correlated, as from n monitoring sites come N observations of the semivariance function. Therefore, generalized least-squares estimation (GLS) is used to estimate (iteratively) $\theta = (\theta_1, \theta_2, ..., \theta_m)$, i.e.

$$\hat{\theta}_{(j)} = \underset{\theta_{(j)}}{\text{Arg min}}[\hat{\gamma} - \gamma(\theta_{(j)})]^T \Sigma^{-1}(\theta_{(j-1)}) [\hat{\gamma} - \gamma(\theta_{(j)})], \quad j = 1, 2, ... \tag{2.4}$$

where $\hat{\gamma} = (\hat{\gamma}_1, ..., \hat{\gamma}_N)^T$, $\gamma(\theta) = [\gamma(h_1, \theta), ..., \gamma(h_N, \theta)]^T$ and $\Sigma(\theta)$ is the covariance matrix of $\hat{\gamma}$. If a Gaussian random field is assumed, Cressie (1985) describes how the entries of $\Sigma(\theta)$ can be updated (Müller, 1998).

2.2 Optimal experimental design for estimation of the semivariance function

Optimal experimental design can be used to maximize the precision of estimation of parameters of regression models, given the number of allowed observations. Müller and Zimmerman (1999) and Bogaert and Russo (1999) showed how the theory of optimal experimental design can be applied to the estimation of the semivariance function. Müller and Zimmerman (1999) clarify the two major differences from the standard optimal design methods for nonlinear regression. The first difference is that the observations of the semivariance function are a result of the spatial configuration of monitoring sites. This means that adding one monitoring site yields n additional pairs of points from which the semivariance function is estimated. Secondly, in optimal design for nonlinear regression models it is usually assumed that the observations are uncorrelated. This is rarely the case for empirical observations of the semivariance function.

Many criteria for optimal designs are functions of the so-called information matrix. Let the monitoring network (design) be denoted as $\xi_n = (s_1, ..., s_n)$, which define $h_1, ..., h_N$. The information matrix corresponding to $\hat{\theta}$ (Equation 2.4) and a monitoring network ξ_n is equal to

$$M[\theta, \xi_n] = J_\theta^T \Sigma^{-1}(\theta, \xi_n) J_\theta \tag{2.5}$$

where J_θ is the well-known Jacobian matrix.

Two ways of calculating the information matrix (2.5) can be found in the literature. One where $\Sigma^{-1}(\theta, \xi_n)$ is approximated by ignoring the off-diagonal elements (Zimmerman and Homer, 1991), mainly in view of the computational advantages. In this way, the correlation between point pairs is neglected. The other way can be found in Müller and Zimmerman (1999) and Bogaert and Russo (1999), where the whole $N \times N$ matrix $\Sigma^{-1}(\theta, \xi_n)$ is included in the calculations.

The so-called D-optimality corresponds to minimization of the determinant of the inverse of the information matrix:

$$\text{D-optimality: } \underset{\xi_n}{\text{Arg min}} \det(M^{-1}[\theta, \xi_n]). \qquad (2.6)$$

Optimizing according to (2.6) results in an optimal design for a certain parameter vector θ. Because of this dependence on the parameter values of the semivariance , the optimal design is a so-called 'locally' optimal design.

3   Case study

In this paper we make use of a case study introduced by Cressie *et al* (1990). This case study considers the Utility Acid Precipitation Study Program (UAPSP) monitoring network located in the eastern and midwestern U.S.A. Annual acid-deposition levels were measured in 1982 and 1983 in a network of 19 U.S. sites ($B_F = \{s_1, ..., s_{19}\}$). There are 11 potential sites ($B_P = \{s_{20}, ..., s_{30}\}$) available for enlarging the monitoring network with one or more sites. The optimal selection of q out of $Q = 11$ sites results in a combinatorial optimization problem, which can be solved by full enumeration with a branch-and-bound algorithm (see Rasch *et al*, 1997). Figure 1 shows the 19 sites of the existing network and the 11 potential sites.

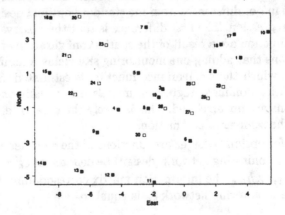

FIGURE 1. The 19 sites of the UAPSP monitoring network (■) and the 11 potential sites (□). Units are in 100 miles.

Zimmerman and Homer (1991) calculate the semivariance function by means of median polish, for the data set considered. The estimated parameter values of a spherical semivariance function were: sill = $1.875 = \hat{\theta}_2$, and range = 236.2 miles = $\hat{\theta}_3$. The nugget was estimated at $\hat{\theta}_1 = 0$, and is removed as a parameter in the estimated semivariance function.

$$\gamma_S(h) = \begin{cases} 1.875 \left\{ \frac{3}{2} \left(\frac{h}{2.362} \right) - \frac{1}{2} \left(\frac{h}{2.362} \right)^3 \right\}, & 0 \leq h \leq 2.362 \\ 1.875 & h > 2.362 \end{cases} \tag{3.1}$$

This estimated function is used as the basic semivariance function.

4　Results

This section of results is split up in three subsections. In the first place, some visualizations are presented. Secondly, results of the branch-and-bound algorithm are discussed. Finally, we will consider the robustness of optimal designs against misspecified parameter values and choice of the model.

4.1　Visualization of the problem

The problem of where to add one additional monitoring site to the 19 existing monitoring sites can be visualized by a contourplot of criterion values on a fine regular grid of points. The criterion value at a certain grid point is calculated by supposing that the additional monitoring site was located on that grid point. The contourplot shows where interesting locations are to add one additional monitoring site, given the existing monitoring network and a certain criterion. Figure 2 presents contourplots for estimation of the semivariance function written in Equation 3.1 in a D-optimal way, neglecting and including correlation between point pairs respectively.

Figure 2 shows, on comparing with Figure 1, that site 26 is the optimal choice to add, when correlation between point pairs is neglected. This corresponds with the result of Zimmerman and Homer (1991). Note that there are relatively few observations in the interval $h = (0, 2.362)$, which is the most interesting part of the semivariance function. Clusters of points result in more observations of the semivariance function at short distances. This tendency can be found in Figure 2. The optimal site to add, changes to site 25 for calculations including the correlation between point pairs, see Figure 2. The surface of the criterion value is less smooth than in the case of neglecting correlation and it tends less to clustering of monitoring sites.

4.2　Results of the branch-and-bound algorithm

Adding only one site to the existing network, can be solved by selection of the site which lowers the criterion value the most. The branch-and-bound

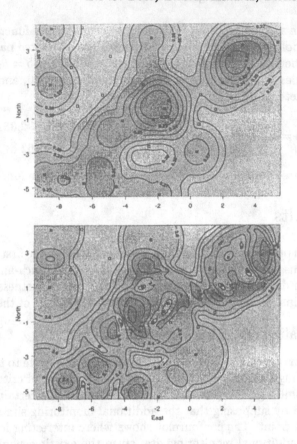

FIGURE 2. Contourplots of the criterion value (D-optimality) for estimation of the (basic) semivariance function (3.1) on a fine grid of points, supposing that an additional monitoring site was located on a grid point. Neglecting correlation between point pairs in (upper graph), including correlation in (lower graph).

algorithm becomes useful when more than one site has to be added to the existing network. To test the branch-and-bound algorithm, 25 sites will be selected from the combined set of 30 monitoring sites: $B_F \cup B_P$. The number of combinations for this problem is equal to 142506.

For both neglecting and including correlation between point pairs, sites with the numbers $\{7, 10, 11, 14, 23\}$ have to be removed out of $B_F \cup B_P$ to obtain an optimal monitoring network of 25 sites. Surprisingly, the computation time for including correlation is less than for neglecting correlation between point pairs, 1054 and 4934 seconds (Pentium II, 266 MHz) respectively. This is due to the fact that including correlation needs less calls (bounding more efficient) of the recursive branch-and-bound algorithm (2885 and 15710 calls). A simple drop algorithm, sequential removal of one site, finds the optimal solution too. However, these kinds of heuristic algorithms can never guarantee that the solution is optimal.

4.3 Robustness analysis of the optimal design

An optimal monitoring network depends on preliminary estimates of the parameter values of the semivariance function. If these preliminary estimates are too different from the true parameter values θ, the optimized monitoring network will not be optimal for the parameter values which have to be estimated (true parameter values). Robustness analysis is applied to see how the optimal designs will change at deviating values of the parameters. The optimal design of the preliminary estimated parameters can be compared with optimal designs which correspond to parameter values deviating from these values. The robustness analyses is applied to the situation of adding 5 from 11 potential sites to the existing monitoring network of 19 sites both including and excluding correlation.

Although the parameter values of the preliminary estimate (by Zimmerman and Homer, 1991) were changed considerably, the optimal combination of sites remains constant for many different parameter values. A doubling of the range (θ_3) results in another choice of stations for the optimized network for the case of including correlation between point pairs. However, if another combination of sites was preferred the criterion values did not differ much. Considering the sill as an overall sill ($\theta_1 + \theta_2 = 1.875$) with rising values for θ_1 had hardly any effect on the optimized networks.

So far, only parameter values of the spherical semivariance function were changed in our robustness analyses. The question arises how robust is the optimal design against another model choice? Therefore, we calculated the optimal designs also for exponential and Gaussian semivariance functions, where the nugget effect was left out of the models. It turned out that the optimal choice of 5 points was equal for all three semivariance functions.

5 Discussion and conclusions

Adding only one site to a monitoring network is easily solved by a simple algorithm. Visualizations of this problem show the difference between neglecting and including correlation between point pairs. Figure 2 shows that extreme clustering of sampling points occurs only when correlation between point pairs is neglected.

Applying a branch-and-bound algorithm (Rasch *et al*, 1997) works well for small cases, as considered in this paper. However, when the size of the combinatorial problem increases, it will soon be too large to solve within a reasonable computation time. Heuristic search algorithms are needed to solve larger problems. For this specific case study, we found that a simple drop algorithm delivers the optimal design for many cases. Only in a few cases was the drop algorithm trapped in a local minimum. For this case study, the branch-and-bound algorithm is efficient for calculating the optimal design including the correlation between point pairs.

Optimal experimental design for estimation of the semivariance function is based on preliminary estimates of parameters of the semivariance function and model choice. Therefore, a robustness study against misspecified parameter values and model choice is advisable. Although the parameter values and the shape of the semivariance function are changed considerably, the optimal monitoring designs and the criterion values do not differ much from each other. The dependence of optimal designs on θ seems not to be a major problem. In our opinion, the application of a more complex criterion, such as an averaged D-optimality, is not necessary.

References

Bogaert, P. and Russo, D. (1999). Optimal spatial sampling design for the estimation of the variogram based on a least squares approach. *Water Resources Research*, **35**, 1275–1289

Cressie, N. (1985). Fitting variogram models by weighted least squares. *J. of the Int. Association for Mathematical Geology*, **17**, 563–586

Cressie, N., Gotway, C.A. and Grondona, M.O. (1990). Spatial prediction from networks. *Chem. and Intelligent Laboratory Systems*, **7**, 251–271

Cressie, N. (1991). *Statistics for spatial data.* Wiley, New York

Fedorov, V.V. (1972). *Theory of optimal experiments.* Academic Press, New York

Isaaks, E.H. and Srivastava, R.M. (1989). *Applied geostatistics.* Oxford University Press, New York

Müller, W.G. (1998). *Collecting spatial data - Optimum design of experiments for random fields.* Physica-Verlag, Heidelberg

Müller, W.G. (1999). Least-squares fitting from the variogram cloud. *Statistics & Probability Letters*, **43**, 93–98

Müller, W.G. and Zimmerman, D.L. (1999). Optimal designs for variogram estimation. *Environmetrics*, **10**, 23-37

Rasch, D.A.M.K., Hendrix, E.M.T. and Boer E.P.J. (1997). Replication-free optimal design in regression analysis. *Comp. Stat.*, **12**, 19–52

Russo, D. (1984). Design of an optimal sampling network for estimating the variogram. *Soil Science Society of America Journal*, **48**, 708–716

Zimmerman, D.L. and Homer, K.E. (1991). A network design criterion for estimating selected attributes of the semivariogram. *Environmetrics*, **2**, 425–441

A-optimal Chemical Balance Weighing Designs with Diagonal Covariance Matrix of Errors

B. Ceranka
K. Katulska

ABSTRACT: The paper deals with the problem of estimating the weights of objects using an A-optimal chemical balance weighing design with the covariance matrix of errors $\sigma^2 \mathbf{G}$, where $\mathbf{G}$ is an $n \times n$ known positive definite diagonal matrix. A lower bound of $tr(\mathbf{X}'\mathbf{G}^{-1}\mathbf{X})^{-1}$ is obtained and a necessary and sufficient condition for this lower bound to be attained is given. The incidence matrices of balanced incomplete block designs have been used to construct A-optimal chemical balance weighing designs.

KEYWORDS: A-optimal chemical balance weighing design.

1 Introduction

The results of n weighing operations performed to determine the individual weights of p objects can be represented by the linear model

$$\mathbf{y} = \mathbf{Xw} + \mathbf{e}, \qquad (1.1)$$

where $\mathbf{y}$ is the $n \times 1$ observation vector, $\mathbf{X} = (x_{ij})$, $i = 1, \ldots, n$, $j = 1, \ldots, p$ is an $n \times p$ matrix with elements $x_{ij} = -1, 1, 0$ according as the j-th object is placed on the left pan, right pan or not included in the i-th weighing, $\mathbf{w}$ is the $p \times 1$ vector of the unknown weights of objects and $\mathbf{e}$ is an $n \times 1$ random vector of errors with $E(\mathbf{e}) = \mathbf{0}_n$ and $E(\mathbf{ee}') = \sigma^2 \mathbf{G}$, where $\mathbf{0}_n$ is the $n \times 1$ vector with zero elements everywhere, $\mathbf{G}$ is an $n \times n$ known positive definite diagonal matrix, $\mathbf{G} = diag(g_1, g_2, \ldots, g_n)$, "$E$" stands for the expectation and $\mathbf{e}'$ denotes the transpose of $\mathbf{e}$. The matrix $\mathbf{X}$ is the design matrix and we refer to it as to the matrix of a chemical balance weighing design with covariance matrix $\sigma^2 \mathbf{G}$.

mODa6, A.C.Atkinson, P.Hackl and W.G.Müller, eds., Physica, Heidelberg, 2001.

The normal equations estimating **w** are of the form

$$\mathbf{X'G^{-1}X\hat{w} = X'G^{-1}y,} \tag{1.2}$$

where (the solution) $\hat{\mathbf{w}}$ is the column vector of the least squares estimates of the true weights.

A chemical balance weighing design is said to be singular or nonsingular, depending on whether the matrix $\mathbf{X'G^{-1}X}$ is singular or nonsingular, respectively. From **G** being a positive definite diagonal matrix it follows that the matrix $\mathbf{X'G^{-1}X}$ is nonsingular if and only if **X** is of full column rank p, i.e. the matrix $\mathbf{X'X}$ is nonsingular.

Now, if $\mathbf{X'X}$ is nonsingular, the least squares estimates of **w** are given by

$$\hat{\mathbf{w}} = \mathbf{(X'G^{-1}X)^{-1}X'G^{-1}y} \tag{1.3}$$

and the covariance matrix of $\hat{\mathbf{w}}$ is

$$\mathrm{Var}(\hat{\mathbf{w}}) = \sigma^2\mathbf{(X'G^{-1}X)^{-1}}.$$

Various aspects of chemical balance weighing designs have been studied by Raghavarao (1975) and by Banerjee (1975) in the case $\mathbf{G = I_n}$, where $\mathbf{I_n}$ is the identity matrix of order n. Katulska (1989) showed that the minimum attainable variance for each of the estimated weights for a chemical balance weighing design with the covariance matrix $\sigma^2\mathbf{G}$, where **G** is a positive definite diagonal matrix, is $\sigma^2/tr(\mathbf{G^{-1}})$, i.e. $\mathrm{Var}(\hat{w}_j) \geq \sigma^2/tr(\mathbf{G^{-1}})$, $j = 1,...,p$. Ceranka and Katulska (1998) considered this same problem under equal correlations of errors, i.e. when $\mathbf{G} = g[(1-\rho)\mathbf{I_n} + \rho\mathbf{1_n1'_n}]$, where g and ρ are some known scalars, the latter denoting a correlation coefficient and $\mathbf{1_n}$ is the $n \times 1$ vector of ones. Wong and Masaro (1984) considered A-optimal chemical balance weighing designs in the case $\mathbf{G = I_n}$.

In this paper we consider the problem of choosing the matrix **X** under which the sum of variances of $\hat{w}_1 \ldots \hat{w}_p$, i.e. $\sigma^2 tr(\mathbf{X'G^{-1}X})^{-1}$ is minimum when **G** is a positive definite diagonal matrix. This design is called A-optimal.

2   A lower bound for $tr(\mathbf{X'G^{-1}X})^{-1}$

Let **X** be an $n \times p$ design matrix of a chemical balance weighing design and **G** be an $n \times n$ known positive definite diagonal matrix. The following result gives a lower bound for $tr(\mathbf{X'G^{-1}X})^{-1}$. With such a lower bound, one gets a feeling for comparing one $tr(\mathbf{X'G^{-1}X})^{-1}$ with another.

Lemma 2.1 *For a positive definite $n \times n$ diagonal matrix **G** and an $n \times p$ matrix **X** of rank p, we have the inequality*

$$tr(\mathbf{X'G^{-1}X})^{-1} \geq \frac{p^2}{tr(\mathbf{X'G^{-1}X})},$$

the equality being attained if and only if $X'G^{-1}X$ *is a scalar multiple of* I_p, *i.e.* $X'G^{-1}X = zI_p$.

Proof. With the trace inner product, we have, by the Cauchy-Schwarz inequality,

$$((X'G^{-1}X)^{-\frac{1}{2}}, (X'G^{-1}X)^{\frac{1}{2}}) \leq \|(X'G^{-1}X)^{-\frac{1}{2}}\| \|(X'G^{-1}X)^{\frac{1}{2}}\|,$$

i.e.

$$p \leq (tr(X'G^{-1}X)^{-1})^{\frac{1}{2}} (tr(X'G^{-1}X))^{\frac{1}{2}},$$

proving the desired result. For the case $G = I_n$ the above lemma was proved by Wong and Masaro (1984).

Theorem 2.1 *For any nonsingular chemical balance weighing design* X *with covariance matrix* $\sigma^2 G$,

$$tr(X'G^{-1}X)^{-1} \geq \frac{p}{trG^{-1}}. \tag{2.1}$$

Proof. Since the matrix X is of full column rank, then from Lemma 2.1 we have

$$tr(X'G^{-1}X)^{-1} \geq \frac{p^2}{tr(X'G^{-1}X)},$$

but

$$tr(X'G^{-1}X) = \sum_{i=1}^{n} g_i^{-1} \sum_{j=1}^{p} x_{ij}^2 \leq p \sum_{i=1}^{n} g_i^{-1} = p\ trG^{-1}.$$

Hence

$$tr(X'G^{-1}X)^{-1} \geq \frac{p}{trG^{-1}}.$$

This proves the theorem.

Definition 2.1 *A nonsingular chemical balance weighing design* X *with covariance matrix* $\sigma^2 G$, *is said to be A-optimal if* $tr(X'G^{-1}X)^{-1}$ *attains the lower bound given by Theorem 2.1, i.e., if*

$$tr(X'G^{-1}X)^{-1} = \frac{p}{tr\ G^{-1}}. \tag{2.2}$$

In other words, an A-optimal design is given by X *satisfying (2.2).*

Now, we investigate the necessary and sufficient conditions under which the matrix X is the matrix of an A-optimal chemical balance weighing design.

Theorem 2.2 *For a positive definite* $n \times n$ *diagonal matrix* G, *the* $n \times p$ *matrix* X *is the matrix of an A-optimal chemical balance weighing design if and only if*

$$X'G^{-1}X = tr(G^{-1})I_p. \tag{2.3}$$

Proof. To prove the necessity part, we observe that the equality in the Cauchy-Schwarz inequality holds if and only if $X'G^{-1}X = zI_p$ and the equality in (2.1) is satisfied if and only if $tr(X'G^{-1}X) = p\,tr(G^{-1})$. These conditions imply that $z = tr(G^{-1})$. Therefore we have (2.3). The proof for the sufficiency part is obvious. Hence the theorem is proved.

It is obvious that we have many interesting possibilities of patterns of G. The constructions of A-optimal chemical balance weighing designs with the covariance matrix $\sigma^2 G$ for each of the forms of G must be investigated separately.

In the present paper we will construct an A-optimal chemical balance weighing design with the covariance matrix $\sigma^2 G$, where G is of the form

$$G = \begin{bmatrix} aI_{n_1} & 0_{n_1}0'_{n_2} \\ 0_{n_2}0'_{n_1} & I_{n_2} \end{bmatrix}, a > 0, \ a \neq 1, n_1 + n_2 = n. \qquad (2.4)$$

This design may be useful in the following situation: there are two kinds of chemical balances, one is of the usual precision and the other is of high precision. If we can use the high precision balance many times, that is, if it is possible to assume that $n_1, n_2 \geq p$ and $n_1, n_2 \equiv 0 \pmod 4$ then there exist Hadamard matrices H_1 of order n_1 and H_2 of order n_2 and we have

Theorem 2.3 *Assume that $n_1, n_2 \geq p$ and $n_1, n_2 \equiv 0 \pmod 4$. Let matrices X_1 and X_2 consist of p columns of Hadamard matrices H_1 and H_2, respectively. Then*

$$X = \begin{bmatrix} X_1 \\ X_2 \end{bmatrix} \qquad (2.5)$$

is an A-optimal chemical balance weighing design with the covariance matrix $\sigma^2 G$, where G is given by (2.4).

Now suppose that the matrix X is likewise partitioned as the matrix G, i.e.

$$X = \begin{bmatrix} 1_{n_1} & X_1 \\ 1_{n_2} & X_2 \end{bmatrix}. \qquad (2.6)$$

Then, from Theorem 2.2, a chemical balance weighing design with X given by (2.6) and the covariance matrix $\sigma^2 G$, where G is given by (2.4), is A-optimal if and only if

(i) $X'_1 1_{n_1} = -aX'_2 1_{n_2}$

and   (2.7)

(ii) $X'_1 X_1 = (n_1 + an_2)I_{p-1} - aX'_2 X_2.$

In the next section we will construct the A-optimal chemical balance weighing design with the covariance matrix $\sigma^2 G$, where G is given by (2.4) under the condition $X_2 = 2N' - 1_b 1'_v$, where N is the incidence matrix of the balanced incomplete block design. If the balanced incomplete block design does not exist, we look for an A-optimal chemical balance weighing design in another way.

3 A-optimal weighing designs

Let N be the incidence matrix of a balanced incomplete block design with the parameters v, b, r, k, λ. The important property of this incidence matrix N is $NN' = (r - \lambda)I_v + \lambda 1_v 1'_v$ and the basic relations between parameters are $vr = bk$ and $\lambda(v - 1) = r(k - 1)$. Suppose further that the matrix X of a chemical balance weighing design is likewise partitioned as the matrix G given by (2.4), i.e.

$$X = \begin{bmatrix} 1_{n_1} & X_1 \\ 1_b & 2N' - 1_b 1'_v \end{bmatrix}. \tag{3.1}$$

In this design we have $p = v + 1$ and $n = n_1 + n_2 = n_1 + b$.

Then from conditions (2.7) we have

Corollary 3.1. A nonsingular chemical balance weighing design X given by (3.1) with the covariance matrix $\sigma^2 G$, where G is of the form (2.4) is A-optimal if and only if

(i) $X'_1 1_{n_1} = a(b - 2r)1_v$

and

(ii) $X'_1 X_1 = (n_1 + ab - 4a(r - \lambda))I_v + a(4r - 4\lambda - b)1_v 1'_v$.

$$\tag{3.2}$$

From (ii) of Corollary 3.1 it is obvious that all diagonal elements of the matrix $X'_1 X_1$ are equal to n_1. This implies that all elements of the matrix X_1 are different from 0. However, from (i) of 3.2 it follows immediately that

$$-n_1 \le a(b - 2r) \le n_1.$$

Now, let $a(b - 2r) = -n_1$, i.e.

$$X = \begin{bmatrix} 1_{n_1} & -1_{n_1} 1'_v \\ 1_b & 2N' - 1_b 1'_v \end{bmatrix}. \tag{3.3}$$

Lemma 3.1 *A design given by X of the form (3.3) is nonsingular.*

Theorem 3.1 *A chemical balance weighing design X given by (3.3) with the covariance matrix $\sigma^2 G$, where G is given by (2.4), is A-optimal if and only if*

$$ab + n_1 = 2ar. \tag{3.4}$$

and

$$r = 2\lambda. \tag{3.5}$$

Proof. For the design matrix $\mathbf{X}$ given by (3.3) and the matrix $\mathbf{G}$ given by (2.4), we have

$$\mathbf{X}'\mathbf{G}^{-1}\mathbf{X} = \begin{bmatrix} b + n_1/a & (2r - b - n_1/a)\mathbf{1}'_v \\ (2r - b - n_1/a)\mathbf{1}_v & 4(r - \lambda)\mathbf{I}_v + (b + 4\lambda - 4r + n_1/a)\mathbf{1}_v\mathbf{1}'_v \end{bmatrix}.$$
$$\tag{3.6}$$

From the conditions (2.3) and (3.6) it follows that a chemical balance weighing design is A-optimal if and only if $2r - b - n_1/a = 0$ and $b + 4\lambda - 4r + n_1/a = 0$. Hence the theorem.

Theorem 3.2 *If a balanced incomplete block design with the parameters* v, b, r, k, λ *exists and the matrix* $\mathbf{X}$ *given by (3.3) is the matrix of an A-optimal chemical balance weighing design with the covariance matrix* $\sigma^2\mathbf{G}$, *where* $\mathbf{G}$ *is given by (2.4) and* $a = 1/s$, *where* s *is a positive integer, then*

$$v = 2t - 1, \quad b = n_1s(2t - 1), \quad r = n_1st, \quad k = t, \quad \lambda = n_1st/2, \tag{3.7}$$

where n_1st *is even and* t *is a positive integer.*

Proof. From (3.4) and (3.5) we have $b = 2r - n_1/a$ and $r = 2\lambda$. From the relations between parameters of balanced incomplete block design, it can easily be verified that $v = 2t - 1$ and $r = n_1st$.
Now, consider the case when $a(b - 2r) = n_1$, i.e.

$$\mathbf{X} = \begin{bmatrix} \mathbf{1}_{n_1} & \mathbf{1}_{n_1}\mathbf{1}'_v \\ \mathbf{1}_b & 2\mathbf{N}' - \mathbf{1}_b\mathbf{1}'_v \end{bmatrix}. \tag{3.8}$$

Lemma 3.2 *The design* $\mathbf{X}$ *given by (3.8) is nonsingular.*

Theorem 3.3 *A chemical balance weighing design* $\mathbf{X}$ *given by (3.8) with the covariance matrix* $\sigma^2\mathbf{G}$, *where* $\mathbf{G}$ *is given by (2.4) is A-optimal if and only if*

$$ab - n_1 = 2ar \tag{3.9}$$

and

$$r = (2a\lambda + n_1)/a. \tag{3.10}$$

Proof. The above results can be proved along similar lines to those used in Theorem 3.1.

Theorem 3.4 *If a balanced incomplete block design with the parameters* v, b, r, k, λ *exists and the matrix* $\mathbf{X}$ *given by (3.8) is the matrix of an A-optimal chemical balance weighing design with the covariance matrix* $\sigma^2\mathbf{G}$, *where* $\mathbf{G}$ *is given by (2.4) and* $a = 1/s$, *where* s *is a positive integer, then*

$$v = 2t + 1, \quad b = n_1s(2t + 1), \quad r = n_1st, \quad k = t, \quad \lambda = n_1s(t - 1)/2, \tag{3.11}$$

where $n_1 s(t-1)$ is even and t is a positive integer.

Proof. The proof of this Theorem is similar to the proof of Theorem 3.2.

Now we are interested in finding an A-optimal chemical balance weighing design when $-n_1 < a(b-2r) < n_1$.

In this case for the matrix X of the form (3.3) or (3.8) the condition (i) of (3.2) is not satisfied. This implies that an A-optimal chemical balance weighing design with matrix X of the form (3.3) or (3.8) does not exist. Therefore we assume that

$$
X = \begin{bmatrix} 1_{b_0} & 1_{b_0} 1'_v \\ 1_{b_1} & 2N'_1 - 1_{b_1} 1'_v \\ 1_b & 2N' - 1_b 1'_v \end{bmatrix}, \tag{3.12}
$$

where $n_1 = b_0 + b_1$, $b_0 \geq 1$, and N_1 denotes the incidence matrix of a balanced incomplete block design with the parameters $v, b_1, r_1, k_1, \lambda_1$.

Lemma 3.3 *The design X given by (3.12) is nonsingular.*

Theorem 3.5 *A chemical balance weighing design X given by (3.12) with the covariance matrix $\sigma^2 G$, where G is given by (2.4) is A-optimal if and only if*

$$
2r_1 - b_1 = a(b-2r) - b_0 \tag{3.13}
$$

and

$$
4(r_1 - \lambda_1) - b_1 = ab - 4a(r - \lambda) + b_0. \tag{3.14}
$$

Theorem 3.6 *Suppose that a balanced incomplete block design with the parameters v, b, r, k, λ exists and $b_0 \neq a(b-2r)$. If the matrix X given by (3.12) is the matrix of an A-optimal chemical balance weighing design with the covariance matrix $\sigma^2 G$, where G is given by (2.4), then*

$$
\begin{aligned}
k_1 &= \{3v - 1 - 2(v-1)a(3r - 2\lambda - b)/[b_0 - a(b-2r)] \pm \sqrt{\Delta}\}/4, \\
b_1 &= v[b_0 - a(b-2r)]/(v - 2k_1), \\
r_1 &= k_1[b_0 - a(b-2r)]/(v - 2k_1), \\
\lambda_1 &= k_1(k_1 - 1)[b_0 - a(b-2r)]/[(v-1)(v - 2k_1)],
\end{aligned} \tag{3.15}
$$

where

$$
\Delta = (v-1)^2 \{2a(3r - 2\lambda - b)/[b_0 - a(b-2r)] - 1\}^2 + 4v.
$$

From Theorem 3.6 we have the following corollaries.

Corollary 3.2. Suppose that a balanced incomplete block design with the parameters $v = s^2, b, r, k, \lambda$ exists, $b = 4(r - \lambda) - b_0/a$ and $b_0 > a(b - 2r)$, where s denotes a positive integer. If the matrix X given by (3.12) is the matrix of an A-optimal chemical balance weighing design with the covariance matrix $\sigma^2 G$, where G is given by (2.4) then

$$k_1 = s(s-1)/2, \quad b_1 = s[b_0 - a(b - 2r)], \tag{3.16}$$
$$r_1 = (s-1)[b_0 - a(b - 2r)]/2, \quad \lambda_1 = (s-2)[b_0 - a(b - 2r)]/4.$$

Corollary 3.3. Suppose that a balanced incomplete block design with the parameters v, b, r, k_1, λ exists, $b = 3r - 2\lambda$ and $b_0 > a(b - 2r)$. If the matrix X given by (3.11) is the matrix of an A-optimal chemical balance weighing design with the covariance matrix $\sigma^2 G$, where G is given by (2.4), then

$$k_1 = (v-1)/2, \quad b_1 = v[b_0 - a(b - 2r)], \tag{3.17}$$
$$r_1 = (v-1)[b_0 - a(b - 2r)]/2, \quad \lambda_1 = (v-3)[b_0 - a(b - 2r)]/4.$$

Theorem 3.7 *Suppose that a balanced incomplete block design with the parameters v, b, r, k, λ exists and $b_0 = a(b - 2r)$. If the matrix X given by (3.12) is the matrix of an A-optimal chemical balance weighing design with the covariance matrix $\sigma^2 G$, where G is given by (2.4), then*

$$k_1 = v/2, \quad b_1 = 2a(v-1)(b - 3r + 2\lambda), \tag{3.18}$$
$$r_1 = a(v-1)(b - 3r + 2\lambda), \quad \lambda_1 = a(v-2)(b - 3r + 2\lambda)/2.$$

References

Banerjee, K.S. (1975). *Weighing Designs for Chemistry, Medicine, Economics, Operations Research, Statistics*. Marcel Dekker Inc., New York.

Ceranka, B. and Katulska, K. (1998). Optimum chemical balance weighing designs under equal correlations of errors. In Atkinson, A.C., Pronzato, L., and Wynn, H.P. (eds.) *MODA 5: Advances in Model-Oriented Data Analysis and Experimental Design*, Physica, Heidelberg, 3-9.

Katulska, K. (1989). Optimum chemical balance weighing designs with non-homogeneity of the variances of errors. *J. Japan Statist. Soc.*, **19**, 95-101.

Raghavarao, D. (1975). Weighing designs: A review article, *Gujarat Statistical Review*, **1**, 1-16.

Wong, C.S. and Masaro, J.C. (1984). A-optimal designs matrices $X = (x_{ij})_{N \times n}$ with $x_{ij} = -1, 0, 1$. *Linear and Multilinear Algebra*, **15**, 23-46.

Replications with Gröbner Bases

A.M. Cohen
A. Di Bucchianico
E. Riccomagno

ABSTRACT: We present an extension of the Gröbner basis method for experimental design introduced in Pistone and Wynn (1996) to designs with replicates. This extension is presented in an abstract regression analysis framework, based on direct computations with functions and inner products. Explicit examples are presented to illustrate our approach.

KEYWORDS: Gröbner basis; orthonormalisation; projection; replicates

1 Introduction

Recently tools from algebraic geometry have been introduced in experimental design. See Pistone and Wynn (1996), Pistone, Riccomagno and Wynn (2000) and Riccomagno (1997). They are particularly useful in the analysis of complex experiments where there is a large number of factors and runs and the structure of the design is not regular, for example there are missing observations from a standard full factorial experiment. Confounding relations among factors and interactions are encoded in the Gröbner bases associated with a design, allowing us to interpret confounding relations of the kind $I = AB$ (where A and B are factors and I is the constant term) for a large class of designs and models.

A major requirement for the application of this technique is that the design has no replicates. However, there are several practical situations where replicates are useful. In the present work we extend the algebraic methods to designs with replicates.

The main idea is to introduce a new variable that counts how many times a point appears in the design. For example, the one-dimensional design with five points $\mathcal{D}^* = \{0, 0, 1, 1, 2\}$ becomes the two-dimensional object $\mathcal{D} = \{(0, 1), (0, 2), (1, 1), (1, 2), (2, 1)\}$. This is encoded in the following set

mODa6, A.C.Atkinson, P.Hackl and W.G.Müller, eds., Physica, Heidelberg, 2001.

of polynomials in two indeterminates

$$x^3 - 3x^2 + 2x, \quad x^2 t - xt - x^2 + x, \quad t^2 - 3t + 2,$$

where x represents the design factor and t counts the number of replicates. The polynomials above can be used to construct a polynomial system of equations whose zeros are the points in $\mathcal{D}$. The zeros of the first polynomial, involving only the x indeterminates, are the distinct points in $\mathcal{D}^*$.

Least squares models are fitted to the data with replications as polynomial interpolators using the Gröbner basis method. In order to accommodate this process, we present a vector space setting for regression analysis in terms of functions on the design points. We perform estimation after orthonormalisation of the model terms (see also Giglio *et al.*, 2000). The traditional sums of squares appear naturally as the lengths of the terms in the orthonormalised model. The coefficients from the non-orthonormalised model are obtained simply by comparing coefficients. A pleasant feature from the computational point of view is that to compute regression coefficients, we do not need to perform matrix inversion as in the standard matrix way of computing regression coefficients. We present several explicit examples to illustrate our method.

2   Basic setup

We start by fixing notation. A design without replicates is a finite subset of $\mathbf{R}^d$. The main idea behind the algebraic geometry approach to experimental design is to view a design as a variety, i.e. the set of common zeroes of a finite set of polynomials. Statistical analysis of data starts with finding a polynomial that interpolates the data at the design points. If points of a design are replicated, then strictly speaking we are dealing with multi-sets rather than ordinary sets. This causes problems for the algebraic geometric approach. Namely there is no polynomial (function) that takes different values at the same point. We overcome this difficulty by introducing an extra variable that counts how many times a point appears in the design

Definition 2.1 *A design $\mathcal{D}$ with replicates is a finite set of points in $\mathbf{R}^d \times \mathcal{L}$, where $\mathcal{L}$ is a finite ordered set (the label set). The associated unreplicated design $\mathcal{D}^*$ is defined by $\mathcal{D}^* = \{a^* \in \mathbf{R}^d \mid \exists \ell \in \mathcal{L} \text{ such that} (a^*, \ell) \in \mathcal{D}\}$. Each element a of $\mathcal{D}$ is of the form $a = (a^*, \ell)$. Thus we may alternatively define $\mathcal{D}^*$ as $\mathcal{D}^* := \{a^* \mid a \in \mathcal{D}\}$.*

Designs without replicates are special designs such that for each $a^* \in \mathcal{D}^*$ there is exactly one $\ell \in \mathcal{L}$ such that $(a^*, \ell) \in \mathcal{D}$. Two designs are isomorphic if their associated unreplicated designs coincide and there is a bijection between the two designs. In general, the unreplicated design $\mathcal{D}^*$ is obtained by projecting $\mathcal{D}$ onto the first d factors. The operation of projection does not take into account the number of replicates. It has a nice algebraic counterpart (see Theorem 4.2 below).

Notation 2.1 *Let $\mathcal{D} \subset \mathbf{R}^d \times \mathcal{L}$ be a design. The set of real-valued functions on $\mathcal{D}$ is denoted by $\mathcal{L}(\mathcal{D})$.*

The inner product on $\mathcal{L}(\mathcal{D})$ given in Definition 2.2 below is directly related to least squares estimation.

Definition 2.2 *If $\mathcal{D}$ is a design, then for all $f, g \in \mathcal{L}(\mathcal{D})$ we define an inner product by $\langle f, g \rangle_{\mathcal{D}} := \sum_{a \in \mathcal{D}} f(a)g(a)$. A norm is defined on $\mathcal{L}(\mathcal{D})$ by $\|f\|_{\mathcal{D}} = \sqrt{\langle f, f \rangle_{\mathcal{D}}}$.*

Note that weighted least squares is easily incorporated in this setup by slightly changing the definition of the inner product $\langle ., . \rangle_{\mathcal{D}}$.

Let $\mathcal{D}$ be a design. Suppose our statistical model is

$$Y(x) = f(x, \theta) + \varepsilon(x), \tag{2.1}$$

where $\theta \in \mathbf{R}^p$ and $\varepsilon(x)$ is a real-valued random variable for all $x \in \mathcal{D}^*$ with $\mathbf{E}\varepsilon(x) = 0$ and $\mathbf{V}\varepsilon(x) = \sigma^2$. Suppose we have observations $Y_1, \ldots, Y_N$ from this model, where Y_i is $Y(a_i^*)$ for $a_i^* \in \mathcal{D}^*$ and replications are allowed, i.e. a_i^* may be equal to a_j^* for $i \neq j$. Then the **least squares estimator** for the parameter vector θ is given by

$$\hat{\theta} = \min_{\theta \in \Theta} \sum_{i=1}^{N} |Y_i - f(a_i^*, \theta)|^2. \tag{2.2}$$

Let $g \in \mathcal{L}(\mathcal{D})$ be the unique function in $\mathcal{L}(\mathcal{D})$ such that $g(a_i) = Y_i$ for all $i = 1, \ldots, N$. Since

$$\hat{\theta} = \min_{\theta \in \Theta} \sum_{i=1}^{N} |Y_i - f(a_i^*, \theta)|^2 = \min_{\theta \in \Theta} \|g - f(., \theta)\|_{\mathcal{D}}^2, \tag{2.3}$$

we see that least squares estimation corresponds to a minimum distance problem in $\mathcal{L}(\mathcal{D})$ with the inner product in Definition (2.2). Note that a function $f(x)$ for $x = (x_1, \ldots, x_d) \in \mathcal{D}^*$ can be naturally extended for $x = (x_1, \ldots, x_d, x_{d+1}) \in \mathcal{D}$ by $(x_1, \ldots, x_d, x_{d+1}) \longmapsto f(x_1, \ldots, x_d)$.

3 Identifiability of linear models

In the sequel we restrict ourselves to linear models, i.e. models such that $f(x, \theta)$ is a linear function of the components of the parameter vector θ

$$Y(x) = \sum_{\alpha \in \mathcal{M}} \theta_\alpha p_\alpha(x) + \varepsilon(x), \tag{3.1}$$

where p_α ($\alpha \in \mathcal{M}$) is an element of $\mathcal{L}(\mathcal{D}^*)$. Clearly the p_α's can be viewed as elements of $\mathcal{L}(\mathcal{D})$. Recall that $\mathcal{L}(\mathcal{D})$ is a vector space over the real numbers and $\mathcal{L}(\mathcal{D})$ is isomorphic to $\mathbf{R}^N$, where N is the number of points in $\mathcal{D}$.

Definition 3.1 *A linear model (3.1) is identifiable by a design $\mathcal{D}$ if the functions p_α ($\alpha \in \mathcal{M}$) are linearly independent elements of $\mathcal{L}(\mathcal{D})$.*

The classical notion of identifiability is equivalent to our definition. Indeed, let $Y = X\theta + \varepsilon$ be a linear model where X is a matrix with p columns. If the design matrix X has rank less than p, then θ is not identifiable since different values of θ yield the same value of $X\theta$. This actually means that the model coincides for different parameter values when restricted to the design points. In other words, the functions on $\mathcal{D}$ that take as values the components of the columns of X are linearly dependent.

For linear models, least squares estimation is the orthogonal projection of g onto span $\{p_\alpha \mid \alpha \in \mathcal{M}\}$. Note that if $\{p_\alpha \mid \alpha \in \mathcal{M}\}$ is an orthogonal subset of $\mathcal{L}(\mathcal{D})$, then elementary linear algebra arguments yield that

$$\widehat{\theta}_\alpha = \frac{\langle g, p_\alpha \rangle_\mathcal{D}}{\langle p_\alpha, p_\alpha \rangle_\mathcal{D}}. \tag{3.2}$$

The functional description of least squares estimation has some advantages over the usual vector space description. It is more natural in our opinion since the model description is also at a functional level. A numerical advantage is that we do not need matrix inversion to compute the coefficient estimates. Indeed, orthogonalisation by the Gram-Schmidt procedure becomes a simple recursive procedure. Note that contrary to the classical use of Gram-Schmidt in the case of $\mathbf{R}^N$, we use Gram-Schmidt in a symbolic way in the space of polynomials. In this polynomial setting rewriting the estimated orthogonalised model in terms of the original model corresponds simply to collecting coefficients.

The functional description given here differs from the abstract setting to linear models initiated by Kruskal (1961). See Drygas (1970) for a self-contained treatment. Specifically, we extensively use computations with polynomials in the next sections. A paper which is closer in spirit to our paper is Neumaier and Seidel (1992), where a design is seen as a normalized measure and optimal designs are derived using arguments in $\mathcal{L}(\mathcal{D})$.

4   A polynomial algebraic representation of $\mathcal{L}(D)$

The set of real-valued functions over a finite set of distinct points can be described using particular classes of polynomials. More precisely let $\mathcal{D}$ be a design in $\mathbf{R}^d \times \mathcal{L}$, let $\mathbf{R}[x_1, \ldots, x_{d+1}]$ be the polynomial ring in $d + 1$ indeterminates with real coefficients and let $\mathrm{Ideal}(\mathcal{D}) \subset \mathbf{R}[x_1, \ldots, x_{d+1}]$ be the set of all polynomials whose zeros include the design points. Then the quotient space $\mathbf{R}[x_1, \ldots, x_{d+1}]/\text{Ideal}(\mathcal{D})$ is a description or representation of $\mathcal{L}(D)$. Moreover vector space bases of $\mathbf{R}[x_1, \ldots, x_{d+1}]/\text{Ideal}(\mathcal{D})$ made of monomials can be determined with Gröbner basis methods. We require the definition of a term ordering.

Definition 4.1 *A term ordering τ on the monomials of $\mathbf{R}[x_1, \ldots, x_d]$ is a total well-ordering such that $x^\alpha \prec_\tau x^\beta$ implies $x^\alpha x^\gamma \prec_\tau x^\beta x^\gamma$ for all $\gamma \neq 0$.*

Theorem 4.1 *Given a design $\mathcal{D} \subset \mathbf{R}^d \times \mathcal{L}$, a term ordering τ and a Gröbner basis $G \subset \mathbf{R}[x_1, \ldots, x_{d+1}]$ for $\mathcal{D}$ with respect to τ, then a vector space basis of $\mathbf{R}[x_1, \ldots, x_{d+1}]/\text{Ideal}(\mathcal{D})$ is given by*

$$Est_{\mathcal{D},\tau} := \{x^\alpha \mid x^\alpha \text{ is not divisible}$$
$$\text{by any of the leading terms of the elements of } G\}$$
$$= \{x^\alpha \mid \alpha \in L_{\mathcal{D},\tau}\}.$$

Moreover, if the set $\{p_\alpha \mid \alpha \in \mathcal{M}\}$ in Model (3.1) is a subset of $Est_{\mathcal{D},\tau}$, then Model (3.1) is identifiable. The set $Est_{\mathcal{D},\tau}$ has exactly N elements where N is the cardinality of $\mathcal{D}$.

Proof. For the first part see for example Cox *et al.* (1996) and for the second and third parts see Pistone, Riccomagno and Wynn (2000). ∎

Note that Theorem 4.1 applies to any design with no replicates, namely to a set of distinct points. Designs defined according to Definition 2.1 are particular examples of sets of distinct points where the "label indeterminate", x_{d+1} distinguishes replicated points. For designs with replicates the trick here is to consider in Model (3.1) only terms of *Est* not involving x_{d+1}.

For statistical inference we need a design, a model, and observations. In a classical screening setup a model is chosen first. However, we may also choose the model after seeing the design (for example the planned design was not completed and there are missing points, see Holliday *et al.*, 1999). In this case, Theorem 4.1 provides a powerful tool in the choice of a regression vector for a linear model of the type in (3.1).

In general different term orderings give different *Est* sets and also typically *Est* includes monomials involving the label indeterminate x_{d+1} which clearly should not be included in Model (3.1). This suggests partitioning *Est*, or equivalently L, in three disjoint parts

$$L_{\mathcal{D},\tau} = L_x^* \cup L_{x_{d+1}} \cup L_{x,x_{d+1}}$$

where L_x^* includes all the elements of $L_{\mathcal{D},\tau}$ that do not involve x_{d+1}, $L_{x_{d+1}}$ includes all the monomials in $Est_{\mathcal{D},\tau}$ which involve only x_{d+1} and $L_{x,x_{d+1}}$ includes the remaining terms. The set $\mathcal{M}$ in Model (3.1) can be chosen to be a subset of L_x^*. The combination of the choice of the term ordering and of the structure of the design determines these three parts.

A reasonable choice for the term ordering is one that eliminates the x_{d+1} variable. For elimination term ordering we refer to Cox *et al.* (1996) and here simply observe that an effect of eliminating x_{d+1} is that the number of monomials in $L_{x_{d+1}}$ is as small as possible.

We conclude this section by showing with polynomial algebra techniques that identifiability is not affected by replications. The elimination of x_{d+1} from Ideal($\mathcal{D}$) corresponds to projecting $\mathcal{D} \subset \mathbf{R}^d \times \mathcal{L}$ onto $\mathbf{R}^d$. For some term orderings the Gröbner basis of Ideal($\mathcal{D}^*$) and $Est_{\mathcal{D}^*}$ can be easily deduced from the Gröbner basis of Ideal($\mathcal{D}$) and $Est_{\mathcal{D}}$.

Theorem 4.2 *1) Ideal($\mathcal{D}$) $\cap$ $\mathbf{R}[x_1, \ldots, x_d]$ = Ideal($\mathcal{D}^*$). 2) Let G be the Gröbner basis of Ideal($\mathcal{D}$) with respect to a term ordering eliminating x_{d+1}. The Gröbner basis of Ideal($\mathcal{D}^*$) is $G \cap \mathbf{R}[x_1, \ldots, x_d]$.*

Proof. 1) Assume $f \in$ Ideal($\mathcal{D}$) $\cap$ $\mathbf{R}[x_1, \ldots, x_d]$. Then for all $a = (a^*, \ell) \in \mathcal{D}$, we have that $0 = f(a) = f(a^*, \ell) = f(a^*)$ as $f \in \mathbf{R}[x_1, \ldots, x_d]$. This implies $f \in$ Ideal($\mathcal{D}^*$). The converse is obvious. 2) See Cox *et al.* (1996). ∎

Clearly Theorem 4.2 applies when instead of x_{d+1} we need to eliminate some other variable. The projection is now on the remaining variables and replications may appear. For example the projection of the 2^2 design at levels ± 1 over the first factor gives ± 1 replicated twice.

5 Examples

The analysis of observations suggested in the paper proceeds as follows. Given a design $\mathcal{D}$, compute $Est_{\mathcal{D}, \tau}$ where τ is a term ordering that eliminates the extra variable t. Orthonormalise the terms of $Est_{\mathcal{D}, \tau}$ that do not involve t. Collect coefficients to determine the parameters of the wanted model from the estimated coefficients in the orthonormalised model.

Example 1 (2^2 full factorial design with centre points)
Consider the 2^2 design at levels ± 1. The standard model associated with it is

$$Y = \theta_0 + \theta_1 x_1 + \theta_2 x_2 + \theta_{12} x_1 x_2.$$

Clearly 1, x_1^2 and x_2^2 are confounded on $\mathcal{D}$. Suppose that we want to test linearity by adding quadratic terms to the model. The simplest way to extend this design such that quadratic terms become identifiable, is to add centre points. We add four observations at $(0, 0)$ and the design becomes

$$\mathcal{D} = \{(-1, -1, 1), (-1, 1, 1), (1, -1, 1), (1, 1, 1),$$
$$(0, 0, 1), (0, 0, 2), (0, 0, 3), (0, 0, 4)\}$$
$$= \{a_i \mid i = 1, \ldots, 8\}.$$

We use a term ordering σ that eliminates the variable t and is a degree reverse lexicographic ordering on x_1 and x_2 (Cox *et al.*, 1996). We obtain

$$Est_D = \{1, x_1, x_2, x_1x_2, x_2^2, t, t^2, t^3\}.$$

An orthonormal basis for the linear span of the terms not involving t is

$$\left\{\frac{1}{\sqrt{8}}, \frac{x_1}{2}, \frac{x_2}{2}, \frac{x_1x_2}{2}, \frac{x_2^2 - \frac{1}{2}}{\sqrt{2}}\right\}.$$

Let g be the polynomial that interpolates the observations $Y_1, \ldots, Y_8$ at the design points, $g(a_i) = Y_i$, $i = 1, \ldots, 8$. The traditional sums of squares correspond to the squares of the inner products. In particular, using Equation (3.2) the average over the centre points minus the average over the full factorial is computed as

$$SS_{\text{pure quadratic}} = \frac{1}{8}(Y_1 + \ldots + Y_4 - (Y_5 + \ldots + Y_8))^2 = \left\langle g, \frac{x_2^2 - \frac{1}{2}}{\sqrt{2}} \right\rangle_D^2.$$

By simple comparison of terms, we read off the coefficients of the required model from the coefficients of the orthonormalised model.

Example 2 (Star composite design with centre points)
If the analysis of the 2^2 full factorial design with centre points indicates that there is curvature, then it is practice to study the quadratic terms. We choose to break the aliasing by augmenting the design with four axial points at $(0, \pm 2)$ and $(\pm 2, 0)$. The new design $\mathcal{D}$ is given below

$$\mathcal{D} = \{(-1, -1, 1), (-1, 1, 1), (1, -1, 1), (1, 1, 1), (-2, 0, 1), (2, 0, 1),$$
$$(0, 2, 1), (0, -2, 1), (0, 0, 1), (0, 0, 2), (0, 0, 3), (0, 0, 4)\}.$$

We use again the term ordering σ used in Example 1. The monomials in Est_D not involving the counting variable are

$$\{1, x_1, x_2, x_1^2, x_2^2, x_1x_2, x_1^3, x_1x_2^2, x_2^4\}.$$

An orthonormal basis for the linear span of these terms by applying the Gram-Schmidt procedure to Est_D in the order above, yields

$$\left\{\frac{\sqrt{3}}{6}, \frac{\sqrt{3}}{6}x_1, \frac{\sqrt{3}}{6}x_2, \frac{\sqrt{6}}{12}(x_1^2 - 1), \frac{\sqrt{3}}{24}(x_1^2 + 3x_2^2 - 4),\right.$$
$$\left.\frac{x_1x_2}{2}, \frac{\sqrt{6}}{12}x_1(3x_2^2 - 1), \frac{\sqrt{6}}{12}x_2(x_2^2 - 3)\right\}.$$

Example 3 (2^{3-1} fractional factorial design with centre points)
Consider the standard 2^{3-1} design with generator $I = ABC$ and four ad-

ditional centre points. The design is

$$\mathcal{D} = \{(1, -1, -1, 1), (-1, 1, -1, 1), (-1, -1, 1, 1), (1, 1, 1, 1),$$
$$(0, 0, 0, 1), (0, 0, 0, 2), (0, 0, 0, 3), (0, 0, 0, 4)\}.$$

Again using the term ordering σ, we obtain $Est_{\mathcal{D}} = \{1, x_1, x_2, x_3, x_3^2, t, t^2, t^3\}$. An orthonormal basis for the linear span of the terms not involving t is

$$\left\{ \frac{1}{\sqrt{8}}, \frac{x_1}{2}, \frac{x_2}{2}, \frac{x_3}{2}, \frac{x_3^2 - \frac{1}{2}}{\sqrt{2}} \right\}.$$

References

Caboara, M. and Robbiano, L. (1997). Families of ideals in statistics. In: Küchlin, W. (ed), *ISSAC'97, Proceedings of the International Symposium on Symbolic and Algebraic Computation, Hawaii*, ACM Press, New York, 404-417.

Cox, D., Little, J. and O'Shea, D. (1996). *Ideals, Varieties, and Algorithms*, 2nd ed. Springer, New York.

Drygas, H. (1970). *The Coordinate-Free approach to Gauss-Markov Estimation*. Springer-Verlag, Berlin.

Giglio, B., Riccomagno, E. and Wynn, H.P. (2000). Gröbner bases in regression. *Journal of Applied Statistics*, **27**, 923-928.

Holliday, T., Pistone, G., Riccomagno, E. and Wynn, H.P. (1999). The application of computational algebraic geometry to the analysis of designed experiments: a case study. *Computational Statistics*, **14**, 213-231.

Kruskal, W. (1961). The coordinate-free approach to Gauss-Markov estimation, and its application to missing and extra observations. In: Neyman, J. (ed), *Proceedings of the 4th Berkeley Symposium on Mathematical Statistics and Probability, Vol. I*, University of California Press, Berkeley, 435-451.

Neumaier, A. and Seidel, J.J. (1992). Measures of strength $2e$ and optimal designs of degree e. *Sankhyā*, **54**, 299-309.

Pistone, G., Riccomagno, E. and Wynn, H.P. (2000). *Algebraic Statistics: Computational Commutative Algebra in Statistics*. Chapman & Hall / CRC Press, London.

Pistone, G. and Wynn, H.P. (1996). Generalised confounding with Gröbner bases. *Biometrika*, **83**, 653-666.

Riccomagno, E. (1997). Algebraic identifiability in experimental design and related topics. Ph.D. thesis, University of Warwick.

Extracting Information from the Variance Function: Optimal Design

D. Downing
V. V. Fedorov
S. Leonov

ABSTRACT: Regression models with the variance function depending on unknown parameters appear in a number of practical problems (variogram fitting and mixed effect models are popular examples). We found that estimation of parameters entering both response and variance functions can be combined in a rather simple way. The proposed estimator belongs to the class of iterated estimators and is numerically very close to the fixed point method, which takes the form of the reweighted least squares method in our setting. The proposed estimators lead to design problems with additive information matrices and therefore can be treated within traditional convex design theory.

KEYWORDS: iterated estimator; locally optimal designs; nonlinear regression; unknown parameters in variance

1   Model, MLE and iterated estimators

Information matrix. Let the observed y have a normal distribution and

$$E[y|x] = \eta(x, \theta), \quad \text{Var}[y|x] = s(x, \theta). \tag{1.1}$$

The score function [cf. Rao (1973), Ch. 6e] of a single normally distributed observation is

$$R(y|x, \theta) = -\frac{1}{2}\frac{\partial}{\partial \theta}\left[\log s(x, \theta) + \frac{\{\eta(x, \theta) - y\}^2}{s(x, \theta)}\right],$$

and the corresponding Fisher information matrix is

$$\mu(x, \theta) = \text{Var}[R(y|x, \theta)]$$

$$= s^{-1}(x, \theta)\frac{\partial \eta(x, \theta)}{\partial \theta}\frac{\partial \eta(x, \theta)}{\partial \theta^T} + \frac{1}{2}s^{-2}(x, \theta)\frac{\partial s(x, \theta)}{\partial \theta}\frac{\partial s(x, \theta)}{\partial \theta^T}. \tag{1.2}$$

mODa6, A.C.Atkinson, P.Hackl and W.G.Müller, eds., Physica, Heidelberg, 2001.

Note that only the second term on the right-hand side of (1.2) distinguishes this case from the traditional regression problem.

If all observations $\{y_i, x_i\}_1^N$ are independent, then the Fisher information matrix for that sample equals

$$M(x_1, \ldots, x_N, \theta) = \sum_{i=1}^{N} \mu(x_i, \theta). \tag{1.3}$$

Estimator consistency. As in Jennrich (1969) and Heyde (1997, Ch.12), one can verify that the maximum likelihood estimator

$$\hat{\theta}_N = \arg\min_{\theta \in \Omega} \sum_{i=1}^{N} \left[\log s(x_i, \theta) + \frac{\{\eta(x_i, \theta) - y_i\}^2}{s(x_i, \theta)} \right], \tag{1.4}$$

is strongly consistent if:

Ω is compact; $x_i \in X$ where X is compact; the functions $\eta(x, \theta)$ and $s(x, \theta) \geq \gamma > 0$ are continuous with respect to θ uniformly on X; the limit

$$\lim_{N \to \infty} N^{-1} \nu_N(\theta, \theta^*) = \nu(\theta, \theta^*) \tag{1.5}$$

exists for all $\theta \in \Omega$,

$$\nu_N(\theta, \theta^*) = \sum_{i=1}^{N} \left[\log s(x_i, \theta) + \frac{\{\eta(x_i, \theta) - \eta(x_i, \theta^*)\}^2 + s(x, \theta^*)}{s(x_i, \theta)} \right],$$

where $\theta^* \in \Omega$ is the vector of true values of the unknown parameters, and the function $\nu(\theta, \theta^*)$ has its unique minimum at $\theta = \theta^*$.

If additionally to the above assumptions θ^* is an internal point of Ω, the functions $\eta(x, \theta)$ and $s(x, \theta)$ are twice differentiable with respect to θ for all $\theta \in \Omega$, and the limit matrix

$$\lim_{N \to \infty} N^{-1} M(x_1, \ldots, x_N, \theta^*) = \bar{M}(\theta^*) \tag{1.6}$$

exists and is regular, then $(\hat{\theta}_N - \theta^*)/\sqrt{N}$ is asymptotically normally distributed with zero mean and dispersion matrix $\bar{M}^{-1}(\theta^*)$. From the last statement it follows that the matrix

$$D(x_1, \ldots, x_N, \hat{\theta}_N) = M^{-1}(x_1, \ldots, x_N, \hat{\theta}_N)$$

may be used as an approximation of the dispersion matrix of θ^*. Note that the proper selection of the series $\{x_i\}_1^N$ is crucial for the consistency and precision of $\hat{\theta}_N$; see also (1.5) and (1.6).

Iterated estimator and the least squares method. We conclude this section with a brief description of an iterative estimation procedure. This procedure may be considered as one of the numerical approaches, and in addition also leads to a better understanding of the statistical nature of (1.4) and various generalizations of the reported results.

Let us assume that

$$\lim_{t\to\infty} \theta_t = \theta_N \tag{1.7}$$

where

$$\theta_t = \arg\min_{\theta\in\Omega} v_N^2(\theta,\theta_{t-1}),$$

$$v_N^2(\theta,\theta') = \sum_{i=1}^{N} \frac{\{\eta(x_i,\theta) - y_i\}^2}{s(x_i,\theta')} + \sum_{i=1}^{N} \frac{\left[s(x_i,\theta) - \{\eta(x_i,\theta') - y_i\}^2\right]^2}{2\,s^2(x_i,\theta')}.$$

If limit (1.7) exists and the above assumptions hold, then

$$\lim_{t\to\infty} \frac{\partial v_N^2(\theta,\theta_t)}{\partial\theta}\bigg|_{\theta=\theta_t} = \frac{\partial v_N^2(\theta,\theta_N)}{\partial\theta}\bigg|_{\theta=\theta_N} = 0. \tag{1.8}$$

At the same time

$$\frac{\partial v_N^2(\theta,\theta')}{\partial\theta}\bigg|_{\theta=\theta'} = \frac{\partial}{\partial\theta} \sum_{i=1}^{N} \left[\log s(x_i,\theta) + \frac{\{\eta(x_i,\theta) - y_i\}^2}{s(x_i,\theta)}\right]. \tag{1.9}$$

Combining (1.4), (1.8) and (1.9), and assuming that the likelihood function has a unique stationary point one may verify that $\theta_N = \hat\theta_N$. Therefore the iterative procedure (1.7) can be used to compute the maximum likelihood estimator if one assures its convergence.

The estimator (1.7) is similar to the iterated estimator proposed by Davidian and Carroll (1987). However they partition the vector θ in two subvectors, with the second one appearing only in the variance $s(x,\theta)$, and perform iterations on each term in $v_N^2(\theta,\theta')$ separately. Moreover, for cases where parameters in $\eta(x,\theta)$ and $s(x,\theta)$ coincide, such a two-stage procedure does not lead to the maximum likelihood estimator.

Introducing $A_N(\theta') = \arg\min_{\theta\in\Omega} v_N^2(\theta,\theta')$ we may present (1.7) as the recursion $\theta_t = A_N(\theta_{t-1})$ for the fixed point point problem $\theta = A_N(\theta)$. Convergence of the recursion is guaranteed [cf. Saaty and Bram (1964), Ch.1.10], if for any $\theta_1,\theta_2 \in \Omega$ there exists a constant K, $0 < K < 1$, such that

$$[A_N(\theta_1) - A_N(\theta_2)]^T [A_N(\theta_1) - A_N(\theta_2)] \le K(\theta_1 - \theta_2)^T(\theta_1 - \theta_2).$$

Observing that $A_N(\theta)$ is nothing but a generalized version of the least squares estimator with non-optimal but predetermined weights and following Jennrich (1969), one can show that $A_N(\theta)$ is strongly consistent, i.e. converges almost surely to θ^* [cf. Rao (1973), Ch.2c]. Using this fact we can show that the probability

$$P[\{A_N(\theta_1) - A_N(\theta_2)\}^T \{A_N(\theta_1) - A_N(\theta_2)\} \le K(\theta_1 - \theta_2)^T(\theta_1 - \theta_2)]$$

tends to 1 as $N \to \infty$. Thus, for large N one may expect that with probability close to 1 the limit (1.7) exists and, consequently, $\theta_N = \hat{\theta}_N$.

If the normality assumption is replaced by the assumption that y has finite moments up to the fourth, then (1.7) can be generalized to comprise the latter case as well. To accomplish that, one has to replace the weights

$$w_2^{-1}(x, \theta') = 2s^2(x, \theta')$$

in the second term of $v_N^2(\theta, \theta')$, see comments to (1.7), by $w_2^{-1}(x, \theta') = \text{Var}[s(x, \theta') - \{\eta(x, \theta') - y\}^2] = \text{E}[\{y - \eta(x, \theta')\}^4 - s^2(x, \theta')]$; the notation $w_1(x, \theta')$ is reserved for the first term,

$$w_1^{-1}(x, \theta') = s(x, \theta').$$

The statistic (1.7) is usually referred to as an "iterated estimator". If the assumption about continuity and differentiability of the function $w_2(x, \theta')$ is added to the assumptions which were used to derive the asymptotic normality of the maximum likelihood estimator, then $(\theta_N - \theta^*)/\sqrt{N}$ is asymptotically normally distributed with mean zero and dispersion matrix

$$D(\theta^*) = \bar{M}^{-1}(\theta^*),$$

where

$$\bar{M}(\theta^*) = \lim_{N \to \infty} N^{-1} \sum_{i=1}^{N} \left[w_1(x_i, \theta) \frac{\partial \eta(x_i, \theta)}{\partial \theta} \frac{\partial \eta(x_i, \theta)}{\partial \theta^T} \right.$$

$$\left. + w_2(x_i, \theta) \frac{\partial s(x_i, \theta)}{\partial \theta} \frac{\partial s(x_i, \theta)}{\partial \theta^T} \right]_{\theta = \theta^*}. \tag{1.10}$$

2 Optimal design

Equivalence theorem. Let

$$\xi_N = \{p_i, x_i\}_1^n, \quad p_i = r_i/N, \quad \sum_{i=1}^{n} r_i = N$$

be a discrete design. In general $r_i > 1$, i.e. we admit repeated observations. Let $\xi(dx)$ be a probability measure on X. Then we can consider   the following optimization problem [cf. Fedorov and Hackl (1997), Ch.2]:

$$\xi^*(\theta) = \arg \min_{\xi \in \Xi} \Psi[M(\xi, \theta)], \tag{2.1}$$

$$M(\xi, \theta^*) = \int_X \xi(dx) \left[w_1(x, \theta) \frac{\partial \eta(x, \theta)}{\partial \theta} \frac{\partial \eta(x, \theta)}{\partial \theta^T} \right.$$

$$+ w_2(x_i, \theta) \frac{\partial s(x_i, \theta)}{\partial \theta} \frac{\partial s(x_i, \theta)}{\partial \theta^T}\Bigg]_{\theta=\theta^*}, \tag{2.2}$$

where Ξ is the set of all probability measures on X, and Ψ is a selected optimality criterion. Design $\xi^*(\theta)$ is usually referred to as a locally optimal design.

From the mathematical point of view, the optimization problem (2.1) belongs to the well studied convex design theory. For instance, for the D–criterion similar to Kiefer's equivalence theorem one can verify the following theorem.

Theorem 2.1 *The following problems are equivalent:*

1. $\max_{\xi \in \Xi} |M(\xi, \theta)|$,

2. $\min_{\xi \in \Xi} \max_{x \in X} \psi(x, \xi, \theta)$,

3. $\max_{x \in X} \psi(x, \xi, \theta) = m$.

In the statement of the theorem, $m = \dim(\theta)$ is the dimension of θ,

$$\psi(x, \xi, \theta) = w_1(x, \theta) d_1(x, \xi, \theta) + w_2(x, \theta) d_2(x, \xi, \theta), \tag{2.3}$$

$$d_1(x, \xi, \theta) = \frac{\partial \eta(x, \theta)}{\partial \theta^T} M^{-1}(\xi, \theta) \frac{\partial \eta(x, \theta)}{\partial \theta}, \tag{2.4}$$

$$d_2(x, \xi, \theta) = \frac{\partial s(x, \theta)}{\partial \theta^T} M^{-1}(\xi, \theta) \frac{\partial s(x, \theta)}{\partial \theta}.$$

Some special cases of the problem (2.1), (2.2) were considered by Atkinson and Cook (1995).

Numerical procedure. Using the above result one can utilize the whole array of methods of optimal design construction [cf. Fedorov and Hackl (1997, Ch. 3), and Gaffke and Mathar (1992)]. In the examples, which are presented in the following section, we have used the first order exchange algorithm:

1. Given ξ_s, find $x_s^+ = \arg\max_{x \in X} \psi(x, \xi_s, \theta)$ and define $\xi_s^+ = \xi_s + \alpha_s \xi(x_s^+)$, where $\xi(x)$ is the design atomized at a single point x, $\alpha_s > 0$.

2. Find $x_s^- = \arg\min_{x \in X_s} \psi(x, \xi_s^+)$, where X_s is the set of support points for the design ξ_s.

3. Construct design $\xi_{s+1} = \xi_s^+ - \alpha_s \xi(x_s^-)$ and return to step 1.

The algorithm converges if, for example,

$$\lim_{s \to \infty} \sum_{i=1}^s \alpha_i = \infty \quad \text{and} \quad \lim_{s \to \infty} \sum_{i=1}^s \alpha_i^2 < \infty.$$

More on sequences $\{\alpha_s\}$ can be found, for example, in Fedorov and Hackl (1997), Ch. 3, Gaffke and Mathar (1992).

3 Examples

In this section we confine ourselves to normally distributed observations exclusively to avoid technical complications.

Quasi-linear heteroscedastic regression. Let $\eta(x,\theta) = \eta(\tau)$, $s(x,\theta) = \sigma^2 s'(\tau)$, $\tau = f^T(x)\gamma$, i.e.

$$\theta = \begin{pmatrix} \gamma \\ \sigma^2 \end{pmatrix}, \quad \frac{\partial \eta(x,\theta)}{\partial \theta} = \frac{\partial \eta}{\partial \tau} \begin{pmatrix} f(x) \\ 0 \end{pmatrix}, \quad \frac{\partial s(x,\theta)}{\partial \theta} = \begin{pmatrix} \sigma^2 \frac{\partial s'}{\partial \tau} f(x) \\ s'(\tau) \end{pmatrix}.$$

Hence

$$\mu(x,\theta) = \begin{pmatrix} \omega(\tau) f(x) f^T(x) & \frac{\partial \ln s'}{\partial \tau} f(x)/\{2\sigma^2\} \\ \frac{\partial \ln s'}{\partial \tau} f^T(x)/\{2\sigma^2\} & \sigma^{-4}/2 \end{pmatrix},$$

where $\omega(\tau) = (\partial \eta/\partial \tau)^2/\{\sigma^2 s'(\tau)\} + (\partial \ln s'/\partial \tau)^2/2$. To allocate observations, one has to analyze the function (cf. Theorem 1)

$$\psi(x,\xi,\theta) = \omega(\tau) f^T(x) D_{\gamma\gamma}(\xi,\theta) f(x)$$
$$+ \sigma^{-2} (\partial \ln s'/\partial \tau)^2 D_{\sigma\gamma}(\xi,\theta) f(x) + \sigma^{-4} D_{\sigma\sigma}(\xi,\theta)/2,$$

where $D(\xi,\theta) = M^{-1}(\xi,\theta)$. All the matrices in the above formula are obviously defined submatrices of $D(\xi,\theta)$. The corresponding design problem belongs to the class of design problems with uncontrolled variables [see Fedorov and Atkinson (1988)], and its solutions depend on the unknown θ. Interestingly enough,

$$\overline{D}_{\gamma\gamma}(\xi,\theta) \leq D_{\gamma\gamma}(\xi,\theta) \leq \underline{D}_{\gamma\gamma}(\xi,\theta),$$

where $\overline{D}_{\gamma\gamma}(\xi,\theta)$ is a standardized dispersion matrix if σ is known [cf. Sebastiani and Settimi (1998)], and $\underline{D}_{\gamma\gamma}(\xi,\theta)$ corresponds to the case when information about γ contained in the variance function is not used, as it was not in Davidian and Carrol (1987), Section 3.1.2 .

Dose-response experiments. In dose response studies, models with the variance function depending on the unknown parameters provide a popular example [see Holford and Sheiner (1981), Bezeau and Endrenyi (1986)]. The response is often modelled by a logistic model

$$\eta(x,\theta) = \theta_1 + \frac{\theta_2 - \theta_1}{1 + (x/\theta_3)^{\theta_4}}, \tag{3.1}$$

where x is a given dose (concentration). The variance function is described by the power model

$$s(x,\theta) = \theta_5 \eta^{\theta_6}(x,\theta) . \tag{3.2}$$

Model (3.1), (3.2) is widely used in ELISA (enzyme-linked immunosorbent assay) or cell-based assays [see Karpinski (1990), Hedayat et al. (1997)].

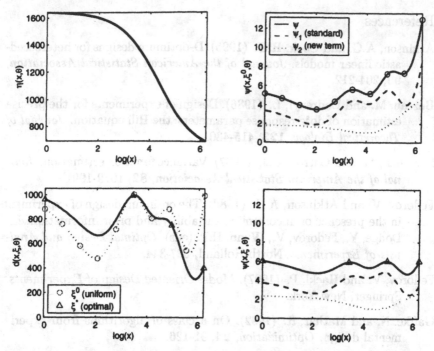

FIGURE 1. Upper left: response. Upper right: normalized variance for uniform design. Lower right: normalized variance for optimal design. Lower left: unnormalized variance (triangles - optimal design, circles - serial dilution)

To illustrate this model, we use the data from a study on the ability of a compound to inhibit the proliferation of cells in a cell-based assay (the compound is not named to shield proprietary information). The vector of unknown parameters was estimated by fitting the data collected from a two-fold dilution design covering a range of concentrations from 0.98 to 500 ng/ml. Thus the design region was set to [-0.02, 6.21] on the log-scale. Figure 1 presents the locally optimal design and variance of prediction for the model (3.1), (3.2), where $\theta = (616, 1646, 75.2, 1.34, 0.33, 0.90)^T$.

The two subplots on the right-side of Figure 1 present the normalized variance of prediction (2.3) for a serial dilution design ξ_0 (i.e. uniform on the log-scale, upper right) and D-optimal design ξ^* (lower right). The solid lines show the function $\psi(x, \xi, \theta)$ while the dashed and dotted lines display the first and second terms on the right-hand side of (2.3), respectively. The unnormalized variance of prediction $d_1(x, \xi, \theta)$ defined in (2.4), is given on the lower-left subplot. It is worth noting that the optimal design in our example is supported at just four points, which is less than the dimension of the parameter space. We also remark that the weights of the support points are not equal. In our example $p = \{0.28, 0.22, 0.22, 0.28\}$.

Acknowledgements. The authors are grateful to the referees for useful comments which helped to improve the presentation of the results.

References

Atkinson, A.C. and Cook, R.D. (1995). D-optimum designs for heteroscedastic linear models. *Journal of the American Statistical Association*, **90**, 204-212.

Bezeau, M. and Endrenyi, L. (1986). Design of experiments for the precise estimation of dose-response parameters: the Hill equation. *Journal of Theoretical Biology*, **123**, 415-430.

Davidian, M. and Carroll, R.J. (1987). Variance function estimation. *Journal of the American Statistical Association*, **82**, 1079-1091.

Fedorov, V. and Atkinson, A.C. (1988). The optimum design of experiments in the presence of uncontrolled variability and prior information. In: Dodge, Y., Fedorov, V., Wynn, H.P. (eds), *Optimal Design and Analysis of Experiments*, North Holland, 327-344.

Fedorov, V. and Hackl, P. (1997). *Model-Oriented Design of Experiments*. Springer, New York.

Gaffke, N. and Mathar, R. (1992). On a class of algorithms from experimental design. *Optimization*, **24**, 91-126.

Hedayat, A.S., Yan, B., and Pezutto, J.M. (1997). Modeling and identifying optimum designs for fitting dose-response curves based on raw optical data. *Journal of the American Statistical Association*, **92**, 1132-1140.

Heyde, C.C. (1997). *Quasi-Likelihood and Its Application*. Springer, New York.

Holford, N.H.G. and Sheiner, L.B. (1981). Understanding the dose response relationship: clinical application of pharmacokinetic-pharmacodynamic models. *Clinical Pharmacokinetics*, **6**, 429-453.

Jennrich, R.I. (1969). Asymptotic properties of nonlinear least squares estimators. *Annals of Mathematical Statistics*, **40**, 633-643.

Karpinski, K.F. (1990). Optimality assessment in the enzyme-linked immunosorbent assay (ELISA). *Biometrics*, **46**, 381-390.

Rao, C.R. (1973). *Linear Statistical Inference and Its Applications*, 2nd Edition. Wiley, New York.

Saaty, T.L. and Bram, J. (1964). *Nonlinear Mathematics*. McGraw-Hill, New York.

Sebastiani, P. and Settimi, R. (1998). First-order optimal designs for nonlinear models. *Journal of Statistical Planning and Inference*, **74**, 177-192.

Model Validity Range in Multicentre Clinical Trials

V. Dragalin
V. V. Fedorov

ABSTRACT: Analyses of multicentre trials consider the estimated treatment effect differences of the individual centres and combine them into an estimate of the overall treatment effect. There has been much debate in the literature concerning the best way to combine these treatment effect differences. We emphasize that first of all one should define the combined response to treatment (CRT), the object that has to be estimated from the results of a multicentre clinical trial. Having the defined target in mind, the least squares estimators of the CRT under three possible models of increasing complexity for multicentre data are derived. We compare these estimators in terms of their mean squared errors. It is shown that the choice of CRT determines not only the best estimator, but also the allocation of patients among the centres that minimizes the MSE.

KEYWORDS: combined response to treatment; model validity range; multicentre trial

1 Combined response to treatment: definitions and models

We consider a randomized two-arm multicentre clinical trial. Assume that we have N centres. The response variable from the kth patient on the jth treatment in the ith centre will be denoted by Y_{ijk}, ($i = 1, \ldots, N$, $j = 1, 2$, $k = 1, \ldots, n_{ij}$).

Let μ_{ij} be the true mean of response on treatment j in centre i and $\delta_i = \mu_{i2} - \mu_{i1}$ be the true treatment effect in centre i. Assume for a moment that we know all μ_{ij}. We have to combine them into a single value which describes the overall treatment effect. In general, this might be a specified function $\psi(\mu) = \psi(\mu_{11}, \ldots, \mu_{N1}, \mu_{12}, \ldots, \mu_{N2})$, which will be called

mODa6, A.C.Atkinson, P.Hackl and W.G.Müller, eds., Physica, Heidelberg, 2001.

the "combined response to treatment" (CRT). In this paper, we confine ourselves only to a linear CRT based on differences $\delta_i = \mu_{i2} - \mu_{i1}$:

$$\delta = \psi(\mu) = \sum_{i=1}^{N} \omega_i \delta_i = \sum_{i=1}^{N} \omega_i \mu_{i2} - \sum_{i=1}^{N} \omega_i \mu_{i1}, \qquad (1.1)$$

where ω_i, $(\omega_i \geq 0, \sum_{i=1}^{N} \omega_i = 1)$, are some given weights for the information, concerning the treatment effect, gained from centre i. As a particular case, one may consider the CRT $\delta = N^{-1} \sum_{i=1}^{N} \delta_i$, i.e., a simple average of treatment differences among centres ($\omega_i = 1/N$).

We consider the following general model for the treatment response

$$Y_{ijk} = \mu_{ij} + \varepsilon_{ijk}, \quad i = 1, \ldots, N, \ j = 1, 2, \ k = 1, \ldots, n_{ij}, \qquad (1.2)$$

where ε_{ijk} are measurement errors which encompass the deviations of a response of the kth patient (in centre i on treatment j) from what is typical for that centre and treatment. Usually, they are assumed to be independent random variables with zero mean and variances $\text{Var}(\varepsilon_{ijk}) = \sigma_{ij}^2$. Further assumptions on homoscedasticity are made, for instance, $\sigma_{ij}^2 = \sigma_i^2$, $(j = 1, 2)$, or $\sigma_{ij}^2 = \sigma^2$. For the sake of simplicity we will assume the latter in the sequel.

The fixed effects (μ_{ij} are assumed to be unknown constants) version of model (1.2) admits different complexity levels:

$$\begin{aligned}
\textbf{Model I:} \quad & Y_{ijk} = \mu + (-1)^j \tau + \varepsilon_{ijk} \\
\textbf{Model II:} \quad & Y_{ijk} = \mu_i + (-1)^j \tau + \varepsilon_{ijk} \\
\textbf{Model III:} \quad & Y_{ijk} = \mu_i + (-1)^j \tau_i + \varepsilon_{ijk}.
\end{aligned}$$

Here τ_i (or τ, if the same across the centres) is the treatment effect in centre i, and μ_i is the centre effect. Model I corresponds to the ANOVA model with treatment as the only fixed effect. Model II corresponds to the ANOVA model with treatment and centre as fixed effects and Model III is the full fixed effects ANOVA model with treatment, centre and treatment-by-centre interaction. Notice, that Model III includes Model II as a particular case ($\tau_i = \tau$), and Model II includes Model I as a particular case ($\mu_i = \mu$), i.e., we have a set of embedded models. From the definitions of δ and $\delta_i = \mu_{i2} - \mu_{i1} = \mu_i + \tau_i - \mu_i + \tau_i = 2\tau_i$, it follows that

$$\delta = \sum_{i=1}^{N} \omega_i \delta_i = 2 \sum_{i=1}^{N} \omega_i \tau_i. \qquad (1.3)$$

2 Estimators and model validity range

We will use a common convention that if a subscript is replaced by a dot, it means that subscript has been summed over. Thus $Y_{ij\cdot} = \sum_{k=1}^{n_{ij}} Y_{ijk}$

and $n_{.j} = \sum_{i=1}^{N} n_{ij}$. The addition of a "bar" indicates that an average is taken, i.e. $\overline{Y}_{ij.} = \frac{1}{n_{ij}} \sum_{k=1}^{n_{ij}} Y_{ijk}$. Although in the equal weights case $\delta = \frac{1}{N} \sum_{i=1}^{N} \delta_i = \bar{\delta}$, we prefer to use the notation δ to be consistent with the general case (1.3).

Let $\Delta_i = \overline{Y}_{i2.} - \overline{Y}_{i1.}$ be the estimator of treatment effect δ_i in centre i. Each of the Models I–III generates a least squares estimator (best linear estimator if the corresponding model is true) for the primary parameter δ. We will denote it by Δ with a Roman index corresponding to the model:

$$\Delta_I = Y_{.2.}/n_{.2} - Y_{.1.}/n_{.1},$$

$$\Delta_{II} = \sum_{i=1}^{N} W_i \Delta_i, \quad \text{with } W_i = \frac{1/\text{Var}(\Delta_i)}{\sum_{k=1}^{N} 1/\text{Var}(\Delta_k)},$$

$$\Delta_{III} = \sum_{i=1}^{N} \omega_i \Delta_i.$$

Since $\text{Var}(\Delta_i) = \sigma^2/n_{i2} + \sigma^2/n_{i1}$, the weights W_i in the estimator Δ_{II} do not depend on σ^2 and

$$W_i = \frac{\left(\frac{1}{n_{i2}} + \frac{1}{n_{i1}}\right)^{-1}}{\sum_{k=1}^{N} \left(\frac{1}{n_{k2}} + \frac{1}{n_{k1}}\right)^{-1}} = \frac{n_{i2} n_{i1}/(n_{i2} + n_{i1})}{\sum_{k=1}^{N} n_{k2} n_{k1}/(n_{k2} + n_{k1})}.$$

Note that estimator Δ_I does not depend on the differences Δ_i and is defined even when a centre has enrolled patients only on one treatment, say either n_{i1} or n_{i2} is equal to zero. Estimators Δ_{II} and Δ_{III} are not defined in such a case, and usually some sort of pooling of the results from such centres is used in practice. Although formally, the estimator Δ_{II} could be defined by omitting that centre from analysis, $\text{Var}(\Delta_i) = \infty$ in such a centre implying that $W_i = 0$.

When $n_{i1} = n_{i2} = n_i$, which we call *balanced randomization* and is approximately the case when blocked randomization is used within each centre, the weights $W_i = n_i/\sum_{k=1}^{N} n_k$ and $\Delta_I = \Delta_{II}$. If furthermore, the n_i are the same for all centres, i.e. an unlikely situation where there is an equal number of patients per treatment per centre, and which we call *balanced enrollment*, then $\Delta_I = \Delta_{II} = \Delta_{III}$.

As soon as both the model for the response and the parameter of interest δ are chosen, the measure of precision of its estimators must be defined. In this study, we stay mainly within the parameter estimation paradigm, in which the mean squared error (MSE)

$$\text{MSE}(\Delta) = E(\Delta - \delta)^2 \tag{2.1}$$

is the most popular measure of an estimator's precision.

First of all, straightforward calculations yield the variance of the three estimators:

$$\text{Var}(\Delta_I) \;=\; \sigma^2 \left(\frac{1}{n_{.2}} + \frac{1}{n_{.1}} \right),$$

$$\text{Var}(\Delta_{II}) \;=\; \sigma^2 \sum_{i=1}^{N} W_i^2 \left(\frac{1}{n_{i2}} + \frac{1}{n_{i1}} \right) = \sigma^2 \left[\sum_{i=1}^{N} \left(\frac{1}{n_{i2}} + \frac{1}{n_{i1}} \right)^{-1} \right]^{-1},$$

$$\text{Var}(\Delta_{III}) \;=\; \sigma^2 \sum_{i=1}^{N} \omega_i^2 \left(\frac{1}{n_{i2}} + \frac{1}{n_{i1}} \right).$$

By the Cauchy-Schwarz inequality, we have

$$\text{Var}(\Delta_I) \leq \text{Var}(\Delta_{II}) \leq \text{Var}(\Delta_{III}). \tag{2.2}$$

Note that $\text{Var}(\Delta_I)$ is minimized when $n_{.1} = n_{.2} = n_{..}/2$, i.e., under balanced total allocation to treatments, $\text{Var}(\Delta_{II})$ is minimized when $n_{i1} = n_{i2}$ in each centre, but not necessarily from centre to centre, i.e., under balanced randomization in each centre, and $\text{Var}(\Delta_{III})$ is minimized when $n_{i1} = n_{i2} = \omega_i n_{..}/2$. In all these cases the minimal value is $4\sigma^2/n_{..}$. Thus, the *optimal design* (in terms of the variance of the CRT estimator) of a trial for estimating (1.3) must be "balanced" and the extent of "balancing" is getting more important as the model complexity increases. The above design problem is a rather special case of the optimal design problem for linear models with a Kronecker product structure (see e.g. Kurotschka, 1978 and Schwabe, 1996).

However, estimator Δ_I is biased under both Model II and Model III and Δ_{II} is biased under Model III. Therefore MSE is a better measure for comparing the estimators. Straightforward calculations yield

$$\text{MSE}(\Delta_I) \;=\; \sigma^2 \left(\frac{1}{n_{.2}} + \frac{1}{n_{.1}} \right) + \left[\sum_{i=1}^{N} \left\{ \left(\frac{n_{i2}}{n_{.2}} - \frac{n_{i1}}{n_{.1}} \right)(\mu_i - \bar{\mu}) \right. \right.$$
$$\left. \left. + \left(\frac{n_{i2}}{n_{.2}} + \frac{n_{i1}}{n_{.1}} - 2\omega_i \right)(\tau_i - \bar{\tau}) \right\} \right]^2,$$

$$\text{MSE}(\Delta_{II}) \;=\; \sigma^2 \sum_{i=1}^{N} W_i^2 \left(\frac{1}{n_{i2}} + \frac{1}{n_{i1}} \right) + 4 \left[\sum_{i=1}^{N} \left(W_i - \omega_i \right)(\tau_i - \bar{\tau}) \right]^2,$$

$$\text{MSE}(\Delta_{III}) \;=\; \sigma^2 \sum_{i=1}^{N} \omega_i^2 \left(\frac{1}{n_{i2}} + \frac{1}{n_{i1}} \right), \tag{2.3}$$

where $\bar{\mu} = \sum_{i=1}^{N} \omega_i \mu_i$ and $\bar{\tau} = \sum_{i=1}^{N} \omega_i \tau_i$. We see that under Model I, Δ_I is an unbiased estimator of δ. Similarly, under both Model I and Model II, Δ_{II} is unbiased, and obviously the estimator Δ_{III} is unbiased under all three models. However, by (2.2) the estimator Δ_{III} can have a significant increase in MSE. Therefore, the estimator Δ_{III} obtained from a more complicated Model III for treatment response may lead to a loss of efficiency when no interaction is present. On the other hand, even when the Model III is true, but the $(\tau_i - \bar{\tau})$ are relatively small, it might happen that estimators Δ_I and Δ_{II} have smaller MSE than Δ_{III}. All bias terms in (2.3) vanish for well designed trials. The required "balancing" is now in a reverse order to that for variance minimization. The simplest model requires the most strict balance, $n_{i2}/n_{.2} = n_{i1}/n_{.1} = \omega_i$. No balancing is required for estimator Δ_{III}, except that both $n_{i1} > 0$ and $n_{i2} > 0$ for all $i = 1, \ldots, N$, but a stronger balancing is required for the simpler models to minimize the bias. Note that we assume here that $\{n_{ij}\}$ are fixed and positive for all centres. However, if the enrollment is random, i.e. $\{n_{ij}\}$ are sampled from a given distribution, there is a possibility that for some centres either n_{i1} or n_{i2} will be zero and Δ_{II} and Δ_{III} must be modified, e.g. $\Delta_{III} = \sum \omega_i \Delta_i$, where summation is over centres with both n_{i1} and n_{i2} positive. But in this case, Δ_{III} will be biased.

As in Fedorov *et al.* (1998), we define a *model validity range* for Model II with respect to Model III, as the set Γ_{II} of all values for the standardized (by σ) deviations of centre specific treatment effects from their mean $(\tau_i - \bar{\tau})/\sigma$, such that $\mathrm{MSE}(\Delta_{II}) < \mathrm{MSE}(\Delta_{III})$ (for the sake of simplicity, it is assumed that all $n_{ij} > 0$):

$$\Gamma_{II} = \left\{ \frac{\tau_i}{\sigma} : \left[\sum_{i=1}^{N} \left(W_i - \omega_i \right) \frac{\tau_i - \bar{\tau}}{\sigma} \right]^2 < \frac{1}{4} \left[\sum_{i=1}^{N} \left(\omega_i^2 - W_i^2 \right) \left(\frac{1}{n_{i2}} + \frac{1}{n_{i1}} \right) \right] \right\},$$

where by definition $\bar{\tau} = \delta/2$. Therefore, the model validity range is the set of all possible values of the standardized treatment-by-centre interaction terms under which one can use a more parsimonious model to get a better estimate of the primary parameter. This set is completely defined by the enrollment $\{n_{ij}\}$.

Similarly, we can determine the model validity range for Model I with respect to Model III:

$$\Gamma_{I} = \left\{ \left(\frac{\mu_i}{\sigma}, \frac{\tau_i}{\sigma} \right) : \left[\sum_{i=1}^{N} \left\{ \left(\frac{n_{i2}}{n_{.2}} - \frac{n_{i1}}{n_{.1}} \right) \frac{\mu_i - \bar{\mu}}{\sigma} + \left(\frac{n_{i2}}{n_{.2}} + \frac{n_{i1}}{n_{.1}} - 2\omega_i \right) \frac{\tau_i - \bar{\tau}}{\sigma} \right\} \right]^2 \right.$$
$$\left. < \sum_{i=1}^{N} \omega_i^2 \left(\frac{1}{n_{i2}} + \frac{1}{n_{i1}} \right) - \left(\frac{1}{n_{.2}} + \frac{1}{n_{.1}} \right) \right\}.$$

Under balanced randomization, i.e., when $n_{i1} = n_{i2} = n_i$, estimators Δ_I and Δ_{II} coincide. Hence

$$\Gamma_{I} = \Gamma_{II} = \left\{ \frac{\tau_i}{\sigma} : \left[\sum_{i=1}^{N} \left(\frac{n_i}{n} - \omega_i \right) \frac{\tau_i - \bar{\tau}}{\sigma} \right]^2 < \frac{1}{2} \left[\sum_{i=1}^{N} \frac{\omega_i^2}{n_i} - \frac{1}{n} \right] \right\},$$

(here, and below, we use $n = n_{.1} = n_{.2}$). Notice that for n_i proportional to ω_i, both parts of the inequality are zero, meaning that, in this special case, the model validity range is empty: the use of the more parsimonious model does not improve the properties of the estimator. However, remember that in this special case all three estimators coincide.

To gain additional insights into the model validity range concept, we assume now that the μ_i and τ_i are sampled from populations with means μ and τ, respectively and variances σ_μ^2 and σ_τ^2, respectively. We would like to emphasise that the fact that μ_i and τ_i are sampled is not used in the construction of the estimators. We need that sampling only to generate some reasonable scenarios for μ_i and τ_i.

To obtain the average MSE, we take the expectation of the MSE with respect to the distribution of τ_i's and μ_i's. We will denote this average MSE as $\overline{\mathrm{MSE}}(\Delta)$. Straightforward calculations yield

$$\overline{\mathrm{MSE}}(\Delta_I) = \sigma^2 \left(\frac{1}{n_{.2}} + \frac{1}{n_{.1}} \right) + \sigma_\mu^2 \sum_{i=1}^{N} \left(\frac{n_{i2}}{n_{.2}} - \frac{n_{i1}}{n_{.1}} \right)^2 + \sigma_\tau^2 \sum_{i=1}^{N} \left(\frac{n_{i2}}{n_{.2}} + \frac{n_{i1}}{n_{.1}} - 2\omega_i \right)^2,$$

$$\overline{\mathrm{MSE}}(\Delta_{II}) = \sigma^2 \sum_{i=1}^{N} W_i^2 \left(\frac{1}{n_{i2}} + \frac{1}{n_{i1}} \right) + 4\sigma_\tau^2 \sum_{i=1}^{N} \left(W_i - \omega_i \right)^2.$$

Similarly, we determine the unconditional model validity range:

$$\overline{\Gamma}_I = \left\{ \left(\frac{\sigma_\mu}{\sigma}, \frac{\sigma_\tau}{\sigma} \right) : \frac{\sigma_\mu^2}{\sigma^2} \sum_{i=1}^{N} \left(\frac{n_{i2}}{n_{.2}} - \frac{n_{i1}}{n_{.1}} \right)^2 + \frac{\sigma_\tau^2}{\sigma^2} \sum_{i=1}^{N} \left(\frac{n_{i2}}{n_{.2}} + \frac{n_{i1}}{n_{.1}} - 2\omega_i \right)^2 \right.$$
$$\left. < \sum_{i=1}^{N} \omega_i^2 \left(\frac{1}{n_{i2}} + \frac{1}{n_{i1}} \right) - \left(\frac{1}{n_{.2}} + \frac{1}{n_{.1}} \right) \right\},$$

$$\overline{\Gamma}_{II} = \left\{ \frac{\sigma_\tau}{\sigma} : \frac{\sigma_\tau^2}{\sigma^2} \sum_{i=1}^{N} \left(W_i - \omega_i \right)^2 < \frac{1}{4} \left[\sum_{i=1}^{N} \left(\omega_i^2 - W_i^2 \right) \left(\frac{1}{n_{i2}} + \frac{1}{n_{i1}} \right) \right] \right\}.$$

To see the practical usefulness of the model validity range we consider the special case of balanced randomization where $n_{i1} = n_{i2} = n_i$. Letting $p_i = n_i/n = W_i$, the inequality used in the definition of $\overline{\Gamma}_{II}(= \overline{\Gamma}_I)$ can be written as

$$\frac{\sigma_\tau^2}{\sigma^2/n} < \frac{1}{2} \sum_{i=1}^{N} \frac{\omega_i^2 - p_i^2}{p_i} \bigg/ \sum_{i=1}^{N} (p_i - \omega_i)^2$$

where $\frac{\sigma_\tau^2}{\sigma^2/n}$ can be seen as the "signal to noise" ratio of the treatment variance to within-centre variance. The model validity range is the interval $[0, R)$ of the "signal to noise" ratios, where the upper bound

$$R = R(N; p_1, \ldots, p_N) = \frac{1}{2} \sum_{i=1}^{N} \frac{\omega_i^2 - p_i^2}{p_i} \bigg/ \sum_{i=1}^{N} (p_i - \omega_i)^2$$

is completely determined by the enrollment $\{p_i\}$.

For illustration let us consider the case of three centres, $N = 3$, and equal weights $\omega_i = 1/3$. The enrollment space (p_1, p_2, p_3) is defined by a portion

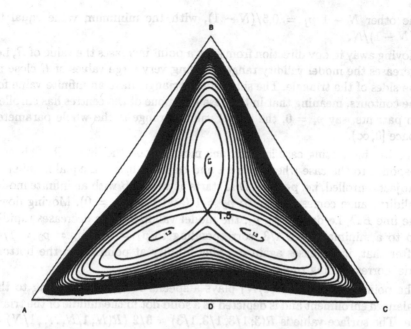

FIGURE 1. Contour plots for the upper bound of the model validity range in the case $N = 3$

of the plane containing the three extreme points $A = (1, 0, 0)$, $B = (0, 1, 0)$ and $C = (0, 0, 1)$, and such that $0 \leq p_i \leq 1$ and $p_1 + p_2 + p_3 = 1$. It is thus a two-dimensional equilateral triangle with the extreme points at the vertices, as depicted in Figure 1. Note that each side of the triangle is a two-centre subspace, with the third centre having zero enrollment. A line parallel to the side BC corresponds to a given value p_1 of the enrollment fraction in the first centre. The side BC itself corresponds to $p_1 = 0$ and moving towards the vertex A with a line parallel to BC corresponds to increasing values of p_1 (up to $p_1 = 1$ at vertex A). Similarly, a line parallel to side AC corresponds to a value of p_2. The intersection of any two lines from these two sets defines a point in the enrollment space (the corresponding $p_3 = 1 - p_1 - p_2$). For example, the marked point in the middle of the triangle corresponds to the balanced enrollment $(1/3, 1/3, 1/3)$.

The contours of the surface $R(3; p_1, p_2, p_3)$ on this simplex are shown in Figure 1. The surface R has three local minima at the enrollment points $(1/2, 1/4, 1/4)$ – the centre of the smallest ellipse on the left, $(1/4, 1/2, 1/4)$ – the centre of the smallest ellipse on the right and $(1/4, 1/4, 1/2)$ – the centre of the smallest ellipse at the top. The value of R at these points is $4/3$. These means, that the smallest model validity range is $[0, 4/3)$ and is achieved under an enrollment when half of the total number of patients are in one centre and the other half is split evenly between the other two centres. This fact is true for general N: the local minima of the surface $R(N; p_1, \ldots, p_N)$ are achieved at N points, each having one $p_i = 0.5$ and the other $N - 1$ $p_j = 0.5/(N - 1)$, with the minimum value equal to $2(N - 1)/N$.

Moving away in any direction from such a point increases the value of R, i.e. increases the model validity range, yielding very large values of R close to the sides of the triangle. The sides of the triangle have an infinite value for the contours, meaning that in the case when one of the centres has enrolled no patients, say $p_i = 0$, the model validity range is the whole parameter space $[0, \infty)$.

Another interesting case is when one moves along the line BD. This corresponds to the case when centre 1 and centre 2 have an equal number of subjects enrolled, i.e. $p_1 = p_3$. One starts at vertex B with an infinite model validity range corresponding to $p_2 = 1$ (and $p_1 = p_3 = 0$). Moving down the line BD, i.e decreasing p_2, the model validity range decreases rapidly up to a minimal value $4/3$, corresponding to $p_2 = 1/2$, $p_1 = p_3 = 1/4$. After that, it increases again reaching infinity at point D on the bottom side, corresponding to $p_2 = 0$, $p_1 = p_3 = 1/2$.

The point $(1/N, 1/N, \ldots, 1/N)$ plays a special role, corresponding to the balanced enrollment and is depicted as a solid dot in the middle of the triangle. The surface value is $R(3; 1/3, 1/3, 1/3) = 3/2$ $(R(N; 1/N, \ldots, 1/N) = N/2$ in general). There are three lines passing through this point, each parallel to one of the sides of the triangle. For example, one parallel to the side AB corresponds to $p_3 = 1/3$. Moving along this line from the bottom side AC towards the side BC, corresponds to increasing p_2 from zero to $2/3$ (of course, corresponding also to decreasing p_1 from $2/3$ to 0): the R value is first decreasing up to the minimum value 1.5 at point $p_1 = p_2 = p_3 = 1/3$ and then increasing again. Therefore, this point is a minimum when one moves along one of these three lines. However, this is not a global minimum. This is in some sense counterintuitive, telling us that even when we have a balanced enrollment (and we know that in that case all estimators coincide) there is a small validity range $[0, N/2)$ for $\frac{\sigma_\tau^2}{\sigma^2/n}$, when estimators Δ_I and Δ_{II} are preferable to Δ_{III}. This is a mathematical artifact obtained by the $0/0$ operation. Notice that in the definition of R, both the numerator and the denominator become zero under the balanced enrollment. However the numerator is an infinitesimal quantity of larger magnitude than the denominator, yielding a non-zero limiting value $R = N/2$.

Hopefully the last fact does not overthrow the major message of this example. Under well-designed experiments (equal centre sizes and an equal number of patients on each treatment), all three models perform well and there is no difference whichever estimator is used. However, as soon as the actual enrollment and randomization become more and more imbalanced (different centre sizes and/or different number of patients on each treatment), the range of possible "signal-to-noise" ratios, under which the parsimonious model may in fact give better estimates, becomes larger and larger.

3   Conclusions

Traditionally (see the papers on multi-centre trials listed in the references), the choice of estimator has been of one of three possibilities, which we have labelled as Type I, II and III. By considering the MSE of all three estimators we can investigate the trade off between variance and bias. We emphasize the importance of an agreed understanding of what we are going to estimate, i.e. a proper choice of CRT. Moreover, we show that the CRT choice also defines the optimal trial design: for the simple average of treatment differences CRT, the optimal design is achieved by a balanced enrollment, i.e. equal numbers of patients per treatment per centre, while for a weighted CRT, the optimal trial design is achieved when the enrollment at each centre is proportional to the centre weight. We have given model validity ranges that indicate, without unblinding the study, which estimator is optimal.

References

Fedorov, V.V., Montepiedra, G. and Nachtsheim, C.J. (1998). Optimal design and the model validity range. *J. Statist. Planning and Inference*, **72**, 215–227.

Fleiss, J. (1986). Analysis of data from multiclinic trials. *Controlled Clinical Trials*, **10**, 237–243.

Gould, A. (1998). Multi-centre trial analysis revisited. *Statistics in Medicine*, **17**, 1779–1797.

Jones, B., Teather, D., Wang, J., and Lewis, J. (1998). A comparison of various estimators of treatment difference for a multi-centre clinical trial. *Statistics in Medicine*, **17**, 1767–1777.

Kurotschka, V. (1978). Optimal design of complex experiments with qualitative factors of influence. *Commun. Statist., Theory Methods*, **A7**, 1363–1378.

Lin, Z. (1999). An issue of statistical analysis in controlled multi-centre studies: How shall we weight the centres? *Statistics in Medicine*, **18**, 365–373.

Schwabe, R. (1996). *Optimum Designs for Multi-Factor Models*. New York: Springer.

Schwemer, G. (2000). General linear models for multicenter clinical trials. *Controlled Clinical Trials*, **21**, 21–29.

Senn, S. (1998). Some controversies in planning and analysing multi-centre trials. *Statistics in Medicine*, **17**, 1753–1765.

Two Models of Nonadaptive Group Testing for Designing Screening Experiments

A.G. D'yachkov
A.J. Macula
D.C. Torney
P.A. Vilenkin

ABSTRACT: We discuss two non-standard models of nonadaptive combinatorial search which develop the conventional disjunct search model of Du and Hwang (1993) for a small number of defective elements contained in a finite ground set or a population. The first model called a *search of defective supersets (complexes)* was suggested in D'yachkov *et al.* (2000c,d). The second model which can be called a *search of defective subsets in the presence of inhibitors* was introduced for the case of an adaptive search by Farach *et al.* (1997) and De Bonis and Vaccaro (1998). For these models, we study the constructive search methods based on the known constructions for the disjunct model from Kautz and Singleton (1964) and from D'yachkov *et al.* (2000a,b).

1 Description of the models

We use the symbol $\triangleq$ to denote definitional equalities.

The standard *disjunct* search model for *designing screening experiments* (DSE, cf. D'yachkov, 1997) has the following form. Let there be a *population* containing t distinguishable *samples*. We identify it with the set $[t] \triangleq \{1, 2, \ldots, t\}$. Assume that the population includes an unknown *defective* subset $p \subset [t]$. We call elements of p *defective samples*.

Our aim is to detect p using a number of *group tests*. Each group test is defined by a subset (*testing group* or *pool*) $G \subset [t]$. The result $r(G, p)$ of this test assumes a binary value 0 (*negative*) or 1 (*positive*) according to

---

mODa6, A.C.Atkinson, P.Hackl and W.G.Müller, eds., Physica, Heidelberg, 2001.

the following rule:

$$r(G, \mathsf{p}) \triangleq \begin{cases} 1, & G_n \cap \mathsf{p} \neq \varnothing, \\ 0, & G_n \cap \mathsf{p} = \varnothing. \end{cases} \tag{1.1}$$

One can see that a group test detects whether the testing group intersects with the defective subset or not.

In the *adaptive* disjunct model each group test G_{n+1} is choosen according to the results of the previous tests $G_1, \ldots, G_n$. This model is considered in Du and Hwang (1993).

We consider the *nonadaptive* disjunct model, in which all pools must be constructed before any test is performed. Thus, an experimenter is not allowed to use the results of the previous tests to construct the next ones. Such models usually occur in the problems of molecular biology (cf. Farach *et al.*, 1997, or D'yachkov *et al.*, 2000b) when different tests can be performed simultaneously as one experiment (one step). Each experiment is expensive, so we are interested in detecting p using only one step, i.e., exactly the nonadaptive search strategy.

We also consider the restriction on the number of defective samples: $|\mathsf{p}| \leq s$, where s is the given positive integer, $s < t$. This condition is usual in the group testing problem. In many cases when the condition does not occur naturally it can be justified by probabilistic arguments. If any sample can be defective with some small probability, then one can choose s so that the inequality $|\mathsf{p}| \leq s$ holds with high probability.

Note, that for practical applications of this model we assume that $s \ll t$. If this condition does not hold, then the present model is not suitable.

Definition 1.1 *A series of N nonadaptive tests $X \triangleq (G_1, G_2, \ldots, G_N)$ which allows us to identify any defective subset $\mathsf{p} \subset [t]$, $|\mathsf{p}| \leq s$, is called a disjunct s-design (Du and Hwang, 1993) of length N and size t. Note that we suppose no error in the results of these tests.*

In the general case to detect a defective subset p using a disjunct s-design one needs to check all possible subsets. Below we introduce an important class of designs called *superimposed codes* for which the decoding algorithm is much more simple.

This model can be generalized in several ways. We can consider the *list decoding procedure* in which one should construct a subset $\mathsf{p}' \subset [t]$ such that $\mathsf{p} \subseteq \mathsf{p}'$ and the size of $\mathsf{p}' \backslash \mathsf{p}$ is not very large. This model was considered in D'yachkov and Rykov (1983) and some other papers. We can also consider designs which correct errors in the results of D'yachkov *et al.* (1989).

In the present paper we introduce two generalizations of the disjunct model of DSE.

1. Nonadaptive search of defective supersets.

Let there be a population $[t] = \{1, 2, \ldots, t\}$. Assume that there exists an unknown *superset* (*complex*) p which is composed of a number of subsets $P \subset [t]$:

$$p \triangleq \{P_1, P_2, \ldots, P_k\}, \quad P_i \subset [t], \quad P_i \not\subset P_j \text{ for } i \neq j. \quad (1.2)$$

To detect p one can use a number of group tests $G \subset [t]$ for which the result $r(G, p)$ is positive if and only if G includes at least one subset $P \subset [t]$ being a member of the complex p, and negative otherwise:

$$r(G, p) \triangleq \begin{cases} 1, & \exists P \in p : P \subseteq G, \\ 0, & \text{otherwise.} \end{cases} \quad (1.3)$$

We consider the following restrictions on complexes p: the number of elements $P \in p$ does not exceed s and the size $|P| \leq \ell$ for any $P \in p$, where s and ℓ are given positive integers, $s + \ell \leq t$.

Definition 1.2 *A series of N nonadaptive tests $X = (G_1, G_2, \ldots, G_N)$ which allows us to identify any such complex p is called a superset (s, ℓ)-design of length N and size t.*

Obviously, for $\ell = 1$ this model is identical to the conventional disjunct model, because each subset $P \in p$ is composed of exactly one element.

One can easily understand the necessity of the additional condition in (1.2): if $P_i \subset P_j$, then we cannot detect the defective property of P_j.

Consider the following simple example of such an application. We have some chemical (or medical) substances. We assume that some combinations of them may be dangerous. To detect these combinations we can perform group tests, i.e., put some of the samples together and test whether the obtained substance is dangerous. The result is positive if the group contains one or more dangerous combinations, that lead to the given model.

2. Nonadaptive search of defective subsets in the presence of inhibitors.

Return to the base disjunct model and assume that along with the defective subset $p \subset [t]$ there exists a subset $I \subset [t]$, $p \cap I = \emptyset$. We call samples from this set *inhibitors*.

Inhibitors make the result of a group test negative despite the existence of defective elements in testing group $G \subset [t]$:

$$r(G, \mathsf{p}, \mathrm{I}) \triangleq \begin{cases} 1, & G \cap \mathsf{p} \neq \varnothing \quad \text{and} \quad G \cap \mathrm{I} = \varnothing, \\ 0, & \text{otherwise.} \end{cases} \qquad (1.4)$$

Definition 1.3 *A series of N nonadaptive tests $X = (G_1, G_2, \ldots, G_N)$ which allows us to identify any defective subset p, $|\mathsf{p}| \leq s$, in the presence of not more then $\imath$ inhibitors, is called an* inhibitory $(s, \imath)$-design *of length N and size t, where $s \geq 1$ and $\imath \geq 0$ are given integers, $s + \imath \leq t$.*

Obviously for $\imath = 0$ this model is identical to the conventional disjunct model of DSE.

The current model arises in applications in which some samples in a population are not defective, but affect the testing result, namely, make it always negative. If a test has a form of some chemical reaction, then it may stop due to the inhibitor. This illustration gives the name to the current model.

The rest of the present paper is organized as follows. In section 2 we consider important classes of designs for the models under consideration. The detecting algorithm for these types of designs are simple. In section 3 we consider a constructive method for such designs. Section 4 contains several examples.

2   Superimposed codes

2.1   Notation

Let $[t]$ be a population of samples. Consider an arbitrary series of N testing groups $X = (G_1, \ldots, G_N)$, $G_n \subset [t]$ for $n = 1, 2, \ldots, N$. Following Du and Hwang (1993), we encode it by the binary $N \times t$ *incidence matrix* $X = \|x_n(u)\|$, where index $n = 1, 2, \ldots N$ denotes a row number and index $u = 1, 2, \ldots t$ denotes a column number. An element $x_n(u)$ of this matrix has the form

$$x_n(u) \triangleq \begin{cases} 1, & \textit{if } u \in G_n, \\ 0, & \text{otherwise.} \end{cases}$$

The n-th row $\mathsf{x}_n \triangleq \big(x_n(1), x_n(2), \ldots, x_n(t)\big)$ of the matrix X encodes the nth test G_n. The uth column $\mathsf{x}(u) \triangleq \big(x_1(u), x_2(u), \ldots, x_N(u)\big)$ is called the uth *codeword*.

Denote by $\mathbf{x} \bigvee \mathbf{y}$ $(\mathbf{x} \bigwedge \mathbf{y})$ the componentwise disjunction (conjunction) of binary vectors $\mathbf{x}$ and $\mathbf{y}$ of the same length. We say that a vector $\mathbf{x}$ *covers* a vector $\mathbf{y}$ if $\mathbf{x} \bigvee \mathbf{y} = \mathbf{x}$. For a matrix X and subset of columns $\tau \subset [t]$ consider the disjunction and conjunction

$$V(X, \tau) \triangleq \bigvee_{u \in \tau} \mathbf{x}(u), \qquad \Lambda(X, \tau) \triangleq \bigwedge_{u \in \tau} \mathbf{x}(u).$$

If $\tau = \varnothing$ then we put $V(X, \varnothing) \triangleq \mathbf{0} = (0, 0, \ldots, 0)$.

2.2 Disjunct model of DSE

Let $\mathbf{p} \subset [t]$ be a defective subset. Denote by $r(X, \mathbf{p})$ the binary vector of results of all N tests: $r(X, \mathbf{p}) \triangleq (r(G_1, \mathbf{p}), r(G_2, \mathbf{p}), \ldots, r(G_N, \mathbf{p}))$. From definition 1.1 one can see that it has the form of the disjunction of codewords:

$$r(X, \mathbf{p}) = V(X, \mathbf{p}) = \bigvee_{u \in \mathbf{p}} \mathbf{x}(u). \tag{2.1}$$

Definition 2.1 *X is a disjunct s-design iff for any two different subsets $\mathbf{p}_1, \mathbf{p}_2 \subset [t]$, $|\mathbf{p}_1| \leq s$, $|\mathbf{p}_2| \leq s$, the result vectors $r(X, \mathbf{p}_1)$ and $r(X, \mathbf{p}_2)$ are different.*

Definition 2.2 *(cf. Kautz and Singleton, 1964) Let s be an integer, $0 < s < t$. A binary $N \times t$ matrix X is called a superimposed s-code of length N and size t if for any subset $\mathbf{p} \subset [t]$, $|\mathbf{p}| \leq s$, and any sample $u \in [t] \backslash \mathbf{p}$ the codeword $\mathbf{x}(u)$ is not covered by the disjunction $r(X, \mathbf{p})$ (2.1). One can see that this is equivalent to the following condition: for the given pair $(\mathbf{p}, u)$ there exists a row number $n \in [N]$ such that $x_n(u) = 1$ and $x_n(u') = 0$ for all $u' \in \mathbf{p}$.*

Lemma 2.1 *(cf. Kautz and Singleton, 1964). Any superimposed s-code is a disjunct s-design.*

Proof. Obviously, for any sample $u \in \mathbf{p}$ the disjunction $r(X, \mathbf{p})$ (2.1) covers the codeword $\mathbf{x}(u)$. If X is a superimposed s-code and $|\mathbf{p}| \leq s$, then from definition 2.2 it follows that for any sample $u \notin \mathbf{p}$ this disjunction does not cover the codeword $\mathbf{x}(u)$. Thus, one can easily detect $\mathbf{p}$ and X is a disjunct s-design. ∎

If X is a disjunct s-design, then the complexity of the trivial algorithm of detecting $\mathbf{p}$ is $\sim \binom{t}{s}$ because we need to perform an exhaustive search over all subsets $\mathbf{p} \subset [t]$, $|\mathbf{p}| \leq s$. The superimposed s-code condition provides the simple decoding algorithm: given the vector $r(X, \mathbf{p})$ and any sample $u \in [t]$ one can easily detect whether u is defective or not. The complexity of this algorithm is $\sim t$.

Superimposed s-codes were introduced in Kautz and Singleton (1964) and studied in many papers such as D'yachkov and Rykov (1983), D'yachkov *et al.* (1989), D'yachkov (1997), D'yachkov *et al.* (2000a,b). Below we consider similar types of designs for the models of searching supersets and searching subsets in the presence of inhibitors.

2.3   Search of defective supersets

Let X be a binary $N \times t$ matrix which encodes a search strategy as described above. Let p be a defective superset (complex) composed of a number of subsets $P \subset [t]$ (1.2). Using the definition (1.3) one can easily prove that the result vector $\mathbf{r}(X, p) \triangleq (r(G_1, p), r(G_2, p), \ldots, r(G_N, p))$ in this model has the form

$$\mathbf{r}(X, p) = \bigvee_{P \in p} \Lambda(X, P) = \bigvee_{P \in p} \bigwedge_{u \in P} \mathbf{x}(u). \qquad (2.2)$$

Definition 2.3 *X is a superset* (s, ℓ)-*design (see Def. 1.2) iff for any two different complexes* p_1 *and* p_2 *composed of not more than s subsets whose sizes do not exceed* ℓ, *the results* $\mathbf{r}(X, p_1) \neq \mathbf{r}(X, p_2)$.

Definition 2.4 *(cf. Mitchell and Piper, 1988). Let s and* ℓ *be positive integers,* $s + \ell \leq t$. *A binary* $N \times t$ *matrix X is called a superimposed* (s, ℓ)-*code if for any subsets* $S, L \subset [t]$, $|S| \leq s$, $|L| \leq \ell$ *and* $S \cap L = \varnothing$ *there exists a row number* $n \in [N]$ *for which* $x_n(u) = 1$ *for any* $u \in L$ *and* $x_n(u') = 0$ *for any* $u' \in S$.

One can see that for $\ell = 1$ this definition coincides with Definition 2.2 of superimposed s-codes. Superimposed (s, ℓ)-codes were first introduced in Mitchell and Piper (1988) for cryptography applications. In the present paper we study them for search problems. Below we introduce some simple results which will be published in a more detailed form in D'yachkov *et al.* (2000d).

Lemma 2.2 *Any superimposed* (s, ℓ)-*code is a superset* (s, ℓ)-*design.*

Proof. Let p be a superset (1.2), where $k \leq s$ and $|P_i| \leq \ell$ for any i. Let X be a superimposed (s, ℓ)-code. Our aim is to detect p given the vector $\mathbf{r}(X, p)$ (2.2).

Any subset $P \subset [t]$, $|P| \leq \ell$, belongs to one of the following two types:

(α) there exists $P_i \in p$ such that $P_i \subseteq P$;

(β) P does not contain any defective subset $P_i \in p$.

Let us show that for any subset P we can detect whether it belongs to the class (α) or (β) given the result $\mathbf{r}(X, p)$. Indeed, if P satisfies (α), then from (2.2) it follows that the vector $\mathbf{r}(X, p)$ covers the conjunction $\Lambda(X, P)$.

Assume that P satisfies condition (β). Then all k sets $P_i \backslash P$ for $1 \leq i \leq k$ are not empty. Take a sample from each of these sets and construct a set $S \subset [t]$ containing all these samples, $|S| \leq k \leq s$. Put $L \triangleq P$, $|L| \leq \ell$. For this pair of sets (S, L) take a row number $n \in [N]$ according to definition 2.4. One can easily prove that the nth components of all vectors $\Lambda(X, P_i)$, $1 \leq i \leq k$, are zeros, and thus the n-th component of $\mathbf{r}(X, \mathbf{p})$ (2.2) is zero. But the n-th component of the vector $\Lambda(X, P)$ is 1. This proves that for the case (β) the vector $\mathbf{r}(X, \mathbf{p})$ does not cover $\Lambda(X, P)$.

The class (α) contains both defective subsets $P_i \in \mathbf{p}$ and subsets P such that $P_i \subsetneq P$ for some i. To separate these cases note that if $P \in \mathbf{p}$, then all subsets of P satisfy (β). And for the second case there exists a subset of P which satisfies (α).

We showed that given the vector $\mathbf{r}(X, \mathbf{p})$ it is possible to detect $\mathbf{p}$, which proves the statement. Moreover, we obtained the algorithm of such detection. Given a subset $P \subset [t]$ this algorithm detects whether $P \in \mathbf{p}$ or not. ∎

If X is a superset (s, ℓ)-design, then the complexity of the trivial algorithm for detecting a complex $\mathbf{p}$ is $\sim \binom{\binom{t}{\ell}}{s}$ because we need to perform an exhaustive search over all supersets $\mathbf{p}$. The superimposed (s, ℓ)-code condition provides the simple decoding algorithm: given the vector $\mathbf{r}(X, \mathbf{p})$ and any subset $P \subset [t]$, $|P| \leq \ell$, one can easily detect whether $P \in \mathbf{p}$ or not. The complexity of this algorithm is $\sim \binom{t}{\ell}$.

2.4 Search of defective subsets in the presence of inhibitors

Let X be a binary $N \times t$ matrix which encodes a search strategy as described above. Let $s \geq 1$ and $\imath \geq 0$ be integers, $s + \imath \leq t$. Denote by $\pi(t, s, \imath)$ the set of all possible pairs $(\mathbf{p}, \mathbf{I})$, where $\mathbf{p}$ is a defective set and $\mathbf{I}$ is a set of inhibitors:

$$\pi(s, \imath, t) \triangleq \{(\mathbf{p}, \mathbf{I}) : \mathbf{p}, \mathbf{I} \subset [t], 1 \leq |\mathbf{p}| \leq s, 0 \leq |\mathbf{p}| \leq \imath, \mathbf{p} \cap \mathbf{I} = \varnothing\}.$$

For a pair of binary symbols $x, y \in \{0, 1\}$ we define the *inhibition of x due to y* operation:

$$x \backslash y \triangleq \begin{cases} 1, & \text{if } x = 1 \text{ and } y = 0, \\ 0, & \text{otherwise.} \end{cases}$$

For a pair of binary vectors $\mathbf{x}$ and $\mathbf{y}$ we denote by $\mathbf{x} \backslash \mathbf{y}$ the componentwise inhibition operation. Note that $\mathbf{x} \backslash \mathbf{y} = \mathbf{0}$ iff $\mathbf{x}$ is covered by $\mathbf{y}$, and $\mathbf{x} \backslash \mathbf{y} = \mathbf{x}$ iff the conjunction $\mathbf{x} \bigwedge \mathbf{y} = \mathbf{0}$.

Let $(\mathbf{p}, \mathbf{I}) \in \pi(s, \imath, t)$ be a pair from the defective set and the inhibitor set. Using the definition (1.4) one can easily prove that the result vector $\mathbf{r}(X, \mathbf{p}, \mathbf{I}) \triangleq (r(G_1, \mathbf{p}, \mathbf{I}), r(G_2, \mathbf{p}, \mathbf{I}), \ldots, r(G_N, \mathbf{p}, \mathbf{I}))$ in this model has the form

$$\mathbf{r}(X,\mathsf{p}) = V(\mathsf{p})\backslash V(I) = \bigvee_{u\in\mathsf{p}} \mathbf{x}(u) \ \Bigg\backslash \ V(I) = \bigvee_{u\in\mathsf{p}} [\mathbf{x}(u)\backslash V(I)] . \qquad (2.3)$$

Definition 2.5 *X is an inhibitory $(s,\imath)$-design (see Def. 3) iff for any two pairs $(\mathsf{p},I),(\mathsf{p}',I') \in \pi(t,s,\imath)$, such that $\mathsf{p} \neq \mathsf{p}'$, the result vectors $\mathbf{r}(X,\mathsf{p},I) \neq \mathbf{r}(X,\mathsf{p}',I')$. Note that this condition allows us to detect the defective subset p but not the inhibitory set I.*

Definition 2.6 *Let $s \geq 1$ and $\imath \geq 0$ be integers, $s+\imath \leq t$. A binary $N \times t$ matrix X is called an inhibitory $(s,\imath)$-code if it is a superimposed $(s+\imath)$-code (see Def. 4).*

Lemma 2.3 *Any inhibitory $(s,\imath)$-code is an inhibitory $(s,\imath)$-design.*

Proof. Let $(\mathsf{p},I) \in \pi(t,s,\imath)$ be a pair of defective and inhibitory sets and X be an inhibitory $(s,\imath)$-code, i.e., a superimposed $(s+\imath)$-code. We should detect p given the vector $\mathbf{r}(X,\mathsf{p},I)$ (2.3).

Let us call a sample $u \in [t]$ $\imath$-*acceptable due to the vector* $\mathbf{r}(X,\mathsf{p},I)$ if there exists a subset $I' \subset [t]$ such that $u \notin I'$, $|I'| \leq \imath$ and the vector $\mathbf{r}(X,\mathsf{p},I)$ covers $\mathbf{x}(u) \setminus V(I')$.

If $u \in \mathsf{p}$, then u is acceptable because all conditions hold for $I' = I$. Assume that $u \notin \mathsf{p}$. Then for any subset $I' \subset [t]$, $u \notin I'$, $|I'| \leq \imath$, consider a pair $(\mathsf{p} \cup I', u)$ and take a row number $n \in [N]$ according to the definition of superimposed $(s+\imath)$-code (Definition 2.2). One can easily prove that the n-th component of the vector $\mathbf{r}(X,\mathsf{p},I)$ is zero and the n-th component of $\mathbf{x}(u) \setminus V(I')$ is 1. So the first vector does not cover the second one. Since this is true for any I', the sample u is not acceptable.

We proved that a sample is defective iff it is $\imath$-acceptable due to $\mathbf{r}(X,\mathsf{p},I)$. Obviously, one can check this given only the result $\mathbf{r}(X,\mathsf{p},I)$. This completes the proof of the statement and also gives the decoding algorithm. Note that one can consider only subsets I' for which $V(I') \bigwedge \mathbf{r}(X,\mathsf{p},I) = 0$. ∎

The complexity of the trivial algorithm for detecting a subset p for an arbitrary inhibitory $(s,\imath)$-design is $\sim \binom{t}{s+\imath} \cdot \binom{s+\imath}{s}$ because we should perform an exhaustive search over all pairs from the set $\pi(t,s,\imath)$. The inhibitory $(s,\imath)$-code condition provides the simple decoding algorithm: given the vector $\mathbf{r}(X,\mathsf{p})$ and a sample $u \in [t]$ we need to check all subsets I'. The complexity of this algorithm is $\sim t \cdot \binom{t}{\imath}$.

3 Concatenated construction for superimposed codes

In the present section we consider a concatenated construction for superimposed (s,ℓ)-codes. For the special case $\ell = 1$ this leads to the construction for superimposed s-codes. For this case a similar method was suggested in Kautz and Singleton (1964) and developed by D'yachkov *et al.* (2000a) and D'yachkov *et al.* (2000b).

Definition 3.1 *(cf. Friedman* et al.*, 1969) Let $q \geq 2$ be an integer and $X = \|x_n(u)\|$ be an $N \times t$ q-ary matrix: $n \in [N]$, $u \in [t]$, $x_n(u) \in [q]$. Let s and ℓ be positive integers, $s + \ell \leq t$. Then X is called a q-ary separating (s, ℓ)-code if for any two subsets $S, L \subset [t]$, $|S| \leq s$, $|L| \leq \ell$, $S \cap L = \varnothing$, there exists a row number $n \in [N]$ such that the corresponding coordinate sets S_n and L_n do not intersect, where*

$$L_n \triangleq \{x_n(u) \,:\, u \in L\} \subset [q], \qquad S_n \triangleq \{x_n(u') \,:\, u' \in S\} \subset [q].$$

Note that for $q = 2$ definition 3.1 does not coincide with definition 2.4 of the binary superimposed (s, ℓ)-code. Binary separating $(2, 2)$-codes were studied before for certain applications by Fredman and Komlos (1984) and Sagalovich (1994).

Lemma 3.1 *Let $X^{(q)}$ be a q-ary separating (s, ℓ)-code of size $t^{(q)}$ and length $N^{(q)}$. Let X' be a binary superimposed (s, ℓ)-code of size q and length N'. Then there exists a binary superimposed (s, ℓ)-code X of size $t = t^{(q)}$ and length $N = N^{(q)} \cdot N'$.*

Proof. Consider the code X obtained by the concatenation of codes $X^{(q)}$ and X', i.e., each q-ary symbol $\theta \in [q]$ in matrix $X^{(q)}$ is replaced by the θ-th codeword from X':

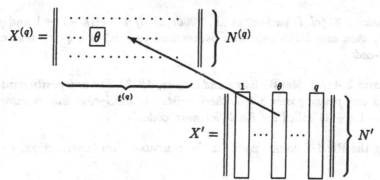

One can easily prove the superimposed (s, ℓ)-code property for X. ∎

A q-ary code $X^{(q)}$ is called an *external code*, and a binary code X' is called an *internal code*. To construct concatenated codes with large sizes we need q-ary external codes with large sizes and binary internal codes with small sizes. Below we discuss some simple methods for constructing them.

Lemma 3.2 *(trivial code) For any positive integers* s, ℓ *and* t, $s + \ell \leq t$, *there exists a superimposed* (s, ℓ)*-code of size* t *and length*

$$N = \min \left\{ \binom{t}{s}; \binom{t}{\ell} \right\}.$$

Proof. To obtain a code of length $N_1 = \binom{t}{s}$ take the binary $N_1 \times t$ matrix the rows of which are all possible binary vectors of length t having exactly s zeros. To obtain a code of length $N_2 = \binom{t}{\ell}$, take the binary $N_2 \times t$ matrix the rows of which are all possible binary vectors of length t having exactly ℓ ones. Obviously, both these matrixes satisfy the (s, ℓ)-code property. ∎

This trivial construction allows us to construct superimposed codes for all possible values of s, ℓ and t. But it is reasonable to use it only for small values of t. Note that for $t = s + \ell$ the trivial code is optimal (has the smallest possible length).

Several methods exist for constructing small superimposed codes for some special values of s and ℓ. Some tables of codes for $\ell = 1$ can be found in D'yachkov *et al.* (2000a) or D'yachkov *et al.* (2000b). The values $s = \ell = 2$ are considered in Sagalovich (1994), D'yachkov *et al.* (2000c,d) and below.

Finally we discuss a method of constructing large q-ary separating codes to be used in concatenated construction. It is based on MDS-codes. Since these codes are well-known we do not consider their properties in details.

Definition 3.2 *(cf. MacWilliams and Sloane, 1977) Any* q*-ary code of size* $t = q^k$, *length* n *and with Hamming distance* $d = n - k + 1$ *is called a maximal distance separable code (MDS-code) with parameters* (q, k, n).

Lemma 3.3 *(cf. D'yachkov* et al., *2000c,d). If* $n \geq s\ell(k-1) + 1$ *and* $q^k \geq s + \ell$, *then any MDS–code with parameters* (q, k, n) *is a* q*-ary separating* (s, ℓ)*-code.*

Lemma 3.4 *(cf. MacWilliams and Sloane, 1977). For any positive integer* λ *and any prime power* $q \geq \lambda$ *there exists an MDS–code with parameters* $(q, \lambda + 1, q + 1)$ *called the Reed–Solomon code.*

Using the Reed–Solomon code for the concatenation construction, we obtain:

Lemma 3.5 *(cf. D'yachkov* et al., *2000c,d) Let* s, ℓ, λ *be positive integers, and* $q \geq s\ell\lambda$ *be a prime power. Assume that there exists a binary superimposed* (s, ℓ)*-code of size* q *and length* N_1. *Then there exists a binary superimposed* (s, ℓ)*-code of size* $t = q^{\lambda+1}$ *and length* $N = N_1(s\ell\lambda + 1)$.

4 Examples

1. The best known superimposed 2-code of size $t = 12$ has length $N = 9$:

$$X = \begin{pmatrix} 0 & 0 & 1 & 1 & 1 & 1 & 0 & 0 & 0 & 0 & 0 & 0 \\ 0 & 0 & 1 & 0 & 0 & 0 & 1 & 1 & 1 & 0 & 0 & 0 \\ 0 & 0 & 1 & 0 & 0 & 0 & 0 & 0 & 0 & 1 & 1 & 1 \\ 0 & 1 & 0 & 1 & 0 & 0 & 1 & 0 & 0 & 1 & 0 & 0 \\ 0 & 1 & 0 & 0 & 1 & 0 & 0 & 1 & 0 & 0 & 1 & 0 \\ 0 & 1 & 0 & 0 & 0 & 1 & 0 & 0 & 1 & 0 & 0 & 1 \\ 1 & 0 & 0 & 1 & 0 & 0 & 0 & 0 & 1 & 0 & 1 & 0 \\ 1 & 0 & 0 & 0 & 1 & 0 & 1 & 0 & 0 & 0 & 0 & 1 \\ 1 & 0 & 0 & 0 & 0 & 1 & 0 & 1 & 0 & 1 & 0 & 0 \end{pmatrix}. \qquad (4.1)$$

It is the first known code for which $N < t$. For all sizes $t_1 < t$ the smallest known codes are trivial. Note that the trivial superimposed s-code of size t is the identity $t \times t$ matrix.

Note also that the matrix (4.1) is an inhibitory $(1, 1)$-code, see Def. 6.

2. Some examples of small superimposed $(2, 2)$-codes are known. For $t = 4$, the optimal code is trivial and has length $N = \binom{4}{2} = 6$. For $t = 5$, the optimal code is also trivial and has length $N = \binom{5}{2} = 10$. For $t = 6$, $t = 7$ and $t = 8$ the optimal code has length $N = 14$. It can be obtained by the concatenated method from the following 3×8 quaternary matrix:

$$C^{(4)} = \begin{pmatrix} 4 & 2 & 3 & 1 & 2 & 4 & 1 & 3 \\ 2 & 4 & 1 & 3 & 2 & 4 & 1 & 3 \\ 1 & 1 & 2 & 2 & 3 & 3 & 4 & 4 \end{pmatrix},$$

which is a separating $(2, 2)$-code. It can be concatenated with the trivial superimposed $(2, 2)$-code of size 4 and length 6. This leads to the superimposed $(2, 2)$-code of size $t = 8$ and length $N = 18$. Examining this code, one can see, that there are two rows in it, which are repeated three times. Removing the copies, we obtain the binary superimposed $(2, 2)$-code of length $N = 14$.

The following table gives several numerical values of the known superimposed $(2, 2)$-codes. Some of them were obtained with the help of V.S. Lebedev.

$t =$	4	5	8	12	16	20	25
$N \leq$	6	10	14	22	28	38	50
$t =$		64	121	512	1331	2^{12}	2^{16}	2^{20}
$N \leq$		70	110	126	198	252	364	476

Some of these codes were known before (cf. Sagalovich, 1994). Most of them are new. D'yachkov *et al.* (2000d) contains an improved table of codes.

References

De Bonis, A. and Vaccaro, U. (1998). Improved Algorithms for Group Testing with Inhibitors. *Information Processing Letters*, **67**, 57-64.

Du, D.-Z. and Hwang, F.K. (1993). *Combinatorial Group Testing and Its Applications*. Singapore: World Scientific.

D'yachkov, A.G. (1997). Designing Screening Experiments. *Lectures in the Bielefeld University*, Bielefeld.

D'yachkov, A.G., Macula, A.J., and Rykov, V.V. (2000a). New Constructions of Superimposed Codes. *IEEE Trans. Inform. Theory*, **46**, 284-290.

D'yachkov, A.G., Macula, A.J., and Rykov, V.V. (2000b). New Applications and Results of Superimposed Code Theory Arising from the Potentialities of Molecular Biology. In *Numbers, Information and Complexity*, Dordrecht: Kluwer, 265-282.

D'yachkov, A.G., Macula, A.J., Torney, D.C., Vilenkin, P.A.,and Yekhanin, S.M. (2000c). New Results in the Theory of Superimposed Codes. *Proc. of the 7-th International Workshop "Algebraic and Combinatorial Coding Theory", ACCT-7*, Bansko, Bulgaria, 126-136.

D'yachkov, A.G., Macula, A.J., Torney, D.C., and Vilenkin, P.A. (2000d). Families of Finite Sets in which No Intersection of ℓ Sets is Covered by the Union of s Others. *J. Combin. Theory, Ser. A* (submitted).

D'yachkov, A.G. and Rykov, V.V. (1983). A Survey of Superimposed Code Theory. *Prob. of Control and Inform. Theory*, **12**, 229-244.

D'yachkov, A.G., Rykov, V.V., and Rashad, A.M. (1989). Superimposed distance codes. *Prob. of Control and Inform. Theory*, **18**, 237-250.

Farach, M., Kannan, S., Knill, E., and Muthukrishnan, S. (1997). Group Testing with Sequences in Experimental Molecular Biology. *Proc. of Compression and Complexity of Sequences, IEEE Computer Society*, 357-367.

Fredman, M.L. and Komlos, J. (1984). On the Size of Separating Systems and Families of Perfect Hash Functions. *SIAM J. Algebraic Discrete Methods*, **5**, 538-544.

Friedman, A.D., Graham, R.L., and Ulman, J.D. (1969). Universal Single Transition Time Asynchronous State Assignments. *IEEE Trans. Comput.*, **18**, 541-547.

Kautz, W.H., and Singleton, R.C. (1964). Nonrandom Binary Superimposed Codes. *IEEE Trans. Inform. Theory*, **10**, 363-377.

MacWilliams, F.J. and Sloane, N.J.A. (1977). *The Theory of Error-Correcting Codes.* Amsterdam: North Holland.

Mitchell, C.J. and Piper, F.C. (1988). Key Storage in Secure Networks. *Discr. Appl. Math.*, **21**, 213-228.

Sagalovich, Yu.L. (1994). Separating Systems. *Probl. of Inform. Theory*, **30**, 105-123.

Optimal Designs for a Continuation-ratio Model

S.K. Fan
K. Chaloner

ABSTRACT: Optimal designs for a trinomial response under the continuation-ratio model are discussed. Examples of locally D-optimal and Bayesian optimal designs for some simple prior distributions are given together with closed form expressions for designs which are approximately optimal.

KEYWORDS: asymptotic optimality; Bayesian design

1 Introduction

In a continuation-ratio model for an ordinal trinomial response, if n_i experimental units are given a dose x_i, the response is trinomial, (y_{1i}, y_{2i}, y_{3i}), where y_{li} are nonnegative integers, $l = 1, 2, 3$ and $\Sigma_l y_{li} = n_i$. The unknown parameters are denoted by θ and the cell probabilities at x_i are $(p_1(\theta, x_i), p_2(\theta, x_i), p_3(\theta, x_i))$ where $\Sigma_l p_l(\theta, x_i) = 1$. The model is

$$\log[p_3(\theta, x)/(1 - p_3(\theta, x))] = a_1 + b_1 x \qquad (1.1)$$
$$\log[p_2(\theta, x)/p_1(\theta, x)] = a_2 + b_2 x. \qquad (1.2)$$

In many applications of this model, $p_1(\theta, x)$ is the probability of no reaction, $p_2(\theta, x)$ is the probability of efficacy, and $p_3(\theta, x)$ is the probability of a severe adverse outcome or death for a dose x in log units (see Thall and Russell, 1998). The model is that of two related logistic regressions: (1.1) is for adverse outcome or not and (1.2) is for cure or no reaction, conditional on no adverse outcome.

Fan and Chaloner (2000) consider a special case, where $b_1 = b_2$. Results are given here for the general case. Define

$$r = b_1/b_2$$
$$d = a_1 - r a_2.$$

mODa6, A.C.Atkinson, P.Hackl and W.G.Müller, eds., Physica, Heidelberg, 2001.

For given a_2 and b_2, when $-d/r$ gets large, the range of x with large $p_2(\theta, x)$ and small $p_3(\theta, x)$ is wider. Figure 1 shows the response probabilities for two cases: the top plot is an example of a small value, $-d/r = 0$, and the bottom plot is that of $-d/r = 15$.

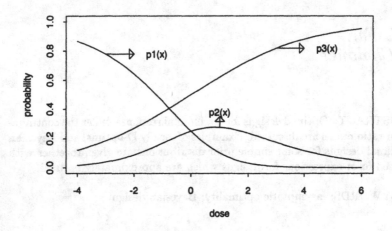

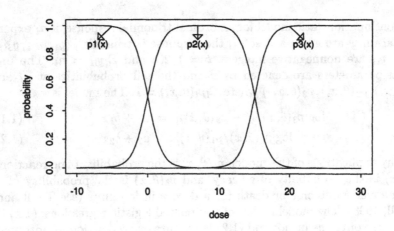

FIGURE 1. Probability plots of probability versus dose: top $(a_2, b_2, a_1, b_1) = (0, 1, 0, 0.5)$, bottom $(a_2, b_2, a_1, b_1) = (0, 1, -15, 1)$.

For a fixed sample size n, the design problem is to choose the number of dose levels, k, the dose levels, $x_1, \ldots, x_k$, and the number of experimental units, n_i, at each x_i, such that $\Sigma n_i = n$, the total sample size. An experimental design will be regarded as a probability measure on the dose domain $\mathcal{X}$.

The design measure η_i at dose x_i is n_i/n. It is assumed that the values $\eta_i, i = 1, \ldots, k$, are positive and sum to one over i but are not otherwise constrained. A design η putting measure η_i at dose x_i for $i = 1, 2, \ldots, k$ will be written as $(\eta_1, \eta_2, \ldots, \eta_k)$ at $(x_1, x_2, \ldots, x_k)$. For a given sample size n the values $n\eta_i$ can be rounded to integers in a systematic way.

Let $\theta = (a_2, b_2, a_1, b_1)$. The Fisher information matrix of a single observation for the continuation-ratio model described by (1.1) and (1.2) can be written as the following matrix of rank 2:

$$
I(\theta, x) = \frac{e^{a_2+b_2x}}{(1+e^{a_2+b_2x})^2(1+e^{a_1+b_1x})}
\begin{bmatrix}
1 & x & 0 & 0 \\
x & x^2 & 0 & 0 \\
0 & 0 & 0 & 0 \\
0 & 0 & 0 & 0
\end{bmatrix} +
$$

$$
\frac{e^{a_1+b_1x}}{(1+e^{a_1+b_1x})^2}
\begin{bmatrix}
0 & 0 & 0 & 0 \\
0 & 0 & 0 & 0 \\
0 & 0 & 1 & x \\
0 & 0 & x & x^2
\end{bmatrix}.
$$

For a design $\eta = (\eta_1, \eta_2, \ldots, \eta_k)$ at $(x_1, x_2, \ldots, x_k)$ and a sample size n, define $M(\theta, \eta) = \sum_{i=1}^{k} \eta_i I(\theta, x_i)$. Then the Fisher information matrix of this design is $nM(\theta, \eta)$. The parameter vector θ will be written as (a_2, b_2, d, r) for convenience but the information matrix $M(\theta, \eta)$, will be used to find optimal designs.

Let the set of probability measures on the dose domain $\mathcal{X}$ be denoted $\mathcal{H}$. A locally D-optimal design is the η in $\mathcal{H}$ maximizing $\phi(\eta) = \log \det M(\theta, \eta)$ for a single value of θ. This criterion averaged over a distribution on the parameters, π, gives Bayesian D-optimality: $\phi(\eta) = E_\pi \log \det M(\theta, \eta)$. Alternatively, the Bayesian c-optimality criterion minimizes the average, over a prior distribution, of the asymptotic variance of the estimate of a specific quantity of interest and local c-optimality corresponds to a point mass prior distribution. These "Bayesian" criteria can be justified as approximations to expected utility, as in Chaloner and Verdinelli (1995).

Criteria such as D-optimality and c-optimality are concave on $\mathcal{H}$ and so the optimality of any design found by numerical optimization can be verified by the General Equivalence Theorem (eg, Whittle, 1971, Chaloner and Larntz, 1989). The theorem requires showing that a directional derivative, at the candidate design, in the direction of all one-point designs is nonpositive and the supremum is 0. Let a one-point design at x be denoted η_x, then the derivative at a design η in the direction of η_x is $F_\phi(\eta, \eta_x) = \lim_{\varepsilon \to 0+} \frac{1}{\varepsilon} [\phi\{(1-\varepsilon)\eta + \varepsilon\eta_x\} - \phi\{\eta\}]$. The theorem is usually applied to univariate response models where the information matrix for η_x is of rank 1, but can easily be extended to multivariate models where the rank is greater than 1.

d value	design point	measure	design point	measure
0	−2.59	0.500	−1.40	0.500
	1.26	0.500	0.374	0.500
1	−3.54	0.500	−1.78	0.500
	0.596	0.500	−0.0347	0.500
3	−8.34	0.253	−2.62	0.500
	−3.57	0.378	−0.863	0.500
	−0.0296	0.369		
	$r = 0.5$		$r = 2$	

TABLE 1. Locally D-optimal designs for $d = 0, 1, 3$ and $r = 0.5, 2$.

Using these directional derivatives, Fan (1999) introduced a concept of "asymptotic optimality" for concave criteria. Consider a sequence of designs $\{\eta_j\}$ for a sequence of prior distributions $\{\pi_j\}$ giving a sequence of criteria $\{\phi_j\}$. Define $f_j = \sup_{x \in \mathcal{X}} F_{\phi_j}(\eta_j, \eta_x)$. Then $\{\eta_j\}$ is said to be asymptotically optimal for $\{\pi_j\}$ if $f_j \to 0$ as j approaches some limit.

Section 2 discusses properties of locally D-optimal designs and Section 3 describes Bayesian D-optimal designs. These optimal designs are found numerically and cannot be expressed in closed form, but some sequences of asymptotically optimal designs are found which can be so expressed.

The quality of designs which are not optimal can be evaluated by their efficiency as defined below.

Definition. The efficiency of a design, as a percentage, is defined to be the sample size required for an experiment using the optimal design to reach the same value of the criterion as an experiment using this design with sample size 100.

2 Locally D-optimal designs

Without loss of generality, for arbitrary d and r, only $a_2 = 0, b_2 = 1$ need be considered. Specifically if $\eta_0^* = (\eta_1^*, \eta_2^*, ..., \eta_k^*)$ at $(x_1^*, x_2^*, ..., x_k^*)$ is locally D-optimal for $\theta = (0, 1, d, r)$ then $\eta^* = (\eta_1^*, \eta_2^*, ..., \eta_k^*)$ at $(\frac{x_1^* - a_2}{b_2}, \frac{x_2^* - a_2}{b_2}, ...,$ $\frac{x_k^* - a_2}{b_2})$ is locally D-optimal for $\theta = (a_2, b_2, d, r)$. In addition, it is assumed that b_1, b_2, and their ratio r are positive, so that $p_3(\theta, x)$ is increasing with x and $p_1(\theta, x)$ is decreasing with x.

The locally D-optimal designs for several combinations of d and r were found numerically, using the software of Clyde (1993), and are listed in Table 1. Note that as the response is multivariate, 2- or 3-point designs can be optimal, even though there are 4 parameters.

Let x^* be the solution of the equation $e^x = (x+1)/(x-1)$: approximately $x^* \doteq 1.543$. For a single logistic regression with intercept 0 and slope 1, the value of x at which the probability of response is $\frac{1}{2}$ is $x = 0$ (that is the LD50 is 0), and the locally D-optimal design puts mass $\frac{1}{2}$ at each of $\pm x^*$ (see eg. Silvey, 1980). For $a_2 = 0, b_2 = 1$ and r fixed, as d becomes large and negative (1.1) and (1.2) correspond to two logistic regression models centered in different regions of $\mathcal{X}$. The LD50 of the logistic regression (1.1) is $-d/r$ and the LD50 for (1.2), efficacy vs. no reaction conditional on no adverse reaction, is 0. It therefore seems reasonable to conjecture that a good design, under local D-optimality, for a large negative value of d, is an equally weighted, 4 point design with the 4 points at the design points for the two separate logistic regressions. This guess leads to asymptotically optimal designs, as is proved in Fan (1999).

Example 1. Let $a_2 = 0, b_2 = 1$ and $r > 0$ be fixed. Define η_d^* as the design measure with mass 0.25 at each of $\{-x^*, x^*, \frac{-d-x^*}{r}, \frac{-d+x^*}{r}\}$. Then, under local D-optimality, as d goes to $-\infty$, the design η_d^* has efficiency that goes to one.

With $a_2 = 0, b_2 = 1$, and $r = 1$, for each $d < 0$ the η_d^* defined above has efficiency at least 91%. In general, for $r \neq 1$, smaller values of r give higher efficiency and larger values give lower efficiency.

3 Bayesian D-optimal designs

Suppose values for a_2 and b_2 are specified and the prior distribution puts probability at several different values of d and r. Let π be the prior distribution putting measure π_i at $\theta_i = (a_2, b_2, d_i, r_i)$, $i = 1, 2, ..., l$, and π_0 be the prior distribution putting the same measure π_i at $\theta_i^0 = (0, 1, d_i, r_i)$, $i = 1, 2, ..., l$. If η_0^* puts $(\eta_1^*, \eta_2^*, ..., \eta_k^*)$ at $(x_1^*, x_2^*, ..., x_k^*)$ and is the Bayesian D-optimal design for prior distribution π_0, then η^* putting $(\eta_1^*, \eta_2^*, ..., \eta_k^*)$ at $(\frac{x_1^* - a_2}{b_2}, \frac{x_2^* - a_2}{b_2}, ..., \frac{x_k^* - a_2}{b_2})$ is the Bayesian D-optimal design for prior π. Only such prior distributions with $a_2 = 0$ and $b_2 = 1$ will therefore be used.

Two numerical examples are given using prior distributions π_1 and π_2. Both distributions have independent marginal distributions for r and d: mass $\frac{1}{3}$ at each of $r = (\frac{1}{2}, 1, 2)$ and, independently, mass $\frac{1}{2}$ at each of $d = (0, d_0)$. For π_1, $d_0 = -10$, and for π_2, $d_0 = -20$. Both, therefore, are uniform discrete distributions with 6 support points. The corresponding Bayesian D-optimal designs are given in Table 2.

A result analogous to Example 1 has not been found using a prior distribution on both d and r. A result has been derived, however, for a 2 point prior distribution on d and $r = 1$. In addition a conjecture has been postulated for other, fixed, values of r.

Suppose $a_2 = 0, b_2 = 1$, and the value of r is given. Let the prior distribu-

Bayesian D-optimal Design			
prior distribution π_1		prior distribution π_2	
design point	measure	design point	measure
−1.800	0.344	−1.89	0.348
−0.304	0.0326	−0.289	0.0534
0.781	0.330	0.885	0.340
4.17	0.0815	9.201	0.0472
5.82	0.0823	10.8	0.0453
9.09	0.0203	18.5	0.0415
11.4	0.0578	21.6	0.0420
17.6	0.0137	36.9	0.0412
22.0	0.0376	43.1	0.0414

TABLE 2. Bayesian D-optimal designs for π_1 and π_2.

tion put mass $\frac{1}{2}$ at each of $d = 0$ and $d = d_0$. Table 3 shows the Bayesian D-optimal designs for $d_0 = -30$ and several different r values. Two design points of weight 0.375 are centered, approximately, around 0, and the other two design points are around $-d/r$. This phenomenon can be explained: when $-d/r$ is large, the conditional logistic regression (1.2) has an LD50 at $x = 0$ and the logistic regression (1.1) has an LD50 of $x = -d/r$. In addition, because information about (1.2) is gained only if an adverse outcome does not occur, more observations must be taken around zero. A reasonable design is therefore a four-point design: two design points are around zero, the other two are around $-d/r$, and the design points around zero should be weighted more than the other two design points. Bayesian D-optimal designs for several different large negative d_0 values in this prior distribution, such as $-20, -40$, have been found to have the same characteristics. The following conjecture therefore arises and has been proved for $r = 1$, but not for other values.

Conjecture 3.1 *Suppose that the prior distribution π puts mass 0.5 at each of $\theta = (0, 1, 0, r)$ and $\theta = (0, 1, d, r)$. Then there exist three constants $l_r, u_r,$ and a_r (they may depend on the value of r) such that the design η_d putting measure 0.375 at each of $x = l_r$ and u_r, and measure 0.125 at each of $x = \frac{-d}{r} + a_r,$ and $\frac{-d}{r} - a_r,$ has efficiency which goes to one, as $d \rightarrow -\infty$.*

For $r = 1$ let $l_r = -1.9$, $u_r = 0.978$, and $a_r = x^* \doteq 1.543$. Then Fan (1999) shows that if, for each d, f_d is the supremum over $x \in \mathcal{X}$ of the directional derivative at η_d in the direction of η_x, then $f_d \rightarrow 0$ as $d \rightarrow -\infty$ as required. In addition, Figure 2 illustrates that even for a small negative d value the efficiency is still very high. The lowest efficiency shown is about 97%.

r value	design point	measure
0.5	−2.15	0.375
	1.39	0.375
	56.9	0.125
	63.1	0.125
1	−1.900	0.375
	0.978	0.375
	28.5	0.125
	31.5	0.125
2	−1.49	0.375
	0.530	0.375
	14.2)	0.125
	15.8	0.125

TABLE 3. Bayesian D-optimal designs for prior distributions with mass $\frac{1}{2}$ on $d = 0$ and $d = d_0 = -30$ for r fixed.

Bayesian D-optimal design for π with $d_1 = -10$					
x_i	−1.34	1.63	4.06	5.76	9.09	11.6
η_i	0.241	0.198	0.156	0.140	0.0464	0.113
x_i	17.5	22.1				
η_i	0.0302	0.0758				
Bayesian D-optimal design for π with $d_1 = 0$					
x_i	−2.19	−0.54	0.734			
η_i	0.419	0.151	0.430			
Bayesian D-optimal design for prior distribution π with $d_1 = 2$					
x_i	−3.75	−1.87	−0.271			
η_i	0.349	0.214	0.437			

TABLE 4. Bayesian D-optimal designs for prior distribution π with $d_1 = -10, 0,$ and 2; x_i = design point; η_i =measure.

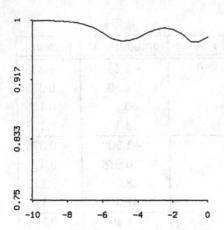

FIGURE 2. (Efficiency/100) vs. d, for the designs $\{\eta_d\}, d < 0$, with r=1.

Now consider a prior distribution π on r which puts measure 1/3 at each of $r = 0.5$, 1, and 2, and $d = d_1$, a fixed constant. Examples of Bayesian D-optimal designs for this prior distribution and three different values of d_1 are given in Table 4.

4   c-optimal designs

The c-optimality criterion minimizes the asymptotic variance of the estimate of a specific quantity of interest, $f(\theta)$. That is $\phi_c(\eta) = -E_\pi c(\theta)^T M(\theta, \eta)^- c(\theta)$ is maximized where $c(\theta)$ is the gradient vector of $f(\theta)$ and $M(\theta, \eta)^-$ is a generalized inverse of $M(\theta, \eta)$. The expectation is either over a prior distribution (Bayesian c-optimality) or over a point mass distribution (local c-optimality). Additional restrictions for estimability may be needed (see Silvey, 1980).

Suppose finding the dose, x_{max}, where $p_2(\theta, x)$ is maximized is such an $f(\theta)$. Then c-optimality for x_{max} appears to be the appropriate criterion but a closed form expression for x_{max} does not exist. The function $p_2(\theta, x)$, is maximized if, and only if

$$b_2(1 + e^{-a_1 - b_1 x}) = b_1(1 + e^{a_2 + b_2 x}).$$

If b_1 and b_2 are positive then the solution exists and is unique since $b_2(1 + e^{-a_1 - b_1 x})$ is strictly decreasing and $b_1(1 + e^{a_2 + b_2 x})$ is strictly increasing. Simulations show that the D-optimal design does not perform well for estimating x_{max} (see Fan, 1999).

Further investigation of designs for estimating x_{max} is in progress. For example if two functions can be found, $l(\theta)$ and $u(\theta)$ such that $l(\theta) <$ $x_{max} < u(\theta)$, then the mixed c-optimal design for estimating $l(\theta)$ and $u(\theta)$ might be a good candidate design. An alternative expression for the asymptotic variance might also be used.

5 Conclusion

Examining designs for these simple prior distributions and approximately optimal designs has led to insight into the design problem for this model.

References

Chaloner, K. and Larntz, K. (1989). Optimal Bayesian design applied to logistic regression experiments. *Journal of Statistical Planning and Inference* **21**, 191-208.

Chaloner, K. and Verdinelli, I. (1995). Bayesian experimental design: a review. *Statistical Science*, **10**, 273-304.

Clyde, M.A. (1993). An object-oriented system for Bayesian nonlinear design using XLISP-STAT. Technical Report, University of Minnesota.

Fan, S. (1999). *Multivariate Optimal Designs*. PhD Thesis, School of Statistics, University of Minnesota.
http://www.stat.umn.edu/~fan/thesis.html

Fan, S. and Chaloner, K. (2000). Limiting optimality for a trinomial response. *Manuscript*

Silvey, D.S. (1980). *Optimal Design*. London: Chapman and Hall.

Thall, P.F. and Russell, K.E. (1998). A strategy for dose-finding and safety monitoring based on efficacy and adverse outcomes in phase I/II clinical trials. *Biometrics* **54**, 251-264.

Whittle, P. (1973). Some general points in the theory of optimal experimental design. *Journal of the Royal Statistical Society*, **B35**, 123-130.

Bayesian Interpolation Schemes for Monitoring Systems

K. Felsenstein

ABSTRACT: Environmental monitoring gives rise to several statistical problems of optimal design. We study the question of interpolation from a Bayesian point of view. Although the considered models are motivated by applications we focus on a theoretical concept rather than the presentation of case studies. We derive Bayesian estimates and risks for variables of interest in monitoring systems under conjugate priors for multivariate normal distributions. Numerical approaches such as MCMC algorithms are discussed in addition.

KEYWORDS: Bayesian interpolation; conjugate normal Wishart distribution; Gibbs sampling; optimal design

1 Introduction

In a system of monitoring incineration plants we deal with different measurements X of elements or substances depending on several cause variables. Let $X(c)$ be a Gaussian process and $c \in C \subseteq \mathbb{R}^l$ be a variable determining the distribution of the measurement $X(c)$. Concerning applications we have a vector c of concentrations of specific substances in waste (input or output) and a resulting quantum $X(c)$ of combustion residual in waste air or slag in mind. On the other hand, the control variable c might represent the time when or the location where $X(c)$ is observed. Since the composition of waste changes within a season time becomes an influence factor. Different types of control variables (such as time and concentrations) are treated separately because of different meanings of a 'distance'.

For a finite choice of k design points $\mathbf{c} = (c_1, \ldots, c_k)$ the mean and covariances of the vector of observations $X(\mathbf{c}) = (X(c_1), \ldots, X(c_k))$ are

$$\mu(\mathbf{c}) = \mathbb{E}X(\mathbf{c})$$

and $\sigma(c_i, c_j) = cov(X(c_i), X(c_j))$. As a class of prior distributions for μ

mODa6, A.C.Atkinson, P.Hackl and W.G.Müller, eds., Physica, Heidelberg, 2001.

and the matrix of covariances Σ we consider the family of conjugate priors. Then the conditional distribution of μ is a normal distribution

$$\mu|\Sigma \sim N(m, \frac{1}{a}\Sigma)$$

and the marginal distribution of the precision matrix $\mathcal{P} = \Sigma^{-1}$ is a Wishart distribution with precision matrix S

$$\Sigma^{-1} \sim W_k(\alpha, S).$$

The mean vector and the covariance matrix follow a Normal-Wishart distribution denoted by $NWis(m, a, \alpha, S)$. Modeling prior beliefs in terms of precisions instead of variances shows methodical and technical advantages. Although not indicated, the prior hyperparameters m, S depend on c.

For fixed c the vector m represents an a priori guess for the mean function and S represents an a priori guess for the covariance matrix. The prior weights $a > 0$ and $\alpha \geq 0$ value the weight of priori knowledge compared to the number of observations.

If $X(c) \in \mathbf{R}^k$ is observed n times and $\overline{X}_n$ denotes the mean vector of n multivariate observations the updating process leads to

$$m^* = \frac{1}{a+n}(a\,m + \overline{X}_n\,n),$$

$$S^* = S + \frac{na}{a+n}(\overline{X}_n - m)(\overline{X}_n - m)^\mathsf{T} + \sum_{i=1}^{n}(X_i - \overline{X}_n)(X_i - \overline{X}_n)^\mathsf{T}$$

and $a^* = a + n$ and $\alpha^* = \alpha + n$ as posterior hyperparameters. Also, the concept of conjugate priors includes the non-informative situation as a limiting case if $a \to \infty$, $\alpha \to \infty$. The non-informative prior density for μ is constant and the density of the precision matrix is

$$\pi(\mathcal{P}) \propto \frac{1}{det(\mathcal{P}))^{(k+1)/2}}.$$

The Bayesian estimates of the parameters are m^* for μ,

$$(\alpha + n)[S + \frac{na}{a+n}(\overline{X}_n - m)(\overline{X}_n - m)^\mathsf{T} + \sum_{i=1}^{n}(X_i - \overline{X}_n)(X_i - \overline{X}_n)^\mathsf{T}]^{-1}$$

for $\mathcal{P}$ and

$$\frac{1}{\alpha+n-k}[S + \frac{n\,a}{a+n}(\overline{X}_n - m)(\overline{X}_n - m)^\mathsf{T} + \sum_{i=1}^{n}(X_i - \overline{X}_n)(X_i - \overline{X}_n)^\mathsf{T}]$$

for Σ respectively.

In principle, there is no substantial difference if we consider n independent observations or one observation $\overline{X}_n$. Therefore, we further assume $n = 1$. The formulation of the model requires a complete specification of the prior covariance structure for the field of control C. For example, the elements of the matrix S are supposed to fulfill

$$\sigma_0(c, c') = \sigma_0 \exp(-\theta_0\, g(\|c - c'\|))$$

where σ_0 and θ_0 are constants and $\|c-c'\|$ is a 'distance' between two design points. Usually, $g(.)$ is a monotone function. We propose a hierarchical prior. It seems practicable to specify a prior distribution $\pi(\theta, \sigma)$ obtaining the target covariance structure as expectation

$$\sigma_0(c, c') = \int \sigma^2 \exp(-\theta\, g(\|c - c'\|))\, d\pi(\theta, \sigma).$$

In the hierarchical model the non-informative case is

$$\pi(\theta) \propto \theta^{-[1+1/\varepsilon]}$$

if $g(t) = t^\varepsilon$ and $1 < \varepsilon < 2$. The application of such a prior leads to an 'optimistic' situation concerning the information about the parameters. Taking the average with respect to the prior of (θ, σ) might lead to narrow predictive distributions due to undersized posterior variances.

2 Estimation of monitoring functions

Exact measurement of several substances in waste air represents a technical challenge. To describe the flow of substances so called "transfer coefficients are introduced. The transfer coefficients have to be calculated from the observations. For example, the variable of interest $X = X_1 X_2$ becomes a product of the measured quantum X_1 and a transfer coefficient X_2. In general, we consider functions H

$$X = H(X_1, X_2, \ldots, X_m),$$

where $X_1, X_2, \ldots, X_m$ are Gaussian variables equipped with Normal-Wishart priors. Regulations and laws for incinerator plants are usually based on threshold values for the mean and the variances of such functions. Thus, there is a need to provide estimates for the mean and variances of X.

A simple function is the product $X = X_1 X_2$. If the monitoring variable X_1 and the transfer coefficient X_2 are independent, the structure of covariance is given through

$$var(X) = \sigma_1^2 \sigma_2^2 + \mu_1^2 \sigma_2^2 + \mu_2^2 \sigma_1^2 \tag{2.1}$$

and

$$cov(X, X') = \sigma_1' \sigma_2' + \mu_1^2 \sigma_2' + \mu_2^2 \sigma_1' \tag{2.2}$$

with corresponding covariances $\sigma_i' = cov(X_i, X_i'), i = 1, 2$.

If independence is abandoned, or more quantities are involved, the covariance structure becomes much more complicated. To keep the model manageable the law of the propagation of errors is applied in practice (see Morf, 1998). Consequently, variances or covariances of the interesting variables can be described as quadratic polynomials or rational functions of components of μ and Σ. Note that the covariances in (2.2) belong to that class of functions.

Since the marginal distribution of $\mathcal{P}$ is a Wishart distribution it follows that quadratic forms (with constant vector d) are χ^2 distributed

$$(d^\mathsf{T} \Sigma d)^{-1} \sim (d^\mathsf{T} n S^* d)^{-1} \chi^2$$

leading to the Bayesian estimates of quadratic forms. It is necessary to consider more generalized 'quadratic forms'.

In view of that we assume that our parameter of interest η consists of sums of quadratic forms of following types,

$$q_1 = \mu^\mathsf{T} M_1 \mu$$

$$q_2 = \mu^\mathsf{T} M_2 \Sigma M_3 \mu$$

$$q_3 = a_1^\mathsf{T} \Sigma M_4 \Sigma a_2$$

$$q_4 = a_3^\mathsf{T} \Sigma M_5 \Sigma^{-1} a_4,$$

where the matrices $M_1, \ldots, M_5$ and the vectors a_1, a_2, a_3 are independent of μ and Σ and of appropriate dimensions. The general Bayesian estimates of quadratic matrix functions are summarized in the following proposition.

Result 1   The Bayesian estimates of quadratic forms $q_1, \ldots, q_4$ under Normal-Wishart prior $NWis(m, a, \alpha, S)$ are

$$\hat{q}_1 = \frac{1}{a^*(\alpha^* - 1 - k)} tr(M_1 S^*) + m^{*\mathsf{T}} M_1 m^*;$$

$$\hat{q}_2 = \frac{m^{*\mathsf{T}} M_2 S^* M_3 m^*}{\alpha^* - 1 - k} + \frac{tr(M_2 S^*[(\alpha^* - 2 - k)M_3 + M_3^\mathsf{T}]S^*)}{a^*(\alpha^* - k)(\alpha^* - k - 1)(\alpha^* - k - 3)}$$
$$+ \frac{tr(M_2 S^*) tr(M_3 S^*)}{a^*(\alpha^* - k)(\alpha^* - k - 1)(\alpha^* - k - 3)};$$

$$\hat{q}_3 = \frac{a_1^\mathsf{T}[(\alpha^* - 2 - k)S^* M_4 S^* + S^* M_4^\mathsf{T} S^* + tr(M_4 S^*)S^*]a_2}{a^*(\alpha^* - k)(\alpha^* - k - 1)(\alpha^* - k - 3)};$$

$$\hat{q}_4 = \frac{a_3^\mathsf{T}[a^* S^* M_5 S^{*-1} - M_5^\mathsf{T}]a_4 - tr(M_5)a_3^\mathsf{T} a_4}{\alpha^* - k - 1}.$$

For the (technical) proofs of the results in this paper we refer to propositions or corollaries in Felsenstein (1996), (2000) and Felsenstein and Pötzelberger (1992).

3 Interpolation

Risk assessments of air pollution requires estimates of concentration levels where no monitoring sites or no vector of concentrations are available. In the practical situation there are observable measurements with points in C and points $(c_0^{(1)}, \ldots, c_0^{(k_0)})$ where it is not possible to observe X. For example, c_0 represents a specific composition of substances in waste but that composition is not available in the actual input. We intend to interpolate $X(c_0)$ for $c_0 = (c_0^{(1)}, \ldots, c_0^{(k_0)})$.

The common vector $(X_0, X) \in \mathbb{R}^{k_0 + k_1}$ consists of the interpolation vector $X_0 = X(c_0)$ and the vector of observable measurements $X = X(c)$ with $c = (c_1, \ldots, c_{k_1})$. A Normal-Wishart distribution for the mean and covariance induces a $k = k_0 + k_1$ dimensional t-distribution for (X_0, X) with precision matrix

$$B = \frac{a}{a+1}(a - k + 1)S^{-1}.$$

The prior covariance matrix S for (X_0, X) is partitioned into

$$S = \begin{pmatrix} S_{00} & S_{01}^\mathsf{T} \\ S_{01} & S_{11} \end{pmatrix}. \tag{3.1}$$

S_{00} consists of prior covariances for the points of interpolation, S_{11} consists of prior covariances for the observable points and S_{01} consists of the covariances between observations and interpolated points. The matrix B is partitioned in the same way as S in (3.1). Also, the conditional distribution of $X_0|X$ which serves as the Bayesian predictive distribution is a t-distribution with $\alpha + 1 - k_0$ degrees of freedom and the mean vector

$$\mathbb{E}(X_0|X) = m_0 - B_{00}^{-1} B_{01}^\mathsf{T}(X - m)$$

where $(m_0, m) \in \mathbb{R}^k$ are the prior means corresponding to (X_0, X). The covariance matrix of that distribution is

$$cov(X_0|X) = \frac{\alpha - k_0 + 1}{\alpha - k_0 - 1} B_{00}^{-1}[\alpha - k + 1 + (X - m)^\mathsf{T}(B_{11} - B_{01}B_{00}^{-1}B_{01}^\mathsf{T})(X - m)].$$

Under the assumption of quadratic loss the Bayesian interpolation vector is $\mathbb{E}(X_0|X) = m_0 - B_{00}^{-1} B_{01}^\mathsf{T}(X - m)$ leading to (minimal) Bayes risk

$$r = \mathbb{E}_X tr(cov(X_0|X)).$$

Since B_{11} is the precision matrix of the (marginal) t-distribution of X the expectation of the trace of the covariance matrix reads

$$r = \frac{\alpha - k + 1}{\alpha - k - 1} tr(B_{00}^{-1}).$$

Therefore, the task to reach an optimal choice of points for observations is to minimize

$$tr(S_{00} - S_{01}^T S_{11}^{-1} S_{01}).$$

Next, we derive the Bayes risk of interpolation of a product $\tilde{X} = X_1 X_2$. The interpolations are $X_1^{(0)} \in \mathbf{R}^{k_0}$ and $X \in \mathbf{R}^{k_1}$ are the observations. A second process $X_2^{(0)} \in \mathbf{R}^{k_0}$ equipped with a Normal-Wishart model is assumed to be independent of X_1 given the observations X. Of course, the prior distributions have to coincide concerning all matters of X. Therefore the prior Normal-Wishart distributions for $(X_1^{(0)}, X)$ and $(X_2^{(0)}, X)$ differ only in the prior Wishart precisions $S^{(1)}$ and $S^{(2)}$ respectively.

For the interpolation of the product we compute the Bayes risk. Without loss of generality, we assume that both variables have mean 0 at the points of interpolation. We denote the elementwise multiplication of matrices by $\star$.

Result 2   The Bayes risk under quadratic loss for the interpolation vector $\tilde{X}_0$ is

$$r = c\, tr([S_{00}^{(1)} - S_{01}^{(1T)} S_{11}^{(1)-1} S_{01}^{(1)}] \star [S_{00}^{(2)} - S_{01}^{(2T)} S_{11}^{(2)-1} S_{01}^{(2)}])$$

where the matrices $S^{(1)}$ and $S^{(2)}$ are partitioned according to (3.1) and the constant c is

$$c = \frac{(\alpha - k_0 + 1)^2 (a+1)^2}{(\alpha - k_0 - 1)^2 a^2} [\alpha - k + 1 + \frac{k_1(2\alpha + k_1\alpha - k_1^2 - k_0 k_1 - 5k_1 - 2k_0}{(\alpha - k - 1)^2(\alpha - k - 3)}].$$

The task of calculating or even minimizing Bayes risk becomes hardly practicable if we consider dependent factors and other functions than products. Therefore, we turn to a simulation model using MCMC algorithms.

The posterior and the predictive distribution are explored via Gibbs sampling. MCMC provides a straightforward method to generate variates from the common distribution of the parameters μ, Σ and the interpolation X_0. The Gibbs sampler as a special MCMC algorithm works in this case as follows. First, pick arbitrary starting values $\eta^{(0)} = (\mu^{(0)}, \Sigma^{(0)}, X_0^{(0)})$ and then successively make random variate drawings from each of the conditional distributions in turn

$$\mu^{(k)} \sim \mu | \Sigma^{(k-1)}, X_0^{(k-1)}, X$$

$$\Sigma^{(k)} \sim \Sigma | \mu^{(k)}, X_0^{(k-1)}, X$$

$$X_0^{(k)} \sim X_0 | \mu^{(k)}, \Sigma^{(k)}, X$$

for $k = 1, 2, \ldots$. The conditional distributions are Normal or Wishart distributions which might be generated from normal variates.

If the chain is simulated long enough the empirical distribution in the k-th step should approximate the distribution of $\eta = (\mu, \Sigma, X_0)$ given the observations X. If the target variable is the mean or covariance of a function of the data process or the loss function it seems to be evident that it is easier to construct and simulate such a Markov chain rather than to analyze the posterior distribution or integrals of it directly. Even for the original conjugate model the Gibbs sampler proves to be useful. The flexibility of the method allows to extend it in several ways.

For example, let us return to the design problem of choosing points $c \in C$ where observations should be taken. In this case we add X as 'parameter' and restart the algorithm. The resulting sample of $\eta = (\mu, \Sigma, X_0, X)$ serves as the basis of a rejection method to select c for $X(c)$ minimizing the Bayes risk.

Of course, the quality of all statistical procedures depends upon the convergence of the Gibbs sampler. Applying results of Smith and Roberts (1993) and Roberts and Rosenthal (1998) we are able to confirm convergence in terms of total variation distance. The total variation distance between probability measures P,Q, means the maximum difference of probabilities over all measurable sets A,

$$\|P - Q\| = \sup_A |P(A) - Q(A)|.$$

Result 3 Let $F^{(k)}$ denote the empirical distribution of $\eta^{(k)} = (\mu^{(k)}, \Sigma^{(k)}, X_0^{(k)})$ generated by Gibbs sampling. If the prior distribution is a conjugate Normal-Wishart or non-informative (together with enough observations) then $F^{(k)}$ converges to the posterior distribution $\pi(.|X)$ of η in the sense of the total variation distance,

$$\|F^{(k)} - \pi(.|X)\| \to 0$$

if $k \to \infty$. For any $\pi(.|X)$ and integrable function H (e.g. the loss function) the moment estimate of H converges to the integral a.s.,

$$\frac{1}{k} \sum_{i=1}^{k} H(\eta^{(i)}) \to \int H(s) \, \pi(s|X) \, ds.$$

This result applies to various other situations of prior distributions. For example, if the priors of μ and Σ are independent (Normal and Wishart) convergence holds.

The problem of establishing rigorous quantitative bounds of convergence remains unsolved in the general case. In Rosenthal (1995) and Roberts and Rosenthal (1998) partially solutions are given. The results are that the rate of convergence is bounded by a geometrical series if certain conditions are fulfilled. The involved constants as well as the conditions are hard to verify and cannot be easily transferred into bounds for practical use.

What remains is purely empirical convergence diagnostic as suggested in Gelman and Rudin(1992). There the statistical properties of the generated samples are used to asses stationarity of the Markov chain. These methods are sensitive and have to be handled with care and statistical precaution.

4   Applications

Finally, we get a glimpse of real data. In Morf (1998) a detailed study of waste management and monitoring in an incineration plant can be found. We take data as well as models out of it to demonstrate the Bayesian method. Suppose the quantity of interest X is the concentration of one of the elements Cu,Zn,Pb,Cd and depends on X_1, the quantity of slag, X_2 the concentration in slag, and X_3 a transfer coefficient. Transfer coefficients describe the flow of a substance into the chemical residuals of combustion. Let $X = H(X_1, X_2, X_3) = cX_1X_2/X_3$. Previous studies provide prior information for mean and correlations of the variables and the results are compared to a non-informative situation. All variables are correlated and the variances are approximated by use of the law of the propagation of errors.

The design points c are vectors of concentrations of other elements, namely S, C, Cl. To solve the problem of optimal interpolation of X or the transfer coefficient for a 'standard' composition

$$c_0 = (C : 250g/kg; S : 2.8g/kg; Cl : 7.1g/kg)$$

we started a numerical calculation. The results of the optimal interpolation using Gibbs sampling with 500 iterations are shown in the following tables.

Concentration (g/kg):	Cu	Zn	Pb	Cd
informative	0.44	0.93	0.62	0.009
non informative	0.41	0.86	0.7	0.01

Transfer coefficient into slag: (%):	Cu	Zn	Pb	Cd
informative	92	45.2	71.8	11.3 .
non informative	91.5	0.81	0.59	10.4

Additionally, we give the coefficients of variation v_c of the prediction distributions.

v_c (%):	Cu	Zn	Pb	Cd
Concentration	29.2	8.1	24.3	18.7 .
Transfer coefficient	5.3	11.5	12.1	20.7

References

Felsenstein, K. (1996). *Bayes'sche Statistik für kontrollierte Experimente.* Vandenhoeck & Ruprecht, Göttingen.

Felsenstein, K. (2000). On Bayes estimators and Bayes risk in conjugate Wishart models. (Submitted for publication).

Felsenstein, K. and Pötzelberger, K. (1992). Ein Bayes'scher Ansatz für die Planung von Meßstellen. *Schriftenr. der Techn. Univ. Wien,* **29,** 89-100.

Gelman, A. and Rudin, D. (1992) Inference from iterative simulation using multiple sequences. *Statistical Science,* **7,** 457-472.

Morf, L. (1998). *Entwicklung einer effizienten Methode zur kontinuierlichen Bestimmung von Stoffflüssen durch eine Müllverbrennungsanlage.* PhD Thesis, Technical University of Vienna.

Roberts, G. and Rosenthal, J. (1998). Markov chain Monte Carlo: Some practical implications of theoretical results. *Canadian J. Statist.,* **26,** 5-31.

Rosenthal, J. (1995). Rates of convergence for Gibbs sampling for variance components models. *Annals of Statistics,* **23,** 740-761.

Smith, A. and Roberts, G. (1993). Bayesian computation via the Gibbs sampler and related Markov chain Monte Carlo methods (with discussion). *Journal of the Royal Statistical Society,* **B55,** 3-24.

Optimality of the Wald SPRT for Processes with Continuous Time Parameter

L.I. Galtchouk

ABSTRACT: The paper deals with the Wald sequential test of two simple hypotheses for processes with continuous time parameter. The observation is a likelihood ratio process, which is a right-continuous local martingale having left-side limits. The Wald sequential test is proved to be optimal in the sense that it minimizes the Kullback-Leibler information under both hypotheses among all tests having no larger error probabilities. The proof is based on the explicit solution of the related optimal stopping problem.

KEYWORDS: Wald sequential test; continuous time observations; martingale; optimal stopping problem

1 Introduction

The classical Wald sequential probability ratio test (SPRT) was introduced for independent identically distributed observations(see Wald, 1947; Wald and Wolfowitz, 1948; Lehmann, 1959 and Ghosh and Sen, 1991). The test is optimal in the sense that it minimizes the mean observation times under both hypotheses among all tests having no larger error probabilities.

After Wald's paper the problem was considered for non identically distributed observations, for dependent observations or, more generally, for processes with continuous time parameter. Dvoretzky, Kiefer and Wolfowitz (1953) studied the Wald test for some homogenuous processes with independent increments. They expected that the Wald test would be optimal in this case, but they did not give the proof.

Shiryaev (1969) has extended Wald's result to the process $X_t = \theta t + W_t$, where $(W_t)_{t \geq 0}$ is the standard Wiener process and the hypotheses are : $H_0 : \theta = 0$ against $H_1 : \theta = 1$.

Later, Lipster and Shiryaev (1974) studied the same problem for the ob-

mODa6, A.C.Atkinson, P.Hackl and W.G.Müller, eds., Physica, Heidelberg, 2001.

servation process $(X_t)_{t \geq 0}$ specified by the stochastic differential equation $dX_t = \theta_t dt + dW_t$, where $(\theta_t)_{t \geq 0}$ is a signal process and the hypotheses are: H_0 : "the signal is present" against H_1 : "the signal is absent". In this case the Wald sequential test minimizes the Kullback-Leibler information under both hypotheses among all tests having no larger error probabilities.

By making use of Shiryaev's method, Yashin (1983) proved optimality of the Wald sequential test in the same sense provided the likelihood ratio process has continuous trajectories.

The Wald sequential test optimality for homogeneous independent increments processes was studied by Irle and Schmitz (1984). Irle (1984) proved the extended optimality of the Wald test for stochastic processes for which the likelihood ratio process attains one of two thresholds at the termination time of the Wald test. Peskir and Shiryaev (2000) have recently studied in detail sequential testing problems for stationary Poisson processes. Some results for processes with jumps are given by Koell (1994).

It must be noted that in the i.i.d. observations case the test minimizes in fact the Kullback-Leibler information because in this case the information is proportional to mean observation time.

This paper deals with the case in which the likelihood ratio process is a local martingale with right-continuous trajectories having left-side limits. This model includes, in particular, both dependent and not identically distributed observations. We prove that the Wald sequential test is optimal in the Kullback-Leibler sense. The proof makes use of the Snell envelope and a convex characterization of the payoff function to obtain the explicit solution of a related optimal stopping problem.

The remainder of this paper is arranged as follows. In Section 2 the optimality property is proved for the Bayesian setting. The main result is proved in Section 3. Section 4 contains some examples and an auxiliary proposition.

2 Optimality of the Wald sequential test in the Bayesian setting

Let $(\Omega, \mathcal{F}, (\mathcal{F}_t)_{t \geq 0}, \mathbf{P}_\pi, \pi \in [0,1])$ be a filtered probability space, where $\mathbf{P}_\pi(\cdot) = \pi \mathbf{P}_1(\cdot) + (1-\pi)\mathbf{P}_0(\cdot)$; $\mathbf{P}_i, i = 0, 1$, be given probability measures on the measurable space $(\Omega, \mathcal{F})$; $(\mathcal{F}_t)_{t \geq 0}$ be the right-continuous filtration completed by all $\mathbf{P}_0 + \mathbf{P}_1$ null sets; π and $1 - \pi$, be *prior probabilities* of hypotheses H_1 : the law is $\mathbf{P}_1, H_0$: the law is $\mathbf{P}_0$, respectively :

$$\mathbf{P}_\pi(H_1) = \pi, \ \mathbf{P}_\pi(H_0) = 1 - \pi.$$

We suppose that $\mathcal{F} = \mathcal{F}_\infty = \sigma(\bigcup_{t \geq 0} \mathcal{F}_t)$. Denote by $\mathbf{P}_{it}$ the restriction of $\mathbf{P}_i$ to the σ-algebra $\mathcal{F}_t, t \geq 0, i = 0, 1$, and suppose that

$$\mathbf{P}_{0t} \sim \mathbf{P}_{1t}, t \geq 0.$$

The equivalence of measures $\mathbf{P}_{it}, i = 0, 1$, implies the existence of non-negative processes

$$m_t = \frac{d\mathbf{P}_{1t}}{d\mathbf{P}_{0t}}, \quad m_t^{-1} = \frac{d\mathbf{P}_{0t}}{d\mathbf{P}_{1t}}, \quad t \geq 0.$$

We call $(m_t), (m_t^{-1})$ *the density processes* or *the likelihood ratio processes*. The processes $(m_t, \mathcal{F}_t, \mathbf{P}_0)_{t \geq 0}$, $(m_t^{-1}, \mathcal{F}_t, \mathbf{P}_1)_{t \geq 0}$ are local martingales. In the sequel we use a modification with right-continuous trajectories having left-side limits.

The goal is to prove that the Wald sequential test is optimal for testing the hypotheses

$$H_0 : \text{the model } (\Omega, \mathcal{F}, (\mathcal{F}_t)_{t \geq 0}, \mathbf{P}_0) \text{ is realized}$$

against

$$H_1 : \text{the model } (\Omega, \mathcal{F}, (\mathcal{F}_t)_{t \geq 0}, \mathbf{P}_1) \text{ is realized}$$

by observations of the process $(m_t)_{t \geq 0}$.

Definition The couple $\delta(\omega) = (\tau(\omega), d(\omega))$ is called *the test* or *the decision rule*, where τ is a $(\mathcal{F}_t)$-stopping time, $d \in \{0, 1\}, d$ is a $\mathcal{F}_\tau$-measurable random variable. The decision $d = i$ means that the hypothesis H_i is accepted, $i = 0, 1$.

For any decision rule $\delta = (\tau, d)$ we define the error probabilities of the first and the second kind

$$\alpha(\delta) = \mathbf{P}_0(d = 1), \quad \beta(\delta) = \mathbf{P}_1(d = 0),$$

respectively.

Let

$$\mathcal{U}(m) = \{x \in \mathbb{R} : \mathbf{P}_0(\exists t \geq 0 : m_t \in \mathcal{U}_\varepsilon(x)) > 0, \text{ for any } \varepsilon > 0\},$$

where $\mathcal{U}_\varepsilon(x) =]x - \varepsilon, x + \varepsilon[$ is an ε-neighbourhood of the point x;

$$\mathfrak{M}_0(G \circ m) = \{\tau : \tau \text{ is a } (\mathcal{F}_t) - \text{stopping time such that}$$

$$\tau < \infty, \mathbf{P}_i - \text{a.s.}, \mathbf{E}_i \mid G \circ m_\tau \mid < \infty, i = 0, 1\},$$

where G is a given function,

$$\Delta_{\alpha, \beta} = \{\delta = (\tau, d) : \alpha(\delta) \leq \alpha, \beta(\delta) \leq \beta, \tau \in \mathfrak{M}_0(\ln m)\},$$

where $\alpha, \beta \in [0, 1]$.

Definition Let $(m_t)_{t \geq 0} = (d\mathbf{P}_{1t}/d\mathbf{P}_{0t})_{t \geq 0}$ be the likelihood ratio process and a, b be two given numbers, $0 < a < m_0 < b < \infty, \mathbf{P}_0-$a.s. The decision rule $\delta_{ab} = (\tau_{ab}, d_{ab})$ is called *the Wald sequential test associated with the thresholds a, b, if*

$$\tau_{ab} = \begin{cases} \inf\{t \geq 0 : m_t \notin]a, b[\}, \\ \infty, \text{if the set}\{\cdot\} = \varnothing, \end{cases}$$

$$d_{ab} = \begin{cases} 1, \text{if } m_{\tau_{ab}} \geq b, \\ 0, \text{if } m_{\tau_{ab}} \leq a. \end{cases}$$

If the hypothesis H_i is true and the decision rule $\delta = (\tau, d)$ is taken, we have to pay the following penalties :

1) the resulting losses $l(i, d)$ of the wrong decision :

$$l(0,0) = l(1,1) = 0, \ l(0,1) = l_0 > 0, \ l(1,0) = l_1 > 0,$$

2) the observed information costs :

$$c \ln(1/m_\tau), \text{ if hypothesis } H_0 \text{ is true,}$$

$$c \ln m_\tau, \text{ if hypothesis } H_1 \text{ is true,}$$

where c is a positive number and (m_t) is the likelihood ratio process.

With every decision rule $\delta = (\tau, d)$ and prior probability π we associate the risk function

$$R(\pi, \delta) = (1 - \pi)l_0\alpha(\delta) + \pi l_1\beta(\delta) + (1 - \pi)c\mathbf{E}_0[\ln(1/m_\tau)] + \pi c\mathbf{E}_1[\ln m_\tau]. \tag{2.1}$$

For fixed $\alpha, \beta \in [0, 1]$, the decision rule $\delta_1 = (\tau_1, d_1) \in \Delta_{\alpha, \beta}$ is called $\pi-Bayesian$, if

$$R(\pi, \delta_1) = \inf_{\delta \in \Delta_{\alpha, \beta}} R(\pi, \delta).$$

Our goal is to find a $\pi-Bayesian$ rule for the risk function $R(\pi, \delta)$.

Lemma 2.1 *Let $R(\pi, \delta)$ be as in (2.1). Then*

$$\inf_{\delta \in \Delta_{\alpha, \beta}} R(\pi, \delta) = c \inf_{\tau \in \mathfrak{M}_0(F \circ m)} \mathbf{E}_0 F \circ m_\tau,$$

where

$$F(x) = \pi x \ln x - (1 - \pi) \ln x + (1 - \pi)w_0 \wedge \pi w_1 x, \ w_i = l_i/c, i = 0, 1. \tag{2.2}$$

The proof of this result is standard(see, for example, Koell, 1994).
We shall need the following results:

Lemma 2.2 *Let* $G = G(x), x \in \mathbb{R}$, *be a continuous function having at least one convex minorant. Then:*

1) for the restriction of G to $\mathcal{U}(m)$ there exists the greatest convex minorant $\underline{G} = \underline{G}(x), x \in \mathcal{U}(m)$;

2) the function $\underline{G}(x), x \in \mathcal{U}(m)$, is linear on each connected component D_i of the set D including the end points, where

$$D = \{x \in \mathcal{U}(m) : \underline{G}(x) < G(x)\};$$

3) equalities

$$\underline{G}'(a_i) = G'(a_i), \underline{G}'(b_i) = G'(b_i)$$

hold for those end points of the component $D_i = (a_i, b_i) \subset D$ which lies in $\mathbb{R} \cap \mathcal{U}(m)$ and at which the function G is differentiable.

This result can be proved by the same arguments as the similar result of Galtchouk and Miroshnichenko (1997).

Let (M_t) be a local martingale; $F = F(x), x \in \mathbb{R}$, be a continuous lower bounded function. Consider the martingale optimal stopping problem

$$\inf_{\tau \in \mathfrak{M}_0(F \circ M)} \mathbf{E}F \circ M_\tau.$$

Denote by $(X_t)_{t \geq 0}$ the Snell envelope for this problem, i.e.

$$X_t = \text{ess} \inf_{\tau \in \mathfrak{M}_0(F \circ M), t \leq \tau} \mathbf{E}[F \circ M_\tau \mid \mathcal{F}_t] \text{ a.s.}$$

There exists a process $(X_t)_{t \geq 0}$ with right-continuous trajectories having left-side limits.

The following properties of the Snell envelope are well-known:

Proposition 2.1 *1) Under the above assumptions concerning the process $(F \circ M_t)$, the process (X_t) is its greatest submartingale minorant.*
Denote

$$\tau_0 = \inf(t \geq 0 : X_t = F \circ M_t)$$

($\tau_0 = \infty$, if the set $(\cdot)$ is empty). Suppose that $\tau_0 \in \mathfrak{M}_0(F \circ M)$. Then
2) the process $(X_{\tau_0 \wedge t})_{t \geq 0}$ is a local martingale,
3) the stopping time τ_0 is optimal :

$$\mathbf{E}F \circ M_{\tau_0} = \inf_{\tau \in \mathfrak{M}_0(F \circ M)} \mathbf{E}F \circ M_\tau.$$

The following result gives a double description for the optimal stopping time in the previous martingale optimal stopping problem.

Theorem 2.1 *Let (M_t) be a local martingale; $F = F(x), x \in \mathbb{R}$, be a continuous lower bounded function; $\underline{F}(x), x \in \mathcal{U}(M)$, be the greatest convex minorant of the restriction of F to $\mathcal{U}(M); (X_t)$ be the Snell envelope of the process $F \circ M$. Denote*

$$D = \{x \in \mathcal{U}(M) : \underline{F}(x) < F(x)\},$$
$$\tau_0 = \inf\{t \geq 0 : X_t = F \circ M_t\},$$
$$\tau_D = \inf\{t \geq 0 : M_t \notin D\},$$
$$\tau_0 = +\infty \ (\text{resp.} \tau_D = +\infty) \text{ if the set} \{\cdot\} = \varnothing.$$

Assume
(i) the set D is bounded connected and nonempty,
(ii) $\tau_D \in \mathfrak{M}_0(F \circ M)$.
Then $\tau_0 = \tau_D$ a.s.

The proof of this theorem is given by Galtchouk (2000).

Theorem 2.2 *Let $F = F(x), x \in \mathcal{U}(m)$, be the function given in (2.2) and $\underline{F} = \underline{F}(x), \ x \in \mathcal{U}(m)$, be its greatest convex minorant. Suppose that the function F is differentiable at the end points of the set $D = \{x \in \mathcal{U}(m) : \underline{F}(x) < F(x)\}$. Then*
1) $D = (a, b)$, where the numbers a, b are defined by the system

$$\begin{cases} \underline{F}'(a) = F'(a), \\ \underline{F}'(b) = F'(b), \end{cases}$$

that is

$$\begin{cases} (b-a)^{-1}[\pi(b\ln b - a\ln a - w_1 a) - (1-\pi)(\ln b - \ln a - w_0)] \\ \qquad\qquad = \pi(\ln a + w_1 + 1) - (1-\pi)/a, \\ (b-a)^{-1}[\pi(b\ln b - a\ln a - w_1 a) - (1-\pi)(\ln b - \ln a - w_0)] \\ \qquad\qquad = \pi(\ln b + 1) - (1-\pi)/b. \end{cases} \quad (2.3)$$

2) For every $w_0 > 0, w_1 > 0, \pi \in (0,1)$, system (2.3) admits the unique solution :

$$a = \frac{1-\pi}{\pi} k^{-1} \frac{w_0 - r(k)}{w_1 + r(k^{-1})}, \quad b = \frac{1-\pi}{\pi} \frac{w_0 - r(k)}{w_1 + r(k^{-1})},$$

where $r(k) = \ln k - k + 1, k = k(w_0, w_1)$ is the unique solution of the equation

$$k = \frac{\left(2r(k) - (w_0 + w_1)\right)^2 - (w_0 - w_1)^2}{\left(2r(k^{-1}) + (w_0 + w_1)\right)^2 - (w_0 - w_1)^2},$$

such that $1 < k(w_0, w_1) < k(w_0 \wedge w_1)$, where $k(w_0 \wedge w_1)$ is the greatest root of the equation

$$r(k) + w_0 \wedge w_1 = 0.$$

Remark 1 *The assumption that the function F given in (2.2) is differentiable at the end points a, b means that these points belong to $\mathcal{U}(m)$ with some neighbourhoods.*

Proof. 1) The function $F(x)$ is convex on the intervals $(0, (1-\pi)w_0/\pi w_1)$ and $((1-\pi)w_0/\pi w_1, \infty)$, because its second derivative is positive on these intervals. The left-hand side and the right-hand side derivatives satisfy the inequality at the point $x = (1-\pi)w_0/(\pi w_1)$

$$F'((1-\pi)w_0/(\pi w_1)-) = \pi[\ln \frac{(1-\pi)w_0}{\pi w_1} - \frac{w_1}{w_0} + 1 + w_1]$$

$$> F'((1-\pi)w_0/(\pi w_1)+) = \pi[\ln \frac{(1-\pi)w_0}{\pi w_1} - \frac{w_1}{w_0} + 1].$$

Hence, the function $F(x)^{\cdot}$ has a beak-like form at the point $x = (1-\pi)w_0/(\pi w_1)$ and $D = (a, b)$. The function $F(x), x \in \mathbb{R}_+$, is differentiable everywhere except the point $x = (1-\pi)w_0/\pi w_1$. By Lemma 2.2, the function F is linear on the interval (a, b) and system (2.3) holds.

2) We write the system (2.3) as

$$\begin{cases} \pi(b \ln \frac{b}{a} - b(1 + w_1) + a) - (1-\pi)(\ln \frac{b}{a} - \frac{b}{a} + (1 - w_0)) = 0, \\ \pi(a \ln \frac{b}{a} - b + a(1 - w_1)) - (1-\pi)(\ln \frac{b}{a} + \frac{a}{b} - (1 + w_0)) = 0. \end{cases}$$

From here by setting $b/a = k$, we obtain

$$\begin{cases} \pi a[k \ln k - k(1 + w_1) + 1] = (1-\pi)[\ln k - k + (1 - w_0)], \\ \pi a[\ln k - k + (1 - w_1)] = (1-\pi)[\ln k + k^{-1} - (1 + w_0)]. \end{cases}$$

Putting $r(k) = \ln k - k + 1$, we have

$$\begin{cases} -\pi ak[r(k^{-1}) + w_1] = (1-\pi)[r(k) - w_0], \\ \pi a[r(k) - k - w_1] = -(1-\pi)[r(k^{-1}) + w_0]. \end{cases} \tag{2.4}$$

The functions $r(k), r(k^{-1})$ are negative, increasing on the interval $(0, 1)$, decreasing on the interval $[1, \infty)$ with the maximum $r(1) = 0$ at the point $k = 1$. The first equation in system (2.4) is well defined on the interval $(k_1(w_1), k_2(w_1))$, where $k_1(w_1) < k_2(w_1)$ are the solutions of the equation

$$r(k^{-1}) + w_1 = 0.$$

Similarly, the second equation of system (2.4) is well defined on the interval $(k_1(w_0), k_2(w_0))$, where $k_1(w_0) < k_2(w_0)$ are the solutions of the equation

$$r(k^{-1}) + w_0 = 0.$$

Denote by $k_1(w_0 \wedge w_1)$ and $k_2(w_0 \wedge w_1)$ the solutions of the equation

$$r(k^{-1}) + w_0 \wedge w_1 = 0,$$

$k_1(w_0 \wedge w_1) < k_2(w_0 \wedge w_1)$. Then system (2.4) is well defined on the interval $k_1(w_0 \wedge w_1) \leq k \leq k_2(w_0 \wedge w_1)$. By dividing the first equation of system (2.4) by the second one we obtain the equation

$$k = \frac{[r(k) - w_0][r(k) - w_1]}{[r(k^{-1}) + w_0][r(k^{-1}) + w_1]} = \frac{\left(2r(k) - (w_0 + w_1)\right)^2 - (w_0 - w_1)^2}{\left(2r(k^{-1}) + (w_0 + w_1)\right)^2 - (w_0 - w_1)^2},$$
$$(2.5)$$

which is well defined on the interval $(k_1(w_0 \wedge w_1), k_2(w_0 \wedge w_1))$.

The value $k = 1$ is the solution of equation (2.5). Denote by $R(k)$ the right-hand side term of the equation (2.5) :

$$R(k) = \frac{\left(2r(k) - (w_0 + w_1)\right)^2 - (w_0 - w_1)^2}{\left(2r(k^{-1}) + (w_0 + w_1)\right)^2 - (w_0 - w_1)^2}.$$

It is easy to see that

$$R'(k) < 0, \text{if } k_1(w_0 \wedge w_1) < k < 1,$$

$$R'(k) = 0, \text{if } k = 1,$$

$$R'(k) > 0, \text{if } 1 < k < k_2(w_0 \wedge w_1),$$

$$R(k) \backsim \left(k/\ln k\right)^2 \uparrow \infty, \text{as } k \uparrow k_2(w_0 \wedge w_1).$$

Therefore, the equation

$$k = R(k)$$

admits the unique solution $k(w_0, w_1)$, such that $1 < k(w_0, w_1) < k_2(w_0 \wedge w_1)$.

From the first equation of system (2.4) and the relation $b = ak$, we find

$$a = \frac{1 - \pi}{\pi} k^{-1} \frac{w_0 - r(k)}{w_1 + r(k^{-1})}, \quad b = \frac{1 - \pi}{\pi} \frac{w_0 - r(k)}{w_1 + r(k^{-1})}, \tag{2.6}$$

where $r(k) = \ln k - k + 1, k = k(w_0, w_1), 1 < k(w_0, w_1) < k(w_0 \wedge w_1), k(w_0 \wedge w_1)$ is the greatest root of the equation $r(k) + w_0 \wedge w_1 = 0$.∎

Theorem 2.3 *Let F be the function given in (2.2); $(m_t)_{t \geq 0}$ be the likelihood ratio process, $m_t = d\mathbf{P}_{1t}/d\mathbf{P}_{0t}, t \geq 0$; the numbers a, b be solutions of system (2.3) and $\delta_{ab} = (\tau_{ab}, d_{ab})$ be the Wald test related to the thresholds a, b.*

Suppose that $a < m_0 < b$ and $\tau_{ab} \in \mathfrak{M}_0(F \circ m)$. Then the decision rule $\delta_{ab} = (\tau_{ab}, d_{ab})$ is π−Bayesian for the risk function $R(\pi, \delta)$ in (1).

Proof. By Lemma 2.1, we have to solve the optimal stopping problem

$$\inf_{\tau \in \mathfrak{M}_0(F \circ m)} \mathbf{E}_0 F \circ m_\tau.$$

By virtue of Theorem 2.1, $\tau_0 = \tau_{ab}$ a.s., where τ_0 is the optimal stopping time given by the Snell envelope for this stopping problem. ∎

3   Main result

The main result of the paper is the following assertion:

Theorem 3.1 *Let $(m_t)_{t \geq 0} = (d\mathbf{P}_{1t}/d\mathbf{P}_{0t})_{t \geq 0}$ be the likelihood ratio process; a, b, be two given numbers such that $0 < a < m_0 < b < \infty, \mathbf{P}_0 - a.s.$; $U_\varepsilon(a), U_\varepsilon(b) \in \mathcal{U}(m)$ for some $\varepsilon > 0$; $\delta_{ab} = (\tau_{ab}, d_{ab})$ be the Wald sequential test associated with the thresholds $a, b; \alpha = \alpha(\delta_{ab}), \beta = \beta(\delta_{ab})$ be errors of the first and the second kinds, respectively, of the test δ_{ab}. Suppose that $\tau_{ab} \in \mathfrak{M}_0(\ln m)$.*
Then, for every $\delta = (\tau, d) \in \Delta_{\alpha, \beta}$, the following inequalities hold

$$\mathbf{E}_0[\ln(1/m_\tau)] \geq \mathbf{E}_0[\ln(1/m_{\tau_{ab}})], \quad \mathbf{E}_1[\ln m_\tau] \geq \mathbf{E}_1[\ln m_{\tau_{ab}}].$$

Remark 2 *The quantities*

$$\mathbf{E}_0[\ln(1/m_\tau)] = \mathbf{E}_0 \left[\ln \frac{d\mathbf{P}_{0\tau}}{d\mathbf{P}_{1\tau}} \right], \quad \mathbf{E}_1[\ln m_\tau] = \mathbf{E}_1 \left[\ln \frac{d\mathbf{P}_{1\tau}}{d\mathbf{P}_{0\tau}} \right]$$

are called the Kullback-Leibler information associated with the measures $\mathbf{P}_{0\tau}, \mathbf{P}_{1\tau}$.
Hence, the Wald sequential test is optimal in the sense that it minimizes both Kullback-Leibler informations in the class $\Delta_{\alpha, \beta}$.

Remark 3 *The condition $U_\varepsilon(a), U_\varepsilon(b) \in \mathcal{U}(m)$ means that the process m visits any small neighbourhood of these points a, b with a positive probability. If it is wrong, for example, at the point a, then one could replace a with $a - \varepsilon$ providing the same probability errors and the same optimal stopping time.*

The proof of the main result. We have seen in Theorem 2.2 that for a given $\pi \in (0, 1)$ and parameters w_0, w_1, there exist the unique thresholds

$$a = \frac{1 - \pi}{\pi} k^{-1}(w_0, w_1) A \circ k(w_0, w_1), \quad b = \frac{1 - \pi}{\pi} A \circ k(w_0, w_1), \quad a \neq b,$$

related to the π−Bayesian test, where $A \circ k(w_0, w_1), k(w_0, w_1)$ are some functions.

We will show that for given thresholds $a, b, 0 < a < b < \infty$ there exist positive parameters w_0, w_1, such that the Wald sequential test associated with the thresholds a, b is π-Bayesian for the risk function

$$R(\pi, \delta) = (1-\pi)w_0\alpha(\delta) + \pi w_1\beta(\delta) + (1-\pi)\mathbf{E}_0[\ln m_\tau^{-1}] + \pi\mathbf{E}_1[\ln m_\tau]. \quad (3.1)$$

Indeed, the system (2.4) related to this risk is linear with respect to the parameters w_0, w_1 for $\pi \in (0,1)$.

Putting $k = b/a$, we find from (2.4)

$$\begin{cases} (1-\pi)w_0 - \pi a k w_1 = (1-\pi)r(k) + \pi a k r(k^{-1}), \\ (1-\pi)w_0 - \pi a w_1 = -(1-\pi)r(k^{-1}) - \pi a r(k). \end{cases} \quad (3.2)$$

The determinant of the coefficients of this system is equal to

$$\pi(1-\pi)a(k-1) > 0, \quad \pi \in (0,1).$$

Therefore, for given numbers $a, b, 0 < a < b < \infty$, system (3.2) admits the unique solution:

$$\begin{cases} w_0 = (1-k)^{-1}[r(k) + kr(k^{-1}) + (r(k) + r(k^{-1}))\pi a k/(1-\pi)], \\ w_1 = (1-k)^{-1}[r(k) + kr(k^{-1}) + (r(k) + r(k^{-1}))(1-\pi)/\pi a], \end{cases} \quad (3.3)$$

where $k = b/a, r(k) = \ln k - k + 1$.

For w_0 and w_1 given in (3.3) the results of the previous section and the condition $\pi \in (0,1)$ show that the thresholds a, b correspond to the test $\delta_{ab} = (\tau_{ab}, d_{ab})$ with

$$\tau_{ab} = \inf\{t \geq 0 : m_t \notin (a,b)\}, \quad \inf\{\varnothing\} = \infty,$$

$$d_{ab} = \begin{cases} 1, & \text{if } m_{\tau_{ab}} \geq b, \\ 0, & \text{if } m_{\tau_{ab}} \leq a. \end{cases}$$

By Theorem 2.3, the test δ_{ab} is π-Bayesian for the risk function (3.1). It means that for any test $\delta = (\tau, d) \in \Delta_{\alpha,\beta}$, we have

$$R(\pi, \delta_{ab}) = (1-\pi)w_0\alpha + \pi w_1\beta + (1-\pi)\mathbf{E}_0[\ln m_{\tau_{ab}}^{-1}] + \pi\mathbf{E}_1[\ln m_{\tau_{ab}}]$$

$$\leq (1-\pi)w_0\alpha(\delta) + \pi w_1\beta(\delta) + (1-\pi)\mathbf{E}_0[\ln m_\tau^{-1}] + \pi\mathbf{E}_1[\ln m_\tau] = R(\pi, \delta),$$

where α and β are the error probabilities of the test δ_{ab}.

This inequality entails the inequality

$$(1-\pi)\mathbf{E}_0[\ln m_{\tau_{ab}}^{-1}] + \pi\mathbf{E}_1[\ln m_{\tau_{ab}}] \leq (1-\pi)\mathbf{E}_0[\ln m_\tau^{-1}] + \pi\mathbf{E}_1[\ln m_\tau],$$

for any $\pi \in [0,1]$, since

$$(1-\pi)w_0\alpha(\delta) + \pi w_1\beta(\delta) \leq (1-\pi)w_0\alpha + \pi w_1\beta, \quad \pi \in [0,1].$$

Therefore, for $\tau \in \mathfrak{M}_0(F \circ m)$, we obtain

$$\mathbf{E}_0[\ln m_{\tau_{ab}}^{-1}] \leq \mathbf{E}_0[\ln m_\tau^{-1}], \quad \mathbf{E}_1[\ln m_{\tau_{ab}}] \leq \mathbf{E}_1[\ln m_\tau]. \blacksquare$$

4 Examples and an auxiliary proposition

Example 1. Let $(\Omega, \mathcal{F}, \mathbf{P}_i)$ be a sample space of a sequence of random variables $(X_n)_{n \geq 1}, i = 0, 1$. Let $\mathcal{F}_n = \sigma(X_1, \ldots, X_n)$. Suppose that $\mathcal{F} = \bigvee \mathcal{F}_n$. Let $\mathbf{P}_{in}$ be the restriction of $\mathbf{P}_i$ to $\mathcal{F}_n$, i.e. $\mathbf{P}_{in}$ is the law of $(X_1, \ldots, X_n)$ under $\mathbf{P}_i, n \geq 1, i = 0, 1$.

Suppose that $\mathbf{P}_{0n} \sim \mathbf{P}_{1n}$ and let

$$m_n = d\mathbf{P}_{1n}/d\mathbf{P}_{0n}, \ n \geq 1.$$

PROBLEM : to test the hypotheses H_0 : the law of the observed sequence (X_n) is $\mathbf{P}_0$ against H_1 : the law is $\mathbf{P}_1$.

We set $\mathcal{F}_t = \mathcal{F}_n, \mathbf{P}_{it} = \mathbf{P}_{in}, i = 0, 1,$ for $t \in [n, n+1)$. Then $m_t = d\mathbf{P}_{1t}/d\mathbf{P}_{0t}$ coincides with m_n for $t \in [n, n+1)$ and the process $(m_t)_{t \geq 0}$ is a local martingale with right continuous paths having left-side limits. By Theorem 3.1, the Wald sequential test is optimal for this problem.

Example 2. Let $(T_n)_{n \geq 1}$ be a sequence of stopping times such that $T_n < T_{n+1}$ on $(T_n < \infty)$ and $T_n \uparrow \infty$ a.s. If N_t is defined as

$$N_t = \sum_{n \geq 1} I_{(T_n \leq t)}, \ t \geq 0,$$

let $(G_t)_{t \geq 0}$ be the filtration generated by (N_t), i.e. $G_t = \sigma(N_s, s \leq t)$. The process (N_t) is called *the point process*.

Let $(A_t^i)_{t \geq 0}$ be a predictable increasing process such that the process

$$(N_t - A_t^i, G_t, \mathbf{P}_i)_{t \geq 0}$$

is a local martingale, $i = 0, 1$. The process $(A_t^i)_{t \geq 0}$ is given by the formula (see Jacod, 1979, proposition 3.41)

$$dA_t^i = \sum_{n=0}^{\infty} \frac{d\mathbf{P}_i(T_{n+1} < t | G_{T_n})}{\mathbf{P}_i(T_{n+1} < t | G_{T_n})} I_{(T_n < t \leq T_{n+1})},$$

where $\mathbf{P}_i(T_{n+1} < t | G_{T_n})$ is the conditional distribution function, $T_0 = 0, G_0$ is trivial.

The process (A_t^i) is called the $\mathbf{P}_i-$compensator of (N_t).

PROBLEM : to test the hypotheses

$$H_0 : \text{the compensator of } (N_t) \text{ is the process } (A_t^0)$$

against

$$H_1 : \text{the compensator of } (N_t) \text{ is the process } (A_t^1)$$

by observation of the process (N_t).

Suppose that $\mathbf{P}_{0t} \sim \mathbf{P}_{1t}, t \geq 0$, where $\mathbf{P}_{it}$ is the restriction of $\mathbf{P}_i$ to G_t. Then the likelihood ratio process

$$m_t = d\mathbf{P}_{1t}/d\mathbf{P}_{0t}$$

is the solution of the equation

$$m_t = m_0 + \int_0^t m_{s-}(Y_s - 1)(1 - \Delta A_s^0)^+ d(N_s - A_s^0),$$

where $x^+ = x^{-1}$, if $x \neq 0$ and $x^+ = 0$, otherwise; (Y_t) is the predictable process such that

$$A_t^1 = \int_0^t Y_s dA_s^0.$$

For the explicit formula of (m_t) see, for example, Jacod (1979).

By Theorem 3.1, the Wald sequential test is optimal for this problem.

We shall give the condition under which the stopping time τ_{AB} is finite. For $0 \leq A < B \leq +\infty$, we denote

$$\tau_{AB} = \begin{cases} \inf\{t \geq 0 : m_t \notin (A, B)\}, \\ \infty, \text{if the set}\{\cdot\} = \varnothing. \end{cases}$$

Lemma 4.1 *Let A, B be given numbers, $0 < A < B < +\infty$. Suppose*

$$\mathbf{P}_{1t} \sim \mathbf{P}_{0t}, 0 \leq t < \infty, \ \mathbf{P}_{1\infty} \perp \mathbf{P}_{0\infty}.$$

Then

$$\mathbf{P}_i(\tau_{AB} < +\infty) = 1, i = 0, 1.$$

Proof. By Kabanov *et al.* (1978), the limits

$$\lim_{t \to \infty} m_t = m_\infty, \lim_{t \to \infty} m_t^{-1} = m_\infty^{-1}$$

exist $\mathbf{P}_0, \mathbf{P}_1$−a.s. In view of the Lebesgue decomposition of $\mathbf{P}_{1\infty}, \mathbf{P}_{0\infty}$ and the property $\mathbf{P}_{1\infty} \perp \mathbf{P}_{0\infty}$, we obtain

$$\mathbf{P}_1(m_\infty = \infty) = 1 \Longleftrightarrow \mathbf{P}_0(m_\infty = 0) = 1,$$

$$\mathbf{P}_0(m_\infty^{-1} = \infty) = 1 \Longleftrightarrow \mathbf{P}_1(m_\infty^{-1} = 0) = 1,$$

whence

$$\mathbf{P}_1(m_\infty = \infty) = 1 = \mathbf{P}_1(m_\infty^{-1} = 0),$$

$$\mathbf{P}_0(m_\infty^{-1} = \infty) = 1 = \mathbf{P}_0(m_\infty = 0).$$

From these equalities we have

$$\mathbf{P}_1(m_\infty = \infty) = 1 \Rightarrow \mathbf{P}_1(\tau_{0B} < \infty) = 1, \qquad (4.1)$$

$$\mathbf{P}_0(m_\infty = 0) = 1 \Rightarrow \mathbf{P}_0(\tau_{A\infty} < \infty) = 1. \tag{4.2}$$

The equality

$$\tau_{AB} = \tau_{A\infty} \wedge \tau_{0B}$$

entails the inclusions

$$(\tau_{A\infty} < \infty) \subseteq (\tau_{AB} < \infty), \ (\tau_{0B} < \infty) \subseteq (\tau_{AB} < \infty).$$

From here and (4.1) and (4.2) it follows that $\mathbf{P}_i(\tau_{AB} < \infty) = 1, i = 0, 1.$ ∎

References

Dvoretzky, A., Kiefer, J., and Wolfowitz, J. (1953). Sequential decision problems for processes with continuous time parameter. Testing hypothesis. *Ann. Math. Statistics*, **24**, 254-264.

Galtchouk, L.I. (2000). Optimality of the Wald SPRT for processes with continuous time parameter: regular case. *Ann. Statistics*, (submitted).

Galtchouk, L.I. and Mirochnichenko, T.P. (1997). Optimal stopping problem for continuous local martingales and some sharp inequalities. *Stochastics and Stochastics Reports*, **61**, 21-33.

Ghosh B.K. and Sen, P.K. (1991). *Handbook of Sequential Analysis*. Marcel Dekker Inc., New York.

Irle, A. (1984). Extended optimality of sequential probability ratio test. *Annals of Statistics*, **12**, 380-386.

Irle, A. and Schmitz, N. (1984). On the optimality of the SPRT for processes with continuous time parameter. *Math. Operationsforschung und Statistik*, **15**, 91-104.

Jacod, J. (1979). Calcul stochastique et problèmes de martingales. *Lecture Notes in Math.*, **714**, Springer-Verlag, Berlin.

Kabanov,Yu., Lipster, R. and Shiryaev, A. (1978). Absolute continuity and singularity of locally continuous probability distributions, I. *Math. Sb.*, **107(149)**, 364-415.

Kabanov,Yu., Lipster, R. and Shiryaev, A. (1979). Absolute continuity and singularity of locally continuous probability distributions, II. *Math. Sb.*, **108(150)**, 32-61.

Karlin S. and Taylor M.M. (1975). *A First Course in Stochastic Processes*. Academic Press, New York.

Koell, Ch. (1994). An optimal procedure for testing hypothesis on cadlag processes. *Sequential Analysis*, **13**, 221-236.

Lehmann, E.L. (1959). *Testing Statistical Hypotheses*. Wiley, New York.

Lipster, R. and Shiryaev, A. (1974). *Statistics of Random Process*. Nauka, Moscow. English translation in two volumes, Springer-Verlag, Berlin.

Peskir, G. and Shiryaev, A.N. (2000). Sequential testing problems for Poisson process. *Ann. of Statistics*, **28**, 837-859.

Shiryaev, A.N. (1969). *Sequential Statistical Analysis (in Russian)*. Nauka, Moscow.

Wald, A. (1947). *Sequential Analysis*. Wiley, New York.

Wald, A. and Wolfowitz, J. (1948). Optimum character of the sequential probability ratio test. *Ann. Math. Statistics*, **19**, 326-339.

Yashin, A. (1983). On a problem of sequential hypothesis testing. *Th. Probability and Applications*, **28**, 157-165.

Efficient Paired Comparison Designs for Utility Elicitation

H. Großmann
U. Graßhoff
H. Holling
R. Schwabe

ABSTRACT: In applications data are often available from the comparison of two alternatives rather than the direct valuation of a single object on its own. Experiments have to be designed for these paired comparisons in a different way than for standard situations. In this note we deal with a problem arising from organizational psychology and present an application of design considerations to the estimation of utility functions.

KEYWORDS: *D*-optimal design; paired comparisons; utility function

1 Introduction

The Productivity Measurement and Enhancement System ProMES (Pritchard, 1990) is a well–known technique in industrial and organizational psychology for measuring and improving organizational productivity. In a recent analysis of a large number of utility functions used there the authors found that a large proportion of these functions can be well approximated by the equation

$$\mu(t; \theta_0, \theta_1, \theta_2) = \theta_0 + \theta_1 t^{\theta_2},$$

where $\theta_2 > 0$ and the stimulus t may vary over the unit interval, $t \in [0, 1]$. Note that this equation is non–linear in the parameter θ_2. By letting $z = -\ln t$ the model can be recognized as a rescaled version of the exponential growth curve associated with the Mitscherlich law of diminishing returns $\theta_0 + \theta_1 e^{-\theta_2 z}$, $z \in [0, \infty)$. Optimal designs for this response have been investigated by Box and Lucas (1959).

mODa6, A.C.Atkinson, P.Hackl and W.G.Müller, eds., Physica, Heidelberg, 2001.

As in various fields of statistical applications including psychological aspects of market research and psychophysical investigations in medicine it often happens in the present context that data are available from the comparison of two (or more) objects rather than the direct valuation of a single object on its own. Additionally, the technique of matching, i.e. of taking differences, is also commonly applied when individual effects occur in paired observations which cannot be modeled appropriately (individual intercepts, correlations, random effects). Evidently, for these *paired comparisons* the experiments should be designed in a different way than for standard experimental situations (see e.g. van Berkum, 1987; Großmann et al., 2000).

Therefore, we consider the problem of designing experiments for estimating the model parameters θ_1 and θ_2 from paired comparison data as follows: For the comparison of two alternative stimuli s and t within an ordered pair (s, t) the response is given by

$$Y(s,t) = \eta(s,t;\theta_1,\theta_2) + Z(s,t),$$

where

$$\eta(s,t;\theta_1,\theta_2) = \mu(s;\theta_0,\theta_1,\theta_2) - \mu(t;\theta_0,\theta_1,\theta_2) = \theta_1(s^{\theta_2} - t^{\theta_2})$$

is the mean response obtained in the comparison. The design region $\mathcal{X} = [0,1]^2$ is given by the Cartesian product as the stimuli presented may be chosen independently. Furthermore, the error distribution $Z(s,t)$ is assumed to belong to a zero mean exponential family and repeated observations are taken to be uncorrelated and homoscedastic.

As usually, in non–linear models (generalized) designs ξ are considered which are defined as finitely supported probability measures on the design region $\mathcal{X}$ (see Kiefer, 1974, for a motivation and discussion of such designs in linear settings). The most popular performance measure of a design ξ is the D–criterion which aims at maximizing the determinant of the information matrix $M(\xi;\theta) = \int_{\mathcal{X}} M(s,t;\theta)\xi(d(s,t))$ where $M(s,t;\theta) = f_\theta(s,t)f_\theta(s,t)^\top$ and $f_\theta(s,t) = \left(\frac{\partial}{\partial\theta_i}\eta(s,t;\theta)\right)_{i=1,2} = \left(s^{\theta_2} - t^{\theta_2}, \theta_1(s^{\theta_2}\ln s - t^{\theta_2}\ln t)\right)^\top$ is the locally linearized regression function. Note that no information is contained in a comparison of a stimulus with itself, $M(t,t;\theta) = 0$. In view of the celebrated equivalence theorem (Kiefer and Wolfowitz, 1960; White, 1973) a D–optimal design ξ^* also minimizes the maximum of the variance function $d(s,t;\xi) = f_\theta^\top(s,t)M(\xi;\theta)^{-1}f_\theta(s,t)$ for the predicted response (G–criterion). This latter criterion is attractive for judging the approximation to the response function.

In contrast to linear design theory, only locally D–optimal designs can be found as the regression function f_θ depends on the unknown parameter θ. It is worthwhile mentioning that in partially linear models the optimal design does not depend on linear parameters. Hence, θ_1 does not affect the D–optimality. For more details on design optimality in non–linear settings we refer to Fedorov and Hackl (1997).

2 Reduction to the canonical form

For every particular parameter value θ the transformation $g_\theta : (s, t) \rightarrow (s^{1/\theta_2}, t^{1/\theta_2})$ induces a correspondence $\xi \rightarrow \xi^{g_\theta}$ on the set of all designs on $\mathcal{X}$. Using this relation we have $|M(\xi; (1, 1))| = \theta_2^2 \theta_1^{-2} |M(\xi^{g_\theta}; \theta)|$ and the left–hand side of this equation is maximized by a design ξ^* if and only if the right–hand side is maximized by the corresponding image under g_θ. Hence, it suffices to consider the design problem for $\theta_1 = \theta_2 = 1$ and the optimal design can be obtained by the inverse transformation g_θ^{-1} of the support. In this sense the transformation g_θ leads to a canonical form in the spirit of Ford, Torsney and Wu (1992) and Sitter and Torsney (1995). In the sequel we will write $M(\xi) = M(\xi; (1, 1))$ for brevity.

Our approach to identify efficient designs is based on geometric ideas and rests on the well–known result that the support of a D–optimal design is included in the set of those points where a minimal circumscribing ellipsoid centered at the origin touches the induced design space (Sibson, 1972; Silvey and Titterington, 1973). This situation is depicted in Figure 1 for the induced design space $\{(s - t, s \ln s - t \ln t); (s, t) \in \mathcal{X}\}$.

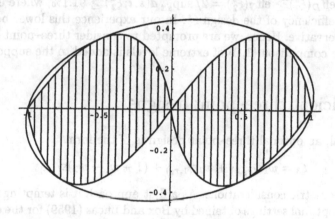

FIGURE 1. Induced design space with circumscribing ellipsoid.

If we ignore comparisons which are obtained by interchanging the alternatives of an ordered pair (s, t), in view of $f_\theta(t, s) = -f_\theta(s, t)$, we can see that the support of a D–optimal design contains two or three points in accordance with Carathéodory's theorem (see Silvey, 1980). Due to the structure of the induced design region these points can be characterized by the three settings $(1, 0)$, $(\tau_1, 0)$ and $(1, \tau_2)$ with $\tau_1, \tau_2 \in [0, 1]$.

3   The best two–point design

In this section we consider two–point designs. As they have minimal support their optimal weights are distributed equally over the two supporting
settings. Hence, by a majorization argument we can confine ourselves to
designs of the form $\xi_2 = \frac{1}{2}\varepsilon_{(\tau_1,0)} + \frac{1}{2}\varepsilon_{(1,\tau_2)}$ where $\varepsilon_{(s,t)}$ denotes the (one–
point) Dirac measure on the setting (s,t). The corresponding value for the
D–criterion is given by $|M(\xi_2)| = \frac{1}{4}((1-\tau_2)\tau_1 \ln \tau_1 + \tau_1\tau_2 \ln \tau_2)^2$. A closed
form solution for the maximization problem does not seem to be feasi-
ble. Numerical optimization yields the optimal settings $\tau_1^* = 0.552$ and
$\tau_2^* = 0.203$ with the determinant $|M(\xi_2^*)| = 0.0484$ for the corresponding
design ξ_2^*.

Checking the condition of the equivalence theorem reveals that ξ_2^* is not
D–optimal within the set of all competing designs since the maximum of
the variance function $d(s,t;\xi_2^*)$ exceeds the number $p = 2$ of parameters.
A closer inspection shows that the maximum is attained at the compar-
ison $(1,0)$ of the extreme settings, $\sup_{s,t} d(s,t;\xi_2^*) = d(1,0;\xi_2^*) = 2.193$.
According to Atwood (1969) a lower bound for the D–efficiency of ξ_2^* is
given by $\mathrm{eff}_D(\xi_2^*) \geq \mathrm{eff}_G(\xi_2^*) = 2/\sup_{s,t} d(s,t;\xi_2^*) \geq 91.1\%$, where $\mathrm{eff}_G(\xi_2^*)$
is the G–efficiency of the design ξ_2^*. In our experience this lower bound is
very conservative. Hence, we are prompted to consider three–point designs
where the comparison $(1,0)$ of extreme levels is added to the support.

4   Efficient three–point designs

In general, an optimal three–point design is of the form

$$\xi_3 = w_1\varepsilon_{(\tau_1,0)} + w_2\varepsilon_{(1,\tau_2)} + (1 - w_1 - w_2)\varepsilon_{(1,0)}$$

due to geometric considerations. As a first approach it is tempting to consider the optimal settings obtained by Box and Lucas (1959) for the directly
observable response, i. e. $\tau_1 = \tau_2 = e^{-1}$ besides the extreme settings 0 and
1. Note that these settings are also optimal in the presence of individual
intercepts (Schwabe, 1995). The optimal weights are found to be the equi-
distribution, $w_1 = w_2 = \frac{1}{3}$, and the corresponding value for the determinant
of the information matrix, 0.0451, does not exceed the corresponding value
for the best two–point design of the previous section. This observation is
in accordance with results by Atkins and Cheng (1999) for correlated observations coming in pairs and by Großmann et al. (2000) in a polynomial
setting.

To facilitate the further calculations we confine ourselves to designs with
equal weights on the non–extreme comparisons, $w_1 = w_2 = w$, i. e.

$$\xi_3 = w\varepsilon_{(\tau_1,0)} + w\varepsilon_{(1,\tau_2)} + (1 - 2w)\varepsilon_{(1,0)}.$$

As in the previous section the optimization problem for the determinant of the information matrix

$$M(\xi_3) = \begin{pmatrix} w(\tau_1^2 + (1-\tau_2)^2) + (1-2w) & w(\tau_1^2 \ln \tau_1 - (1-\tau_2)\tau_2 \ln \tau_2) \\ w(\tau_1^2 \ln \tau_1 - (1-\tau_2)\tau_2 \ln \tau_2) & w((\tau_1 \ln \tau_1)^2 + (\tau_2 \ln \tau_2)^2) \end{pmatrix}$$

does not seem to be amenable to a full analytical solution. However, a necessary condition for the optimal weight w in terms of the levels τ_1 and τ_2 is given by

$$w = -\frac{(\tau_1 \ln \tau_1)^2 + (\tau_2 \ln \tau_2)^2}{2[(\tau_1 \ln \tau_1)^2(\tau_2^2 - 2\tau_2 - 1) - 2\tau_1^2 \tau_2(\tau_2 - 1)\ln \tau_1 \ln \tau_2 + (\tau_1^2 - 2)(\tau_2 \ln \tau_2)^2]}.$$

Using this equation in the calculation of $|M(\xi_3)|$, optimal levels $\tau_1^* = 0.506$ and $\tau_2^* = 0.242$ are found numerically with the optimal common weight $w^* = 0.417$. The determinant of the information matrix for the corresponding design ξ_3^* equals $|M(\xi_3^*)| = 0.0493$. The inspection of the variance function $d(s, t; \xi_3^*)$ for the design ξ_3^*, which is shown in Figure 2, yields $\sup_{s,t} d(s, t; \xi_3^*) = d(\tau_1^*, 0; \xi_3^*) = 2.0015$ and thus reveals the near optimality of ξ_3^*. In fact, as can be seen from $\text{eff}_D \geq 2/\sup_{s,t} d(s, t; \xi_3^*) \geq 99.9\%$ the design is almost D-optimal.

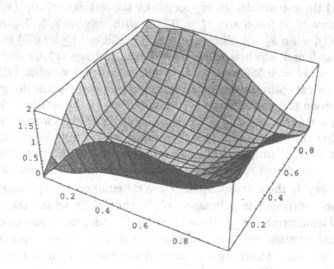

FIGURE 2. Variance function for the design ξ_3^*.

Note that in the present setting of paired comparisons the variance function vanishes on the main diagonal, $d(t, t; \xi) = 0$. Hence, it is preferable to draw the surface of the variance function $d(s, t; \xi)$ rather than the sometimes more appealing information function $1/d(s, t; \xi)$.

In view of the efficiency of the design ξ_3^* the best two–point design ξ_2^* can be re–evaluated. By calculating the relative efficiency with respect to ξ_3^* a D–efficiency $\mathrm{eff}_D(\xi_2^*) \geq 99.0\%$ is obtained. Hence, the two–point design ξ_2^* is a good competitor when an even number of observations are to be taken.

For practical applications the weights of a design have to be transformed to numbers of observations to be allocated to the corresponding settings. Apart from those fortunate situations where the weights are a multiple of $1/n$, where n is the sample size in the experiment, those weights have to be rounded. For example, in the present setting the optimal weight $w^* = 0.417$ may apparently be replaced by $w' = 2/5$. This results in a design ξ_3' with $|M(\xi_3')| = 0.0492$ and a D–efficiency which is still as high as $\mathrm{eff}_D(\xi_3') \geq 99.9\%$. Thus the design ξ_3' can be used as a highly efficient surrogate for the D–optimal design when the number n of observations is a multiple of 5.

5 Discussion

We considered the model equation $Y(s,t) = \theta_1(s^{\theta_2} - t^{\theta_2}) + Z(s,t)$ of paired comparisons when the utility follows a logarithmically rescaled exponential growth curve. In this situation we have obtained a highly efficient ($\geq 99.9\%$) three–point design supported by the settings $(\tau_1^*, 0)$, $(1, \tau_2^*)$ and $(1, 0)$, where $\tau_1^* = 0.506$ and $\tau_2^* = 0.242$, with weights $2/5$, $2/5$ and $1/5$, respectively, when $\theta_2 = 1$. Similarly, a highly efficient ($\geq 99.0\%$) two–point design was found which is supported by the settings $(\tau_1^*, 0)$ and $(1, \tau_2^*)$, where, now, $\tau_1^* = 0.552$ and $\tau_2^* = 0.203$, with equal weights $1/2$. Moreover, numerical optimization shows that the best three–point design under the restriction of equal weights $1/3$ is supported by the settings $(0.462, 0)$, $(1, 0.280)$ and $(1, 0)$ resulting in an efficiency of 98.5%. It is worthwhile mentioning that the routine approach of running the Fedorov–Wynn algorithm (Fedorov, 1972; Wynn, 1970) in SAS (version 8.00) with procedure OPTEX on the induced design space yields essentially the same efficient designs.

For arbitrary θ_2 the efficient settings are determined by replacing τ_1^* and τ_2^* by their corresponding images $(\tau_1^*)^{\theta_2}$ and $(\tau_2^*)^{\theta_2}$ under the reversed canonical transformation $g_\theta^{-1}(t) = t^{\theta_2}$ while the weights remain unchanged. This transformation equivariance is not shared by other popular design criteria like IMSE which aims at minimizing the average of the variance function for the prediction.

Acknowledgments: The authors would like to thank Prof. R. D. Pritchard for providing a copy of the ProMES database and an anonymous referee for helpful comments. Part of this work was supported by the *Deutsche Forschungsgemeinschaft* under grant Ho 1286/2-1.

References

Atkins, J.E. and Cheng, C.-S. (1999). Optimal regression designs in the presence of random block effects. *J. Statist. Plann. Inference*, **77**, 321-335.

Atwood, C.L. (1969). Optimal and efficient designs of experiments. *Ann. Math. Statist.*, **40**, 1570-1602.

Box, G.E.P. and Lucas, H.L. (1959). Design of experiments in non–linear situations. *Biometrika*, **46**, 77-90.

Fedorov, V.V. (1972). *Theory of Optimal Experiments*. Academic Press, New York.

Fedorov, V.V. and Hackl, P. (1997). *Model–Oriented Design of Experiments. Lecture Notes in Statistics*, **125**. Springer, New York.

Ford, I., Torsney, B. and Wu, C.J.F. (1992). The use of a canonical form in the construction of locally optimal designs for non–linear problems. *J. R. Statist. Soc.*, **B54**, 569-583.

Großmann, H., Holling, H., Graßhoff, U. and Schwabe, R. (2000). Efficient designs for paired comparisons with a polynomial factor. In: Atkinson, A.C., Bogacka, B. and Zhigljavsky, A.A. (eds.), *Optimum Design 2000* Kluwer, Dordrecht (to appear).

Kiefer, J. (1974). General equivalence theory for optimum designs (approximate theory). *Ann. Statist.*, **2**, 849-879.

Kiefer, J. and Wolfowitz, J. (1960). The equivalence of two extremum problems. *Can. J. Math.*, **12**, 363-366.

Pritchard, R.D. (1990). *Measuring and Improving Organizational Productivity: A Practical Guide*. Praeger, New York.

Schwabe, R. (1995). Designing experiments for additive nonlinear models. In: Kitsos, C.P. and Müller, W.G. (eds.), *MODA 4 - Advances in Model–Oriented Data Analysis*. Physica, Heidelberg, 77-85.

Sibson, R. (1972). Contribution to discussion of papers by H.P. Wynn and P.J. Laycock. *J. R. Statist. Soc.*, **B34**, 181-183.

Silvey, S.D. (1980). *Optimal Design*. Chapman & Hall, London.

Silvey, S.D. and Titterington, D.M. (1973). A geometric approach to optimal design theory. *Biometrika*, **60**, 21-32.

Sitter, R.R. and Torsney, B. (1995). *D*-optimal designs for generalized linear models. In: Kitsos, C.P. and Müller, W.G. (eds.), *MODA 4 - Advances in Model–Oriented Data Analysis*. Physica, Heidelberg, 87-102.

van Berkum, E.E.M. (1987). *Optimal Paired Comparison Designs for Factorial Experiments*, *CWI Tract*, **31**.

White, L.V. (1973). An extension of the general equivalence theorem to nonlinear models. *Biometrika*, **60**, 345-348.

Wynn, H.P. (1970). The sequential generation of *D*-optimum experimental designs. *Ann. Math. Statist.*, **41**, 1655-1664.

Optimal Design for the Testing of Anti-malarial Drugs

L.M. Haines
G.P.Y. Clarke
E. Gouws
W.F. Rosenberger

ABSTRACT: In dose-response experiments involving anti-malarial drugs the number of maturing parasites is recorded but the number of parasites originally present in the blood sample is unknown. This situation, commonly referred to as Wadley's problem, can be modelled by means of a particular Poisson distribution. In this paper designs for which the parameters of interest or functions of those parameters are estimated as precisely as possible are developed. In particular locally D-, D_s- and c-optimal designs are constructed and compared and the D_s-optimal design is used to provide a bench mark for appraising existing procedures.

KEYWORDS: D- and c-optimality; Elfving's set; Wadley's Problem

1 Introduction

The incidence of malaria has increased alarmingly over the last decade. An estimated 1 million people in the world die from the disease each year and a further 2 billion are at risk (WHO, 1998). The development of anti-malarial drugs is clearly of vital importance and the efficient and effective screening of such drugs is, in turn, essential. Dose-response experiments for anti-malarial drugs involve the recording of the number of maturing parasites in a given blood sample but with the added complication that the number of parasites originally present in that sample is unknown (Gouws, 1995). This situation, referred to more generally as Wadley's problem, can be modelled by assuming that the number of parasites present follows an appropriate distribution, such as the Poisson or negative binomial, that the number of maturing parasites for a fixed number originally present is

mODa6, A.C.Atkinson, P.Hackl and W.G.Müller, eds., Physica, Heidelberg, 2001.

binomial and that the probability of a parasite dying is a suitable function of the log dose (Wadley, 1949; Morgan, 1992, pp. 105-107; Gouws, 1995). Attention then focuses on estimating parameters such as the LD50 and the LD95 which characterize the performance of the drug.

The aim of the present study is to invoke the classical theory of optimal design in order to construct designs for dose-response experiments for antimalarial drugs which in some sense provide good estimates of the quantities of interest.

2 Preliminaries

Suppose that the total number of parasites in a given blood sample follows a Poisson distribution with mean τ. Suppose further that the number of parasites maturing after exposure to a log dose $x \in \mathbb{R}$, with the number of parasites originally present assumed fixed, is binomial with the probability of a parasite dying given by the logistic function

$$p = p(x, \alpha, \beta) = \frac{e^{(x-\alpha)/\beta}}{1 + e^{(x-\alpha)/\beta}}$$

where $\beta > 0$ and thus of a parasite maturing by $q = 1 - p$. Then it is straightforward to show that the number maturing follows a Poisson with mean parameter $\tau(1 - p)$ and that this model is in fact a generalized nonlinear model. Furthermore the information matrix for $\theta = (\alpha, \beta, \tau)$ based on a single observation x can be derived following Dobson (1990, p. 41) and expressed as $M(x, \theta) = g(x, \theta)g(x, \theta)'$ where

$$g(x, \theta) = \begin{bmatrix} \dfrac{\sqrt{\tau}p\sqrt{q}}{\beta} \\ \dfrac{(x - \alpha)\sqrt{\tau}p\sqrt{q}}{\beta^2} \\ \sqrt{\dfrac{q}{\tau}} \end{bmatrix}.$$

Consider now an approximate design which is a probability measure μ putting weights w_i on the support points x_i for $i = 1, \ldots, n$. Then the information matrix for μ is given by $M(\mu, \theta) = \sum_{i=1}^{n} w_i M(x_i, \theta)$ and clearly depends on the unknown parameters in θ. In the present study it is assumed that reasonable guesses for the parameters α, β and τ are available and that locally optimal designs which are designs based on these values and which maximize an appropriate convex function of the information matrix are to be constructed (Chernoff, 1953). In particular data for the action of the drug halofantrine on a Gambian strain of the malaria parasite, and involving concentrations of the drug of 0, 1, 2, 4, 8, 16, 32 and 64 μm/l, were analysed by Gouws (1995) and maximum likelihood estimates of $\hat{\alpha} = 0.6286$, $\hat{\beta} = 0.2857$ and $\hat{\tau} = 5760$ obtained. These estimates are taken here to be appropriate guesses for the true parameter values.

3 Optimal designs

3.1 *D*- and D_s-optimal designs

Designs for which the determinant of the information matrix is maximized are termed *D*-optimal designs and clearly accommodate precise estimation of all the parameters in a model. In the present case it is readily seen that such designs are independent of the value of the parameter τ and also, following the arguments of Ford *et al.* (1992), that the design problem can be expressed in canonical form. Thus the *D*-optimal design for parameters values α, β and τ can be obtained from that for $\alpha = 0, \beta = 1$ and $\tau = 1$ by transforming the design points according to the relation $x = \alpha + \beta z$ and by retaining the same weights for the transformed points. Specifically for $\alpha = 0$ and $\beta = 1$ the *D*-optimal design puts equal weights on the support points, $-\infty$, -0.3265 and 2.3664 and thus the corresponding design for $\alpha = 0.6286$ and $\beta = 0.2857$ puts equal weights on the points $-\infty$, 0.5353 and 1.3047. This latter design is thus based on the doses 0, 3.42 and 20.17 μm/l, and clearly the requirement of a point at zero dose relates to the precise estimation of the unknown count parameter τ. Note that the global optimality of this design, and indeed of all designs reported in the present study, was confirmed by invoking the appropriate Equivalence Theorem (see for example Atkinson and Donev, 1992).

In fact for dose-response experiments on anti-malarial drugs interest centres on the parameters α and β, with the mean count τ regarded as a nuisance parameter. It is thus more appropriate to consider the criterion of D_s-optimality for which the determinant of the inverse of the variance matrix for the parameter estimates of α and β eliminating τ is maximized. This criterion can be expressed succinctly as $|M(\mu, \theta)|/M_{33}(\mu, \theta)$, where $M_{33}(\mu, \theta)$ is element $(3, 3)$ of the information matrix $M(\mu, \theta)$, and the resultant D_s-optimal design for $\alpha = 0.6286$ and $\beta = 0.2857$ has support points $-\infty$, 0.6095 and 1.3871 with attendant weights 0.2276, 0.3089 and 0.4635 respectively. The corresponding doses are 0, 4.069 and 24.384 μm/l and together with the associated weights reflect the fact that emphasis is on the estimation of α and β rather than of τ.

3.2 *c*–optimal designs

In dose-response experiments the precise estimation of the dose of a drug which kills a specified proportion of subjects, such as the LD50 or the

LD95, is invariably of interest. In the present example the dose which is lethal for a given proportion of parasites γ corresponds to the γth percentile of the underlying logistic distribution and is formulated explicitly as $x_g = \alpha + \beta\Gamma_g$, where $\Gamma_g = \ln\{\gamma/(1-\gamma)\}$. The approximate asymptotic variance of the maximum likelihood estimator of a lethal dose x_g can in turn be expressed as $c^T M^-(\mu, \theta)c$, where $c = \{1, \Gamma_g, 0\}$ and A^- denotes the generalized inverse of the matrix A. In order to achieve precise estimation of x_g, designs minimizing this variance are therefore sought. Note that the optimality criterion invoked here is a particular case of c-optimality with the last element of c equal to 0.

It is immediately clear that c-optimal designs for the present example are independent of the nuisance parameter, τ, and furthermore that the construction of such designs for linear combinations of the parameters α and β can be expressed in the canonical form of Ford et al. (1992). It thus follows that the c-optimal design for a lethal dose x_g and parameter values α and β can be obtained directly from the corresponding design for $\alpha = 0$ and $\beta = 1$ by a straightforward linear transformation of the support points. A particularly awkward feature of c-optimality is that the information matrix of the optimal design can be, and indeed often is, singular. In such cases the algebra, and specifically the implementation of the Equivalence Theorem, is somewhat intractable (Silvey, 1980, p. 49) and in addition the designs are difficult to compute numerically (Atkinson and Donev, 1992, p. 110, 113). However for models with two or three parameters the geometric approach to constructing c-optimal designs developed by Elfving (1952) is both powerful and attractive and is thus invoked here.

Specifically consider the induced design space or design locus defined by $\mathcal{G} = \{g(x, \theta) : x \in \mathbb{R}\}$ and describing a curve in 3-dimensional space (Box and Lucas, 1959). Then the Elfving set, $\mathcal{R}$, is the convex hull of the set $\mathcal{G} \cup -\mathcal{G}$, where $-\mathcal{G}$ denotes the set obtained by reflecting $\mathcal{G}$ through the origin, and the c-optimal design for a linear combination of the parameters, $c^T\theta$, is described by the point at which a ray through the origin in the direction of the vector c intersects the boundary of the Elfving set, $\partial\mathcal{R}$. Since interest here centres on c-optimal designs for lethal doses, and thus for linear combinations of the parameters α and β only, attention can be restricted to the intersection of the Elfving set with the xy plane, denoted $\mathcal{R}_{xy}$, and vectors c through the origin lying in that plane. It is convenient at this stage to consider specifically the canonical form of the design problem, and thus the case of $\alpha = 0$ and $\beta = 1$, and to introduce the parametric representation of $g(x, \theta)$ specified by

$$g(u) = \left\{\tau u(1 - \tau u^2), \tau u(1 - \tau u^2)\ln\left(\frac{1}{\tau u^2 - 1}\right), u\right\}$$

where $u = 1/\sqrt{\tau(1 + e^x)}$ and $0 \leq u \leq 1/\sqrt{\tau}$. Then the induced design space can be expressed as $\mathcal{G} = \{g(u) : 0 \leq u \leq 1/\sqrt{\tau}\}$ and the set $\mathcal{C}$, generated by the intersection of lines joining points in $\mathcal{G}$ and $-\mathcal{G}$ with the xy-plane can be specified succinctly as

$$x(u, u_1) = uu_1(u_1 - u)\tau^2$$
$$y(u, u_1) = \frac{uu_1\tau}{(u + u_1)} \left\{ (1 - u^2\tau) \ln\left(\frac{1}{\tau u^2} - 1\right) - (1 - u_1^2\tau) \ln\left(\frac{1}{\tau u_1^2} - 1\right) \right\}$$

for u and $u_1 \in [0, 1/\sqrt{\tau}]$. Parametric curves defined by this formulation for selected u_1 values and $0 < u < \frac{1}{\sqrt{\tau}}$ are shown in Figure 1(a) and indicate clearly the attractive form of the set $\mathcal{C}$.

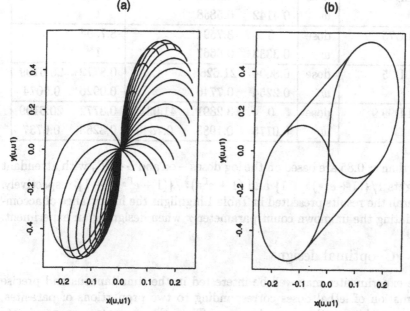

(a)   **(b)**

FIGURE 1. (a) Parametric curves describing the set $\mathcal{C}$. (b) Boundaries of $\mathcal{C}$ and its convex hull.

Furthermore a little reflection reveals that the convex hull of $\mathcal{C}$ coincides with the set $\mathcal{R}_{xy}$. The boundaries of the sets $\mathcal{C}$ and $\mathcal{R}_{xy}$ are shown in Figure 1(b) and it now follows from algebraic and numerical calculations relating to this geometry that c-optimal designs for lethal doses are based on either two or three support points. Some representative examples of these designs for $\alpha = 0.6286$ and $\beta = 0.2857$ are summarized in Table 1, together with the corresponding designs obtained by assuming that the parameter τ, describing the number of parasites present in the original sample, is known.

More specifically it is readily shown that c-optimal designs for the parameter τ assumed Poisson distributed and lethal doses corresponding to the proportion γ less than ≈ 0.40477 and greater than ≈ 0.9985 are based on the same 3 support points but with appropriately varying weights. Otherwise these designs comprise two points of support and in particular for γ

TABLE 1. c-optimal designs for a range of lethal doses with $\alpha = 0.6286$ and $\beta = 0.2857$

Lethal dose		Poisson τ			Fixed τ	
LD5	dose	0	3.2891	41.9311	0.8772	20.6109
	w	0.2704	0.4196	0.3100	0.9074	0.0926
LD25	dose	0	3.2891	41.9311	2.0641	
	w	0.3595	0.4975	0.1430	1	
LD50	dose	0	4.2521		4.2521	
	w	0.4142	0.5858		1	
LD75	dose	0	8.7594		8.7594	
	w	0.3333	0.6667		1	
LD95	dose	0.8604	21.6268		0.8772	20.6109
	w	0.2254	0.7746		0.0926	0.9074
LD99.9	dose	0	3.2891	41.9311	0.8772	20.6109
	w	0.0175	0.1983	0.7843	0.3263	0.6737

less than ≈ 0.85 are based on the log doses $-\infty$ and $\alpha + \beta\Gamma_g$ with attendant weights $1/\{(1 + e^{\Gamma_g})^{\frac{1}{2}} + 1\}$ and $(1 + e^{\Gamma_g})^{\frac{1}{2}}/\{(1 + e^{\Gamma_g})^{\frac{1}{2}} + 1\}$ respectively. Overall the results presented in Table 1 highlight the importance of accommodating the unknown count parameter τ when designing an experiment.

3.3 C–optimal designs

The experimenter may well be interested in the simultaneous and precise estimation of lethal doses corresponding to two proportions of parasites, say γ_1 and γ_2, rather than just one. Then the asymptotic approximate variance matrix for the maximum likelihood estimates of these lethal doses is given by $CM^-(\mu, \theta)C^T$, where

$$C = \begin{bmatrix} 1 & \Gamma_1 & 0 \\ 1 & \Gamma_2 & 0 \end{bmatrix}$$

with $\Gamma_1 = \ln(\gamma_1/(1-\gamma_1))$ and $\Gamma_2 = \ln(\gamma_2/(1-\gamma_2))$, and designs which in some sense minimize this variance matrix are sought. Clearly the design obtained by minimizing the determinant of the matrix $CM^-(\mu, \theta)C^T$ coincides with the D_s-optimal design. It is therefore interesting to consider C-optimal designs for which the trace of $CM^-(\mu, \theta)C^T$ is minimized (Atkinson and Donev, 1992, p. 113-114). The C-optimality criterion is scale dependent and indeed it is usual to scale the rows of the matrix C by the asymptotic variances of the associated c-optimal designs (Atkinson $et\ al.$, 1993). In the present case the C-optimal design for the lethal doses LD50 and LD95, with appropriate scaling, was found to comprise the support points $-\infty, 0.6247,$

and 1.3624 corresponding to doses $0, 4.2135$ and 23.0369 and with associated weights $0.2349, 0.3155$, and 0.4496 respectively and is very similar to the D_s-optimal design. Indeed both of these designs were in turn found to be similar to the C-optimal designs for a range of γ_1 and γ_2 values.

4 Conclusions

The main aim of this paper has been to construct designs for dose-response experiments with anti-malarial drugs for which the parameters of interest are in some sense estimated as precisely as possible. The c-optimal designs for doses lethal to specified proportions of the parasites are mathematically interesting and challenging to derive. However in the example considered here the c-optimal designs for proportions lying between 40.48% and 99.85% are based on only two points of support and thus have little practical appeal. The D_s-optimal design which ensures the precise estimation of the location and scale parameters describing the logistic tolerance distribution would seem to be attractive in its own right and in addition is similar to, and thus representative of, both the D-optimal design for all parameters in the model and a range of C-optimal designs for pairs of lethal doses. The design is based on three points of support and while it cannot therefore be used to appraise the lack-of-fit of the model, it does provide a valuable bench mark for other designs. Thus for example it is interesting to observe that the design used in laboratory procedures and comprising the doses $0, 1, 2, 4, 8, 16, 32$, and 64 $\mu m/l$ has an efficiency of 78.61% relative to the D_s-optimal design and to further note that this efficiency is reassuringly high.

The present study is essentially a preliminary one and raises a number of interesting questions. First it should be emphasized that the designs considered in this study are locally optimal and results are presented for specific values of the parameters. This restriction can be addressed by constructing Bayesian optimal designs for which a prior distribution is placed on the parameters of the model. More generally the analysis of the data on the drug halofantrine indicated considerable overdispersion in the chosen model and also demonstrated that adding a constant to the raw doses improves the fit of the model. It would thus be interesting to examine optimal designs for models which accommodate these features and in particular for the model with the parameter τ assumed negative binomial which is known to account for overdispersion. Finally from a design point of view the study highlights the fact that c-optimality is a difficult criterion to handle both algebraically and numerically and emphasizes the need for more tractable general results in this regard. Even for the simple case considered here the calculations based on the 3-dimensional geometry underpinning Elfving's Theorem were intricate and example-specific. The above ideas are currently being investigated.

Acknowledgments: Professors Haines and Clarke would like to thank the University of Natal and the National Research Foundation of South Africa for financial support. Professor Rosenberger's research is supported by grant R29-DK51017-05 from the National Institute of Diabetes and Digestive and Kidney Diseases. This work was begun while he was visiting the Statistics and Biometry Group at the University of Natal Pietermaritzburg. He thanks the University for its hospitality.

References

Atkinson, A.C., Chaloner, K., Herzberg, A.M., and Juritz, J. (1993). Optimum experimental designs for properties of a compartmental model. *Biometrics*, **49**, 325-337.

Atkinson, A.C. and Donev, A.N. (1992). *Optimum Experimental Designs*. Clarendon Press, Oxford.

Box, G.E.P. and Lucas H.L. (1959). Design of experiments in non-linear situations. *Biometrika*, **46**, 77-90.

Chernoff, H. (1953). Locally optimal designs for estimating parameters. *Annals of Mathematical Statistics*, **24**, 586-602.

Dobson, A.J. (1990). *An Introduction to Generalized Linear Models*. Chapman and Hall, London.

Elfving, G. (1952). Optimum allocation in linear regression theory. *Annals of Mathematical Statistics*, **23**, 255-262.

Ford, I., Torsney, B., and Wu, C.F.J. (1992). The use of a canonical form in the construction of locally optimal designs for non-linear problems. *Journal of the Royal Statistical Society* B, **54**, 569-583.

Gouws, E. (1995). *Drug Resistance in Malarial Research : The Statistical Approach*. M.Sc. thesis, University of Natal, South Africa.

Morgan, B.J.T. (1992). *Analysis of Quantal Response Data*. Chapman and Hall, London.

Silvey, S.D. (1980). *Optimal Design: An Introduction to the Theory for Parameter Estimation*. Chapman and Hall, London.

Wadley, F.M. (1949). Dosage-mortality correlation with number treated estimated from a parallel sample. *Annals of Applied Biology*, **36**, 196-202.

World Health Organization (1998). *Malaria*. Fact Sheet Number 94, WHO, Geneva.

Optimal Adaptive Designs for Delayed Response Models: Exponential Case

J. Hardwick
R. Oehmke
Q. F. Stout

ABSTRACT: We propose a delayed response model for a Bernoulli 2-armed bandit. Patients arrive according to a Poisson process and their response times are exponential. We develop optimal solutions, and compare to previously suggested designs.

KEYWORDS: multi-arm bandit; sequential sampling; design of experiments; clinical trial; ethics; algorithms; parallel processing

1 Introduction

Adaptive designs are effective mechanisms for flexibly allocating experimental resources – particularly in clinical trials. Unfortunately, optimal *fully* sequential designs require immediate responses and cannot be applied when responses are delayed. In this paper, we seek to optimize an objective function for a problem in which there are two populations and the responses, which may be delayed, are independent Bernoulli random variables.

Perhaps the simplest model to consider is one in which observations are delayed a fixed amount of time. Such models have been considered by several researchers, including Bandyopadhyay and Biswas (1996), Douke (1994), Ivanova and Rosenberger (2000), although the optimal design was only recently obtained in Hardwick, Oehmke and Stout (2001). Far more complex, however, is the problem in which the response times follow arbitrary distributions. Such models are too difficult to optimize exactly.

Taking a less general approach, here we consider the model in which patients arrive via a Poisson process and their response times follow independent exponential distributions. We assume that the arrival rate and the mean response times are known, and the goal is to optimize total patient

mODa6, A.C.Atkinson, P.Hackl and W.G.Müller, eds., Physica, Heidelberg, 2001.

successes during the experiment. We can model this problem as a 2-armed bandit (2AB) with delayed response. Recall that the objective of a bandit problem is to allocate resources to different experimental "arms" in such a way that the total return from the experiment is optimized.

There has been some work done on the related problem of maximizing patient survival times in a 1-armed bandit (1AB) model. In the 1AB there are actually two arms, but the attributes of one of them are completely known. In Eick (1988), the author addresses the extent to which geometric response delays affect standard behavioral characteristics of the 1AB, where the survival rate of one arm is known and the goal is to maximize total survival time by allocating patients to either the known or unknown therapy. Some of these results have been extended in Wang (2000).

In the next section, we develop models for the delayed response bandit and present the requisite dynamic programming equations. In Section 3, we present a delayed version of the randomized play-the-winner rule (RPW). In Section 4, we compare the delayed bandit and RPW rules with each other and to the optimal non-delayed solution generated by the 2-armed bandit algorithm. The last section, Section 5, is a discussion.

2   Models with exponential delay

Suppose that patients arrive according to a Poisson process with rate λ_s. As they arrive, they are assigned either to arm (treatment) 1 or 2. Patient responses are Bernoulli with success rates π_1 and π_2. Prior distributions on the π_i are $\text{Be}(a_i, b_i)$, $i = 1, 2$, respectively. The response time for a patient on arm i is exponential with mean λ_i, $i = 1, 2$. Response times are independent among themselves and independent of arrival times and of actual responses. The experiment will allocate a total of n patients.

If a patient arrival occurs at time t, the patient is allocated to arm 1 or 2 based on data up until t. This includes past arrival times, response times and the responses, as well as the priors. A sufficient statistic is $\langle s_1(t), f_1(t), u_1(t); s_2(t), f_2(t), u_2(t) \rangle$, where $s_i(t)$, $f_i(t)$ are the number of successes and failures on arm i and $u_i(t)$ is the number outstanding on arm i at time t, $i = 1, 2$. Because the problem is stationary in time, we can drop the time notation. Thus a policy is a function that depends on the priors and n and maps $\langle s_1, f_1, u_1; s_2, f_2, u_2 \rangle$ to $\{1, 2\}$. Optimal solutions are policies that are optimized for a given objective function. As noted, the objective here is to maximize total patient successes during the experiment, and the problem thus has the form of a two armed bandit with delay. We call this optimization problem the *delayed 2-armed bandit*, D2AB. However, our approach also works for general objective functions.

It is well-known that such optimization problems can be solved via dynamic programming. However, computational space and time grow exponentially in the number of arms, and the delay complicates this further. The state space involves all possible variations of its components, as long as all are nonnegative and their sum is no greater than n; i.e., the state space corresponds to all possible sufficient statistics. There are $\binom{n+6}{6} = \Theta(n^6)$ states in the D2AB, and the delayed k-arm bandit will have $\binom{n+3k}{3k}$ states. This is in contrast to the $\Theta(n^4)$ states in the standard 2AB, and $\binom{n+2k}{2k}$ states in the standard k-arm Bernoulli bandit.

To apply dynamic programming, one needs to know the value of each terminal state, i.e., those states which can be directly evaluated without recourse to recursion. These are the states for which $s_1 + f_1 + s_2 + f_2 = n$. Ultimately, the goal is to determine the value, V, of the initial state $\langle 0, 0, 0; 0, 0, 0 \rangle$.

There are various ways to tackle this problem, but finding one that is computationally feasible is a keystone of the solution. Perhaps the most natural approach is the one in which time is marked by patient arrivals, because these are the only times when action is taken and decisions are needed. Unfortunately, this formulation is too hard to solve computationally, taking $\Theta(n^{10})$ time. For further details, see Hardwick et al. (2001).

A second approach marks time by *events*, where an event is either a subject arrival or a response from one of the arms. Because we are using continuous time, we can assume that only one event occurs at a time. Let $P_1(u_1, u_2)$, $P_2(u_1, u_2)$, $P_s(u_1, u_2)$ represent the probability that the next event is an observation on arm 1, an observation on arm 2, or a subject arrival, respectively. Fortunately, P_1, P_2 and P_s have a simple form:

$$P_s(u_1, u_2) = \frac{\lambda_s}{\lambda_s + u_1 \cdot \lambda_1 + u_2 \cdot \lambda_2} \quad \text{and} \quad P_i(u_1, u_2) = \frac{u_i \cdot \lambda_i}{\lambda_s + u_1 \cdot \lambda_1 + u_2 \cdot \lambda_2}.$$

Let $\pi_i(s_i, f_i)$ denote the probability that an observation on arm i will be a success, given that s_i successes and f_i failures have been previously observed on the arm. Also, let $\hat{y}$ represent component y increased by one and $\sigma + \hat{y}$ be state σ with component y increased by one. Then the dynamic programming equation for determining the value of state $\sigma = \langle s_1, f_1, u_1; s_2, f_2, u_2 \rangle$ is:

$$\begin{aligned}
V(\sigma) = \quad & P_1(u_1, u_2) * \Big[\pi_1(s_1, f_1) \cdot V(\sigma + \hat{s_1} - \hat{u_1}) \\
& \qquad\qquad + (1 - \pi_1(s_1, f_1)) \cdot V(\sigma + \hat{f_1} - \hat{u_1}) \Big] \\
& + P_2(u_1, u_2) * \Big[\pi_2(s_2, f_2) \cdot V(\sigma + \hat{s_2} - \hat{u_2}) \\
& \qquad\qquad + (1 - \pi_2(s_2, f_2)) \cdot V(\sigma + \hat{f_2} - \hat{u_2}) \Big] \\
& + P_s(u_1, u_2) * \max \{ V(\sigma + \hat{u_1}), \ V(\sigma + \hat{u_2}) \}
\end{aligned}$$

Here, the allocation choice is handled in the last term, where if there is a subject arrival then we just determine to which arm we allocate. Initially this just means that the arm has one more unobserved allocation. The advantage of this approach is that it requires only $\Theta(n^6)$ time. While still formidable, this can be achieved for useful sample sizes. For example, problems of size $n = 200$ have been optimized using a parallel computer. See Oehmke, Hardwick and Stout (2001) for a discussion of the parallelization process and optimizations to improve performance.

3   A randomized play-the-winner rule

Exact evaluations of arbitrary, sub-optimal allocation designs are possible via slight modifications to the algorithm in Oehmke et al. (2001). One popular such rule is the randomized play the winner (RPW) rule which first appeared in Wei and Durham (1978). In this urn model, there are initial balls representing the treatment options. Patients are assigned to arms according to the type of ball drawn at random from the urn. Sampling is with replacement, and balls are added to the urn according to the last patient's response. Using RPW, the proportion of allocations to the better arm converges to one.

One advantage of urn models like RPW is the natural way in which delayed observations can be incorporated into the allocation process. When a delayed response eventually comes in, balls of the appropriate type are added to the urn. Since sampling is with replacement, any delay pattern can be accommodated. We call this design the *delayed RPW rule* (DRPW). The same approach was used in Ivanova and Rosenberger (2000), in which responses occurred with a fixed delay. In Bandyopadhyay and Biswas (1996) the authors consider a slightly altered version of this rule for a related best selection problem.

4   Results of comparisons

We have carried out exact analyses of the exponential delay model for both the D2AB and DRPW. In these preliminary analyses, we take $n = 100$. For the DRPW we initialize the urn with one ball for each treatment. If a success is observed on treatment i then another ball of type i is added to the urn, while if a failure is observed then another ball of type $3 - i$ is added.

For comparative purposes, we look at base and best case scenarios. The best fixed in advance allocation procedure is the base case, i.e., the optimal solution when no responses will be available until after all n patients have been allocated. To maximize successes one should allocate all patients to the treatment with the higher expected success rate. We denote the expected number of successes in the base case by $E_b[S]$. Here, we consider only uniform priors on the treatment success rates π_1 and π_2, in which case any fixed allocation is best. For these priors, $E_b[S] = n/2$.

λ_1	λ_2						
$\downarrow$	10^{-5}	10^{-4}	10^{-3}	10^{-2}	10^{-1}	10^0	10^1
10^{-5}	50.1						
10^{-4}	51.2	51.2					
10^{-3}	55.4	55.4	55.8				
10^{-2}	59.3	59.4	59.9	61.5			
10^{-1}	60.9	61.0	61.6	63.1	64.1		
10^0	61.3	61.3	61.9	63.5	64.5	64.8	
10^1	61.3	61.3	62.0	63.5	64.6	64.8	64.9

TABLE 1. Bandit: E[S] as (λ_1, λ_2) vary, $n = 100$, $\lambda_s = 1$, uniform priors

We encounter the best possible case when all responses are observed immediately (full information). In this situation, DRPW is simply the regular RPW and the D2AB is the regular 2-armed bandit. Recall that the regular 2-armed bandit optimizes the problem of allocating to maximize total successes. Letting $E_{opt}[S]$ represent expected successes in the best case, we have $E_{opt}[S] = 64.9$ for our example. Using the difference $E_{opt}[S] - E_b[S]$ as a scale for improvement, we can think of the values on this scale, (0, 14.9), as representing the "extra" successes over the best fixed allocation of 100 observations. We take $R(\delta) = (E_\delta[S] - E_b[S])/(E_{opt}[S] - E_b[S])$ to be the *relative improvement* over the base case for any allocation rule δ. While $R(\delta)$ also depends on n and the prior parameters, these are omitted from the notation.

Note that $R(D2AB) \to 1$ and $R(DRPW) \to 1$ as $n \to \infty$. However, this asymptotic behavior gives little information about the values for practical sample sizes. Hence, their behavior must be determined computationally.

Tables 1 and 2 contain the expected successes for the D2AB and the DRPW rules, respectively. Patient response rates, λ_1 and λ_2, vary over a grid of values between 10^{-5} and 10^1, and the patient arrival rate is fixed at 1. Note that, for both rules, when $\lambda_1 = \lambda_2 = 10^{-5}$, $E[S] \approx 50$. When $\lambda_1 = \lambda_2 = 10$, the delayed bandit rule gives E[S]=64.9 as one would expect. Note that in the best case scenario for the DRPW, $E[S] = 57.9$, which gives an R of 0.53. With the RPW, we can expect to gain only 7.9 successes as compared to the 14.9 for the optimal bandit.

Moving away from the extreme points, consider the case when λ_1, λ_2 and λ_s are all the same order of magnitude. The D2AB rule is virtually unaffected, with an R value of 0.99. This is true because, on average, there is only one patient unobserved (but allocated) throughout the trial. For the DRPW, also, R(DRPW) is only slightly smaller than R(RPW) = 0.52. Both rules seem quite robust to mild to moderate delays in adaptation. It is only when *both* response rates are at least three orders of magnitude below the arrival rate that results begin to degrade seriously. When $\lambda_1 = \lambda_2 = 10^{-3}$,

λ_1	λ_2						
$\downarrow$	10^{-5}	10^{-4}	10^{-3}	10^{-2}	10^{-1}	10^0	10^1
10^{-5}	50.0						
10^{-4}	50.2	50.4					
10^{-3}	51.6	51.7	52.6				
10^{-2}	54.8	54.8	54.9	55.7			
10^{-1}	56.5	56.5	56.5	56.7	57.3		
10^0	56.9	56.9	56.9	57.1	57.6	57.8	
10^1	57.0	57.0	57.0	57.2	57.6	57.8	57.9

TABLE 2. RPW: E[S] as (λ_1, λ_2) vary, $n = 100$, $\lambda_s = 1$, uniform priors

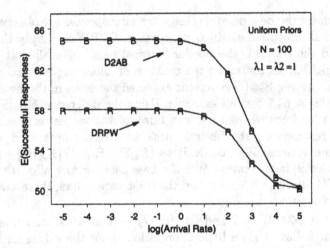

FIGURE 1. Expected successes for D2AB and DRPW, $\lambda_1 = \lambda_2 = 1$

for example, R(D2AB) is only 0.40, and R(DRPW) is a dismal 0.17. It is also interesting to note that even when the response rate is only 1/100th the arrival rate, the D2AB does better than the RPW with immediate responses. Figure 1 illustrates R(D2AB) and R(DRPW) when the response rates are both one but the arrival rate varies between 10^{-5} and 10^5.

When we consider scenarios in which only one treatment arm supplies information to the system, we see an interesting result. For example, using uniform priors, when $\lambda_1 = \lambda_s = 1$ but $\lambda_2 = 10^{-5}$, the relative improvement is 0.76 for the D2AB and 0.47 for the DRPW. This is an intriguing result for the DRPW since its R-value is 89% of the best possible RPW value. Still, one clearly prefers the D2AB since we only get a 24% loss over the optimal solution while excluding half the information.

One way to view this problem independently from the allocation rules is to examine the expected number of allocated but unobserved patients when a new patient allocation decision must be made. As noted, when the response delay rate is 1, at any point in time one expects only a single observation to be delayed, and the impact on performance is minimal. When $\lambda = 0.1$, once approximately 20 patients have been allocated there is a consistent lag of about 10 patients. Connecting this value to the results in Tables 1 and 2, one finds that a loss of roughly 10% of the total information at the time of allocation of the last patient (and a significantly higher loss rate for earlier decisions), corresponds to a loss of only about 5% in terms of the improvement available from each rule.

When the response rate is about 100 times slower that the arrival rate, asymptotically there will be approximately 100 unobserved patients at any point in time. Fortunately, for a sample size of 100, one is quite far from this asymptotic behavior, and approximately 37% of the responses have been observed by the time the last allocation decision must be made. This allows the D2AB to achieve 77% of the relative improvement possible, while the DRPW rule attains only 38%.

While for space reasons this paper has only analyzed problems in which both treatments have uniform priors, similar results hold for more general priors.

5 Conclusions

Because there has been scant research addressing optimal adaptive designs with delayed responses, there are numerous outstanding problems in the area. One might argue that fully optimal designs aren't necessary in practice if good ad hoc options are available. However, without a basis of comparison it is difficult to know how good an ad hoc option is, since asymptotic analyses give only vague information about their behavior for practical sample sizes. Examining the properties of optimal designs can also lead to the development and selection of superior sub-optimal alternatives.

An important concern is the design's robustness. For example, one can evaluate robustness with respect to departures from prior specifications and from the assumption of exponential response times. One way to improve robustness might be to use prior distributions on the response rate parameters. We are also interested in operating characteristics such as the distribution of the objective function, number of allocations to each arm, how the allocations vary with increasingly delay, etc. Some of these issues are examined in Hardwick et al. (2001).

Recall that the goal of this paper is to develop exactly optimal delayed response designs that allow for the use of *any* objective function, not just the bandit objective of maximizing reward (successes). The algorithm presented in Oehmke et al. (2001) has this capability, and in future work we will examine its performance for other objectives. For example, some researchers have considered two-stage models in which the first stage is adaptive and in the second stage all patients are assigned to the arm judged to be best at the end of the first stage. In this situation, the optimal first stage allocation will be nudged closer to equal allocation to insure a better decision for the second stage.

To summarize our findings, we have developed optimal designs for a clinical trial model with Bernoulli observations and exponentially delayed response and patient arrival times. We found that under fairly broad circumstances, the delayed response design performed extremely well compared with the optimal non-delayed algorithm. We also found that the most commonly proposed ad hoc rule for such problems, the DRPW rule, performed significantly less well than the optimal delayed design, which suggests that there is need for better ad hoc strategies.

Acknowledgments: This work was partially supported by National Science Foundation grants DMS-9504980 and DMS-0072910. Parallel computing facilities were provided by the University of Michigan's Center for Parallel Computing.

References

Bandyopadhyay, U. and Biwas, A. (1996), Delayed response in randomized play-the-winner rule: a decision theoretic outlook, *Calcutta Statist. Assoc. Bul.* **46**, 69–88.

Douke, H. (1994), On sequential design based on Markov chains for selecting one of two treatments in clinical trials with delayed observations, *J. Japanese Soc. Comput. Statist.* **7**, 89–103.

Eick, S. (1988), The two-armed bandit with delayed responses, *Ann. Statist.* **16**, 254–264.

Hardwick, J., Oehmke, R. and Stout, Q.F. (2001), Optimal adaptive designs for delayed response models, EECS Technical Report, University of Michigan.

Ivanova, A and Rosenberger, W. (2000), A comparison of urn designs for randomized clinical trials of $k > 2$ treatments, *J. Biopharm. Statist.* **10**, 93–107.

Oehmke, R., Hardwick, J. and Stout, Q.F. (2001), Scalable algorithms for adaptive statistical designs. To appear in *Scientific Programming*.

Wang, X. (2000) A Bandit Process with Delayed Responses, *Statistics and Probability Letters* **48**, 303–307.

Wei, L.J. and Durham, S. (1978), The randomized play-the-winner rule in medical trials, *J. Amer. Statist. Assoc.* **73**, 830–843.

Non-D-optimality of the Simplex Centroid Design for Regression Models Homogeneous of Degree p

R.-D. Hilgers

ABSTRACT: Models with regression functions homogeneous of degree p are suitable in some situations to describe the data derived from mixture experiments. The uniformly weighted simplex centroid design, supported by the barycenters corresponding to the regression functions may be an intuitive design. The aim of the investigation is to show the non-D-optimality of the design.

KEYWORDS: additivity; D-optimal design; homogeneous regression functions; mixture experiments

1 Introduction

Since the introduction of the simplex centroid design by Scheffé (1963), the design is frequently used to estimate the parameters of different types of regression models for mixture experiments. A special class of such models defined on the q-dimensional simplex

$$\mathcal{U}_q = \left\{ \mathbf{x} \in \mathbb{R}^q : 0 \leq x_r \leq 1, \ 1 \leq r \leq q, \ \sum_{r=1}^{q} x_r = 1 \right\}$$

can be described by the following equation:

$$\eta(\mathbf{x}) = \sum_{\ell=1}^{\nu} \sum_{1 \leq s_1 < \cdots < s_\ell \leq q} \vartheta_{s_1,\ldots,s_\ell} h_{s_1,\ldots,s_\ell}(\mathbf{x}), \quad \nu < q. \qquad (1.1)$$

According to response surface methodology applied to the data of mixture experiments, Snee (1971, p. 160) mentioned, that in most applications, the model (1.1) with degree ν equal to one or two is used in particular in

mODa6, A.C.Atkinson, P.Hackl and W.G.Müller, eds., Physica, Heidelberg, 2001.

cases where the number of components is large. Note, that the number of parameters grows very fast with ν as well with increasing q, leading to a high number of experiments. However, in practical situations, cost considerations often restrict the number of experiments. This may be the reason for application of the intuitive simplex centroid design, supported by the barycenters corresponding to the regression functions in the model. The most popular models are polynomial based models, where the functions $h_{s_1,\ldots,s_\ell}(\mathbf{x})$ are defined as product of the respective components $x_{s_1},\ldots,x_{s_\ell}$, c. f. Scheffe (1963). However, other authors considered different regression functions, c. f. Becker (1968, 1978). The latter models can be embedded in the following general definitions for the regression functions of model (1.1).

(A1) The regression functions should be homogeneous of degree $p, p \in \mathbb{N}$, by means of

$$h_{s_1,\ldots,s_\ell}(\alpha\mathbf{x}) = \alpha^p h_{s_1,\ldots,s_\ell}(\mathbf{x}) \text{ for all } \alpha \in \mathbb{R}, \mathbf{x} \in \mathbb{R}^q .$$

(A2) The functions $h_{s_1,\ldots,s_\ell}(\mathbf{x})$ should contain no further parameters and depend on the components $x_{s_1},\ldots,x_{s_\ell}$ of $\mathbf{x}$ only. Moreover, $h_{s_1,\ldots,s_\ell}(\mathbf{x})$ should be zero, whenever at least one of the variables $x_{s_1},\ldots,x_{s_\ell}$ vanishes.

(A3) The set of regression functions should be permutation invariant in the sense that with $h_{s_1,\ldots,s_\ell}(\mathbf{x})$ all functions $h_{\pi(s_1),\ldots,\pi(s_\ell)}(\pi(\mathbf{x}))$ are included that result from permutations π of all q components, $s_1 < \ldots < s_\ell, \pi(s_1) < \ldots < \pi(s_\ell), 1 \leq \ell \leq \nu$.

Denote by $\mathbf{e}_r$ the r-th q-dimensional unitvector so that $\mathbf{e}_1^t = (1,0,\ldots,0); \mathbf{e}_2^t = (0,1,0,\ldots,0)$ and so on. Then a direct consequence of (A1) and (A3) is, that at $\mathbf{z}^m = (1/m)\sum_{s=1}^{m}\mathbf{e}_{u_s}$, a barycenter of order m, the function $h_{s_1,\ldots,s_\ell}$ equals either zero or can be calculated by

$$h_{s_1,\ldots,s_\ell}(\mathbf{z}^m) = \left(\frac{1}{m}\right)^p h_{s_1,\ldots,s_\ell}\left(\sum_{s=1}^{m}\mathbf{e}_{u_s}\right) = \left(\frac{1}{m}\right)^p c_\ell,$$

if $z_{s_r} \neq 0, 1 \leq r \leq \ell$. Moreover it is assumed that the constants c_ℓ do not vanish $(h_{s_1,\ldots,s_\ell}\left(\sum_{r=1}^{\ell}\mathbf{e}_{s_r}\right) = c_\ell \neq 0)$.

In section 2 we will address the question, that the minimum support simplex centroid design, supported by the barycenters corresponding to the regression functions only, is not D-optimal to estimate the parameters of the regression model (1.1). The results will be discussed in section 3.

2   Non D-optimality

The design under consideration is the uniformly weighted simplex centroid design denoted by ξ_ν which is supported by the barycenters of depth one up to ν. Let

$$A(\nu,p) = \sum_{\ell=1}^{\nu} \binom{\nu+1}{\ell} \left(\frac{\ell}{\nu+1}\right)^{2p}$$

Lemma 2.1 *Let $p > 0$ and $\nu < q$ in model (1.1) and the regression functions fulfill the assumptions (A1) - (A3). If*

$$A(\nu,p) > 1, \tag{2.1}$$

then the design ξ_ν is not D–optimal.

Proof. We will show, that under (2.1), in contradiction to the equivalence theorem of Kiefer and Wolfowitz (1960), the generalized variance exceeds the value k, the number of parameters in model (1.1). Thus ξ_ν is not D–optimal.

So assume ξ_ν is D–optimal. For $1 \le \ell \le \nu$ let $\{s_1,\ldots,s_\ell\} \subseteq \{1,\ldots,q\}$ and define

$$g_{s_1,\ldots,s_\ell}(\mathbf{x}) = \frac{\ell^p}{c_\ell} \left[h_{s_1,\ldots,s_\ell}(\mathbf{x}) + \sum_{r=1}^{\nu-\ell}(-1)^r \frac{c_\ell}{c_{\ell+r}} \sum_{\substack{t_1,\ldots,t_r \notin \{s_1,\ldots,s_\ell\} \\ 1 \le t_1 < \cdots < t_r \le q}} h_{s_1,\ldots,s_\ell,t_1,\ldots,t_r}(\mathbf{x}) \right],$$

where the summation is zero if it is taken over an empty index set.

Note that the functions $g_{s_1,\ldots,s_\ell}$ equal zero in all barycenters having at least one vanishing component indexed by $s_1,\ldots,s_\ell$. In particular this holds true for all barycenters of depth less than ℓ. Further $g_{s_1,\ldots,s_\ell}$ equals one in the barycenter of depth ℓ with components indexed by $s_1,\ldots,s_\ell$ and zero in the other barycenters of depth ℓ. Consider now a barycenter $\mathbf{z}^m$, of depth $m, \ell < m \le \nu$. If at least one of the components $s_1,\ldots,s_\ell$ equals zero it follows that the function $g_{s_1,\ldots,s_\ell}$ equals zero. If, on the other hand, all components $s_1,\ldots,s_\ell$ do not vanish, it is easy to see that $\binom{m-\ell}{r}$ of the functions $h_{s_1,\ldots,s_\ell,t_1,\ldots,t_r}(\mathbf{x})$ do not vanish, so that $g_{s_1,\ldots,s_\ell}$ equals zero, because

$$g_{s_1,\ldots,s_\ell}(\mathbf{z}^m) = \frac{\ell^p}{c_\ell} \left[\left(\frac{1}{m}\right)^p c_\ell + \sum_{r=1}^{m-\ell}(-1)^r \frac{c_\ell}{c_{\ell+r}} \binom{m-\ell}{r} \left(\frac{1}{m}\right)^p c_{\ell+r} \right]$$

$$= \left(\frac{\ell}{m}\right)^p \left[1 + \sum_{r=1}^{m-\ell}(-1)^r \binom{m-\ell}{r} \right]$$

$$= \left(\frac{\ell}{m}\right)^p \left[\sum_{r=0}^{m-\ell}(-1)^r \binom{m-\ell}{r} \right] = 0.$$

Thus, the set of functions $g_{s_1,\ldots,s_\ell}(\mathbf{x})$ constitute an orthonormal set of functions which span the same function space as the regression functions, c. f. Kiefer (1961), Hilgers (1993).

In a barycenter $z^{\nu+1}$ of depth $\nu + 1 \leq q$ the function $g_{s_1,\ldots,s_\ell}$ is either zero or

$$g_{s_1,\ldots,s_\ell}(z^{\nu+1}) = \left(\frac{\ell}{\nu+1}\right)^p \left[\sum_{r=0}^{\nu-\ell}(-1)^r\binom{\nu+1-\ell}{r}\right]$$

$$= (-1)^{\nu-\ell}\left(\frac{\ell}{\nu+1}\right)^p. \tag{2.2}$$

There are $\binom{\nu+1}{\ell}$ of the functions $g_{s_1,\ldots,s_\ell}$ which do not vanish in $z^{\nu+1}$. Thus, the generalized variance in that point divided by k is

$$d(z^{\nu+1};\xi_\nu)/k = \sum_{\ell=1}^{\nu} \sum_{1\leq s_1<\cdots<s_\ell\leq q} g^2_{s_1,\ldots,s_\ell}(z^{\nu+1})$$

$$= \sum_{\ell=1}^{\nu} \binom{\nu+1}{\ell}\left(\frac{\ell}{\nu+1}\right)^{2p} \tag{2.3}$$

and exceeds one, contrary to the D–optimality of ξ_ν. ∎

Remark:

1. It is easy to see that $A(\nu,p)$ is decreasing in p for fixed ν and increasing in ν for fixed p.

2. Non D-optimality for model (1.1) with regression function homogeneous of degree zero ($p = 0$), follows from $A(\nu,0) > 1$ for all $\nu \geq 1$.

3. Note, that $A(\nu,1)$ is the second factorial moment of the binomial distribution with $p = 1/2$, c.f. Johnson and Kotz (1969, p. 51, 4.2). This implies the non D–optimality for all models (1.1) with regression function homogeneous of degree one and $\nu > 1$, c.f. Hilgers (2000).

4. A sufficient condition, for $A(\nu,p)$ to be greater than 1 is:

$$\nu \geq 2p.$$

Obviously for $\nu \geq \ell$ and $m > 1$ it follows that $(\nu + 1)/\ell < (\nu + 1 - m)/(\ell - m)$. Thus for $\nu \geq 2p$:

$$A(\nu,p) > \binom{\nu+1}{2p}\left(\frac{2p}{\nu+1}\right)^{2p}$$

$$= \frac{\nu+1}{2p}\frac{2p}{\nu+1}\cdot\frac{(\nu+1)-1}{2p-1}\frac{2p}{\nu+1}\cdots\frac{(\nu+1)-2p+1}{1}\frac{2p}{\nu+1}$$

$$> 1.$$

However, $A(\nu,p)$ may exceed 1 even for smaller values of ν than $\nu \geq 2p$ which can be seen by the numerical results in Table 1.1.

TABLE 1. Minimal values of ν, with $A(\nu, p) > 1$

p	2	3	4	5	6	7	8	9	10
ν	4	5	5	6	7	7	8	9	9

3 Comments

Becker (1978, p. 198) introduced a model which is homogeneous of degree 0 for $q = 3$ components satisfying the condition $(A1 - A3)$:

$$y(\mathbf{x}) = \vartheta_0 + \vartheta_1 \frac{x_1}{x_1 + x_2} + \vartheta_2 \frac{x_2}{x_2 + x_3} + \vartheta_3 \frac{x_3}{x_3 + x_1}$$

$$+ \vartheta_{12} \frac{x_1 x_2}{(x_1 + x_2)^2} + \vartheta_{13} \frac{x_1 x_3}{(x_1 + x_3)^2} + \vartheta_{23} \frac{x_2 x_3}{(x_2 + x_3)^2}$$

$$+ \vartheta_{123} \frac{x_1 x_2 x_3}{(x_1 + x_2)(x_1 + x_3)(x_2 + x_3)},$$

where $(x_{s_1} \cdots x_{s_\ell})/(x_{s_1} + \cdots + x_{s_\ell})^{\ell-1} = 0$, if $x_{s_1} = \cdots = x_{s_\ell} = 0$.

However, I have not found an application of the model in the literature up to now. In contrast to regression models homogeneous of degree 0 as well as models homogeneous of degree $p > 1$, the following three types of regression functions suggested by Becker (1968) were used in different scientific areas, c.f. Chick (1984), Cornell (1978), D'Agostino (1985), Johnson (1981), Novik (1983), Sahrmann (1987), Sobolev (1976).

$$h_{s_1,\ldots,s_\ell}(x_{s_1}, \ldots, x_{s_\ell}) = \begin{cases} \min\{x_{s_1}, \ldots, x_{s_\ell}\} & (H_1), \\ \sqrt[\ell]{x_{s_1} \cdots x_{s_\ell}} & (H_2), \\ \dfrac{x_{s_1} \cdots x_{s_\ell}}{(x_{s_1} + \cdots + x_{s_\ell})^{\ell-1}} & (H_3), \end{cases}$$

where $(x_{s_1} \cdots x_{s_\ell})/(x_{s_1} + \cdots + x_{s_\ell})^{\ell-1} = 0$, if $x_{s_1} = \cdots = x_{s_\ell} = 0$.

Lemma 2.1 generalizes the results on non D-optimality of the simplex centroid design for $\nu = 2, q = 3$ in Becker (1978, p. 204), for the models H_1, H_2 and H_3 with $\nu = 2 < q$ in Liu and Neudecker (1997, p. 60) and for the minimum polynomial with $\nu < q$ in Hilgers (2000). Remember also the D-optimality for the tic polynomials ($4 \leq \nu < q$), c.f. Scheffé (1963) and Atwood (1969). However, one can argue that the simplex centroid design works quite well for homogeneous models with $\nu < q$, although non D-optimality is shown. To this end consider the G-efficiency of the simplex centroid design defined by, $e^G(\xi) = k/ \sup_{\mathbf{x} \in \mathcal{U}_q} d(\mathbf{x}; \xi)$, c. f.

Pukelsheim (1993,p. 132) and use

$$d(\mathbf{z}; \xi_\nu)/k = \sum_{\ell=1}^{\nu} \sum_{1 \leq s_1 < \cdots < s_\ell \leq q} g^2_{s_1,\ldots,s_\ell}(\mathbf{z}) \, .$$

Clearly $d(\mathbf{z}; \xi_\nu) = k$ at all supporting points of ξ_ν, i. e. barycenters of depth less or equal to ν. Thus evaluate the generalized variance only in barycenters $\mathbf{z}^{\nu+m}$, of depth $(\nu + m), 0 \leq m \leq q - \nu$. Similar arguments to those used to derive (2.3) and the equation

$$\sum_{\ell=1}^{\nu} (-1)^\ell \binom{\nu + m}{\ell} = (-1)^\nu \binom{\nu + m - 1}{\nu} - 1 \, .$$

lead to

$$g^2_{s_1,\ldots,s_\ell}(\mathbf{z}^{\nu+m}) = \left(\frac{\ell}{\nu + m} \right)^{2p} \left(\frac{\nu - \ell + m - 1}{m - 1} \right)^2 \, .$$

Thus the generalized variance divided by k equals

$$
\begin{aligned}
d(\mathbf{z}^{\nu+m}; \xi_\nu)/k &= \sum_{\ell=1}^{\nu} \sum_{1 \leq s_1 < \cdots < s_\ell \leq q} g^2_{s_1,\ldots,s_\ell}(\mathbf{z}^{\nu+m}) \\
&= \sum_{\ell=1}^{\nu} \binom{\nu + m}{\ell} \left(\frac{\ell}{\nu + m} \right)^{2p} \left(\frac{\nu - \ell + m - 1}{m - 1} \right)^2 . \quad (3.1)
\end{aligned}
$$

Note that (3.1) is decreasing in p, so that upper bounds can be given for $p = 0$, c.f. table 2. All efficiencies are remarkably low and thus the simplex centroid design is not an appealing design. It can be argued that the low efficiencies reflect the fact that addidional experimental points should be included in the design. Consequently, no good explanation can be given, why, for fixed q, the efficiency bounds are smaller for intermediate values of ν, than for larger smaller values. However, it is not at all clear whether the set of barycenters yields a suitable set of supporting points.

The non-D-optimality property of the simplex centroid design shown in Lemma 2.1 confronts with the problem of optimal design. It is easy to verify that the simplex centroid design is D-optimal to estimate the parameters of Becker's models H_1, H_2 and H_3 with $\nu = 2 = q$, c. f. Hilgers (1991). However, this does not hold for the whole family of regression functions which are homogeneous of degree one. Consider for example the simple case $\nu = q = 2$. Obviously if $\nu = 2 = q$ and $h_{12}(x_1, x_2) = \sqrt{x_1 x_2} \left(2 - \sin^2 \left(2\pi \frac{x_1 x_2}{(x_1 + x_2)^2} \right) \right)$ model (1.1) is homogeneous of degree one, but the generalized variance exceeds 1 for $x_1 = \frac{1}{6}$ and $x_2 = \frac{5}{6}$ in contradiction to the D-optimality of the simplex centroid design.

TABLE 2. Upper bounds for the G-efficiency of ξ_ν for model (1.1) with regression functions homogeneous of degree 0

q	ν							
	2	3	4	5	6	7	8	9
3	0.1667							
4	0.0455	0.0714						
5	0.0182	0.0105	0.0333					
6	0.0090	0.0027	0.0031	0.0161				
7	0.0051	0.0009	0.0006	0.0010	0.0079			
8	0.0032	0.0004	0.0001	0.0001	0.0004	0.0039		
9	0.0021	0.0002	0.0000	0.0000	0.0000	0.0001	0.0020	
10	0.0015	0.0001	0.0000	0.0000	0.0000	0.0000	0.0001	0.0010

Another way to answer the question is by numerical methods which are not within the scope of the present paper.

Acknowledgments: I am indebted to the constructive comments of two anonymous referees.

References

Atwood, C.L. (1969). Optimal and efficient designs of experiments. *Annals of Statistics*, **40**, 1570-1602.

Becker, N.G. (1968). Models for the response of a mixture. *Journal of the Royal Statistical Society*, **B30**, 349-358.

Becker, N.G. (1978). Models and designs for experiments with mixtures. *Australian Journal of Statistics*, **B 3**, 195-208.

Chick, L.A. and Piepel, G.F. (1984). Statistically designed optimization of a glass composition. *Journal of the American Ceramc Society*, **67**, 763-768.

Cornell, J.A. and Gorman, J.W. (1978). On the detection of an additive blending component in multicomponent mixtures. *Biometrics*, **34**, 251-263.

D'Agostino, G. and Castagnetta, L. (1985). Computer-aided mobile phase optimization and chromatogram simulation in high-performance liquid chromatography. *Journal of Chromatography*, **338**, 1-23.

Hilgers, R.-D. (1991). Optimale Versuchsplanung in Mischungs-Mengen-Experimenten. *PhD Thesis, University of Dortmund*.

Hilgers, R.-D. (1993). A useful set of multiple orthogonal polynomials on the q-simplex and its application to D-optimal designs. In: Müller, W.G., Wynn, H.P., and Zhigljavski, A.A. (eds.), *Model-Oriented Data Analysis*, Physica-Verlag, Heidelberg, 91-104.

Hilgers, R.-D. (2000). D-optimal design for Becker's minimum polynomial, *Statistics and Probability Letters*, **49**, 175-179.

Johnson, N.L. and Kotz, S. (1969). *Discrete Distributions*. Wiley, New York.

Johnson, T.M. and Zabik, M.E. (1981). Response surface methodology for analysis of protein interactions in angel food cakes. *Journal of Food Science*, **46**, 1226-1230.

Kiefer, J.C. (1961). Optimum designs in regression problems II. *Annals of Mathematical Statistics*, **32**, 298-325.

Kiefer, J. and Wolfowitz, J. (1960). The equivalence of two extremum problems. *Canadian Journal of Mathematics*, **12**, 129-132.

Liu, S.Z. and Neudecker, H. (1997). Experiments with mixtures: Optimal allocation for Becker's models. *Metrika*, **45**, 53-66.

Novik, F.S. (1983). Synthesis of a model on a simplex. *Industrial Laboratory USSR*, **49**, 78-81.

Pukelsheim, F. (1993). *Optimal Design of Experiments*. Wiley, New York.

Sahrmann, H.F., Piepel, G.F., and Cornell, J.A. (1987). In search of the optimum Harvey Wallbanger recipe via mixture experiment techniques. *The American Statistican*, **41**, 190-194.

Scheffé, H. (1958). Experiments with mixtures. *Journal of the Royal Statistical Society*, **B20**, 344-360.

Scheffé, H. (1963). The simplex centroid design for experiments with mixtures. *Journal of the Royal Statistical Society* **B25**, 235-263.

Snee, R.D. (1971). Design and analysis of mixture experiments. *Journal of Quality Technology*, **B3**, 159-169.

Sobolev, N. W. (1976). Construction of simplicial linear models of composition property diagrams. *Industrial Laboratory USSR*, **42**, 103-108.

New Upper Bounds for Maximum-Entropy Sampling

A. Hoffman
J. Lee
J. Williams

ABSTRACT: We develop new upper bounds for the constrained maximum-entropy sampling problem. Our partition bounds are based on Fischer's inequality. Further new bounds combine the use of Fischer's inequality with previously developed bounds. We demonstrate this in detail by using partitioning to strengthen spectral bounds. Computations suggest that these bounds may be useful in finding optimal solutions by branch-and-bound.

KEYWORDS: branch-and-bound; Fischer's inequality; matching.

1 Introduction

The constrained maximum-entropy sampling problem (*CMESP*) is an important problem in the design of experiments (see Shewry and Wynn, 1987; Sebastiani and Wynn, 2000 and Lee, 1998). It arises in designing monitoring networks (see Zidek *et al.*, 2000). Let C be an $n \times n$ real symmetric positive definite matrix with row and column indices $N = \{1, 2, \ldots, n\}$. For $S, T \subset N$, we let $C[S, T]$ denote the submatrix of C with row indices S and column indices T. We let $\det C[S, S]$ denote the determinant of the principal submatrix $C[S, S]$, with the convention that $\det C[\varnothing, \varnothing] = 1$. Let s be an integer satisfying $0 < s \leq n$. Let M be a finite index set. Let a_{ij} and b_i be real numbers, for $i \in M$ and $j \in N$. The CMESP is

$$z := \max_{\substack{S \subset N: \\ |S| = s}} \ln \det C[S, S] ; \tag{1.1}$$

$$\text{subject to } \sum_{j \in S} a_{ij} \leq b_i , \ \forall \, i \in M . \tag{1.2}$$

In the design of experiments, C is a covariance matrix of a set of Gaussian

mODa6, A.C.Atkinson, P.Hackl and W.G.Müller, eds., Physica, Heidelberg, 2001.

random variables, and, up to constants, $\ln \det C[S, S]$ is the entropy of the set of random variables associated with S. So the CMESP is to find a most-informative s-subset of a set of n candidate Gaussian covariates, subject to some side constraints.

The CMESP is already NP-Hard when $M = \emptyset$. Ko *et al.* (1995) introduced exact algorithms based on the branch-and-bound framework. The key ingredient in such a method is an upper bound on z. Ko *et al.* introduced the *spectral bound*

$$v := \sum_{l=1}^{s} \ln \lambda_l (C) \;, \tag{1.3}$$

where λ_l denotes the lth greatest eigenvalue. Taking advantage of the side-constraints (1.2), Lee (1998) introduced the *Lagrangian spectral bound*

$$v(A, b) := \min_{w \in \mathfrak{R}_+^M} v(A, b, w) \;, \tag{1.4}$$

where

$$v(A, b, w) := \sum_{l=1}^{s} \ln \lambda_l (D_w C D_w) + \sum_{i \in M} w_i b_i \;, \tag{1.5}$$

and

$$D_w := \operatorname{diag}_j \left\{ \exp \left\{ -\frac{1}{2} \sum_{i \in M} w_i a_{ij} \right\} \right\}. \tag{1.6}$$

Lee demonstrated that $v(A, b, w)$ is convex in w, and he described descent methods for calculating a minimizing w in (1.4). Other upper bounds are based on convex programming; see Anstreicher *et al.* (1999) and Lee (2000). Lower bounds for fathoming are determined by heuristic search methods. Branching consists of fixing an index j out of the solution (deleting row and column j of C) or fixing an index j in to the solution (pivoting in C on C_{jj}, decrementing s by one, and decreasing each b_i by a_{ij}).

In Anstreicher *et al.* (1999), we described how different bounds can be calculated by considering the equivalent *complementary problem* (when $\det C \neq 0$):

$$z = \ln \det C + \max_{\substack{N \setminus S \subset N: \\ |N \setminus S| = n - s}} \ln \det C^{-1}[N \setminus S, N \setminus S] \;;$$

$$\text{subject to} \sum_{j \in N \setminus S} (-a_{ij}) \leq b_i - \sum_{j \in N} a_{ij} \;, \; \forall \, i \in M \;.$$

We can calculate a bound with respect to choosing the $n - s$ element set $N \setminus S$ for this problem, and then just add $\ln \det C$ to that bound. The spectral bound for the original and complementary problems are identical, but this is not the case for other bounds.

In Section 2, we develop new upper bounds based on Fischer's inequality. All upper bounds are valid in the non-Gaussian case as well, since, for fixed covariance matrix C, the Gaussian maximizes entropy. In Section 3, we demonstrate how to combine the use of Fischer's inequality with previously developed bounds. We demonstrate this in detail by strengthening the spectral bounds. Computational experiments suggest that these bounds may be useful in solving problems to optimality in a branch-and-bound framework.

2 Partition bounds

We base new upper bounds on the following inequality of Fischer. In what follows a partition of a set S is just a set of disjoint sets (or "blocks") $S_1, S_2, \ldots, S_s$ whose union is S. Note that we allow empty parts for convenience.

Lemma 2.1 (Fischer, 1908, pp. 36-7; also see Horn and Johnson, 1985, pp. 478-9, Theorem 7.8.3). *Let B be a square symmetric positive-semidefinite matrix with rows and columns indexed from the cardinality s set S^*. Let $S_1, S_2, \ldots, S_s$ be a partition of S^*. Then*

$$\det B \leq \prod_{k=1}^{s} \det B[S_k, S_k] . \tag{2.1}$$

A sequence of refinements of the partition yields a nondecreasing sequence of upper bounds on $\det B$.

Let $\Pi = \{\pi_1, \pi_2, \ldots, \pi_s\}$ be a "partition" of s; that is, a multiset of nonnegative integers such that $\sum_{k=1}^{s} \pi_k = s$. We introduce the *partition bound*

$$\psi(\Pi) := \quad \max \sum_{k=1}^{s} \ln \det C[S_k, S_k] ; \tag{2.2}$$

$$\text{subject to } S_k \subset N , \ \forall \, k = 1, 2, \ldots, s ; \tag{2.3}$$

$$|S_k| = \pi_k , \ \forall \, k = 1, 2, \ldots, s ; \tag{2.4}$$

$$S_k \cap S_{k'} = \varnothing , \ \forall \, 1 \leq k < k' \leq s ; \tag{2.5}$$

$$\sum_{k=1}^{s} \sum_{j \in S_k} a_{ij} \leq b_i , \ \forall \, i \in M . \tag{2.6}$$

We note that when $\Pi = \{s, 0, 0, \ldots, 0\}$, we have $z = \psi(\Pi)$. Next, we establish that the partition bound is in fact an upper bound on z.

Proposition 2.1 $\quad z \leq \psi(\Pi)$.

Proof. Suppose that S^* is an optimal solution to (1.1–1.2). So $z = \ln \det C[S^*, S^*]$. Choose any partition $S = \{S_1, S_2, \ldots, S_s\}$ of S^* satisfying (2.4); the conditions (2.3,2.5) are obviously satisfied. Moreover,

$$\sum_{k=1}^{s} \sum_{j \in S_k} a_{ij} = \sum_{j \in S^*} a_{ij} \,,$$

so (2.6) is satisfied. Therefore, S is a feasible solution to the program (2.2–2.6). Now, applying Fischer's inequality (2.1) to $B = C[S^*, S^*]$, and considering the monotonicity of the logarithm, the result follows. $\square$

Next, we focus on situations where we can compute $\psi(\Pi)$ efficiently — either in a practical or theoretical sense. The simplest such situation is based on the finest partition, where we take $\Pi = \{1, 1, \ldots, 1\}$. In this case, we can recast the *diagonal bound* $\psi_1 := \psi(\Pi)$ as the optimal value of the following ILP (i.e, integer linear program):

$$\psi_1 = \max \sum_{j \in N} (\ln C_{jj}) \, x_j \,;$$

$$\text{subject to } \sum_{j \in N} x_j = s \,;$$

$$\sum_{j \in N} a_{ij} x_j \le b_i \,, \; \forall \, i \in M \,;$$

$$x_j \in \{0, 1\}, \; \forall \, j \in N \,.$$

General methods of ILP can be applied to solve this bounding program (see Nemhauser and Wolsey, 1988). When $M = \varnothing$, we obtain

$$\psi_1 = \sum_{l=1}^{s} \ln C_{[l]} \,,$$

where $C_{[l]}$ is the lth greatest diagonal element of C. So when $M = \varnothing$, we can calculate ψ_1 efficiently in the theoretical sense.

Example 1 *Let n be even, $s := n/2$ and $M := \varnothing$. Let the nonzeros of C consist of the $n/2$ diagonal blocks*

$$\begin{pmatrix} C_{2l-1,2l-1} & C_{2l-1,2l} \\ C_{2l,2l-1} & C_{2l,2l} \end{pmatrix} := \begin{pmatrix} 1 & 1 \\ 1 & 1 \end{pmatrix} ,$$

for $l = 1, 2, \ldots, n/2$. We have $z = \psi_1 = 0$ and $v = \ln 2^{n/2}$. So here the diagonal bound is much better than the spectral bound.

Example 2 *Let $C := \varepsilon I_n + 1_{n \times n}$, with $\varepsilon > 0$, and $M := \varnothing$. Clearly $\psi_1 = \ln (1 + \varepsilon)^s$, which tends to 0 as $\varepsilon \to 0^+$. Also $v = \ln (n + \varepsilon)\varepsilon^{s-1}$ and $z = \ln (s + \varepsilon)\varepsilon^{s-1}$, which both tend to $-\infty$ as $\varepsilon \to 0^+$. So here the spectral bound is much better than the diagonal bound.*

Another situation that we can exploit is when $\Pi = \{2, 2, \ldots, 2, 0, 0, \ldots, 0\}$ for even s and $\Pi = \{2, 2, \ldots, 2, 1, 0, 0, \ldots, 0\}$ for odd s. Again, we can recast this *matching bound* as the optimal value of an ILP. We let N^2 be the set of all subsets T of N satisfying $|T| = 2$. For all $T \subset N^2$, we define binary variables y_T. The form of the program depends on the parity of s. If s is even, then our bound $\psi_2 := \psi(\Pi)$ is

$$\psi_2 = \max \sum_{T \in N^2} (\ln \det C[T, T]) \, y_T \, ;$$

$$\text{subject to} \sum_{T \in N^2} y_T = s/2 \, ;$$

$$\sum_{\substack{T \in N^2: \\ j \in T}} y_T \leq 1 \, , \forall \, j \in N \, ;$$

$$\sum_{T \in N^2} \left(\sum_{j \in T} a_{ij} \right) y_T \leq b_i \, , \, \forall \, i \in M \, ;$$

$$y_T \in \{0, 1\}, \, \forall \, T \in N^2 \, .$$

General ILP methods may be applied to solve this bounding program. The CMESP is already quite difficult at $n = 100$, so this approach might be quite reasonable. We note that when $M = \varnothing$, there is a theoretically-efficient algorithm for solving this program. We define a complete graph with vertex set N and edge set N^2. For edge $T \in N^2$, we assign weight $\ln \det C[T, T]$. Then we find a maximum-weight "matching" (i.e., a set of pairwise disjoint edges) having cardinality $s/2$, hence the moniker "matching bound".

If s is odd, we define additional binary variables x_j, for $j \in N$. In this case, we calculate our bound by solving the ILP

$$\psi_2 = \max \sum_{j \in N} (\ln C_{jj}) \, x_j + \sum_{T \in N^2} (\ln \det C[T, T]) \, y_T \, ;$$

$$\text{subject to} \sum_{j \in N} x_j = 1 \, ;$$

$$\sum_{T \in N^2} y_T = (s - 1)/2 \, ;$$

$$x_j + \sum_{\substack{T \in N^2: \\ j \in T}} y_T \leq 1 \, , \forall \, j \in N \, ;$$

$$\sum_{j \in N} a_{ij} x_j + \sum_{T \in N^2} \left(\sum_{j \in T} a_{ij} \right) y_T \leq b_i \, , \, \forall \, i \in M \, ;$$

$$x_j \in \{0, 1\}, \, \forall \, j \in N \, ; \quad y_T \in \{0, 1\}, \, \forall \, T \in N^2 \, .$$

Again, when $M = \varnothing$, we can recast this program using matchings. We start with the same graph as before, but now we incorporate an additional vertex 0. We join vertex 0 to each other vertex $j \in N$. Then, for each $j \in N$, we give weight $K + \ln C_{jj}$ to the edge $\{0, j\}$, for sufficiently large K. Then, we again find a maximum-weight matching having cardinality $s/2$. For large enough K, the optimal weight matching will include exactly one of the edges meeting vertex 0. The optimal weight for this matching is $K + \psi_2$.

Example 3 *Continuing with the matrix from Example 2, we calculate $\psi_2 = \ln (2\varepsilon + \varepsilon^2)^{s/2}$ which tends to $-\infty$ as $\varepsilon \to 0^+$. So here the matching bound is much better than the diagonal bound. Furthermore, holding ε constant (at some small positive value), and letting n increase, we can make the matching bound do much better than the spectral bound as well.*

Although ψ_2 is harder to calculate than ψ_1, since $\psi_2 \leq \psi_1$, it may be worth the extra effort, in the context of branch-and-bound, to calculate ψ_2.

We performed some computational experiments using environmental monitoring data; see Ko *et al.* (1995). Our results focus on the quality of the bounds and not on the computational effort. In the tables, the "bars" indicate bounds applied to the complementary problem. The first problem has $n = 48$ and no side constraints. In Table 1 we display the gaps (i.e., upper bound minus optimal entropy) $\psi_k - z$ and $\bar{\psi}_k - z$. This gives an indication of the behavior of the partition bound ψ_k and the complementary partition bound $\bar{\psi}_k$ as k increases. For small values of k, we can observe how ψ_k (resp. $\bar{\psi}_k$) does better than $\bar{\psi}_k$ (resp. ψ_k) when s is small (resp. large) relative to n. Unfortunately, the improvement in the bound is often rather slight as k increases, while the difficulty in calculating the bound grows very quickly. We note that experience has shown that a gap of up to perhaps 3 indicates that a problem might be solved to optimality by branch-and-bound within a reasonable amount of time.

Table 2 compares these partition bounds to previous bounds for a data set having $n = 124$; "Id,Di,Tr" refer to particular parameter choices for some convex-programming bounds; see Anstreicher *et al.* (1999) for details. We note that for s not too small nor too large, these problems are beyond our current capability to solve, so we have tabulated bounds rather than gaps. For this data set, ψ_2 (resp. $\bar{\psi}_2$) offers no significant improvement over ψ_1 (resp. $\bar{\psi}_1$). Furthermore, for small (resp. large) values of s, ψ_1 (resp. $\bar{\psi}_1$), which is very cheap to compute, is competitive with the other bounds.

3 Spectral partition bounds

Let $\mathcal{N} = \{N_1, N_2, ..., N_n\}$ denote a partition of N (we are allowing empty parts for convenience). For $k = 1, 2, \ldots, n$, let $\Lambda(N_k)$ be the multiset of $|N_k|$ eigenvalues of $C[N_k, N_k]$. Let $\Lambda(\mathcal{N})$ denote the multiset union of the n elements from the sets $\Lambda(N_k)$. For a multiset $\Lambda(\cdot)$, $\Lambda_l(\cdot)$ denotes the lth

$\psi_k - z$	$s=12$	$s=24$	$s=36$	$\bar{\psi}_k - z$	$s=12$	$s=24$	$s=36$
$k=1$	2.960525	6.346888	11.544246	$k=1$	4.180991	2.372326	0.716902
$k=2$	2.683129	6.208118	11.335711	$k=2$	4.180985	2.372316	0.716723
$k=3$	2.298198	5.915404	10.871107	$k=3$	4.180968	2.372173	0.712107

TABLE 1. Gaps ($n = 48$)

$s=$	10	20	30	40	50	60	70	80	90	100	110	120
$\psi_1 =$	44.2	81.5	116.2	149.1	180.6	210.4	237.6	262.8	285.3	303.5	316.9	326.8
$\psi_2 =$	44.2	81.5	116.2	149.1	180.6	210.4	237.6	262.8	285.3	303.5	316.9	326.8
$\bar{\psi}_1 =$	290.1	293.1	291.4	286.1	278.0	267.1	253.9	237.7	217.3	191.6	160.3	122.2
$\bar{\psi}_2 =$	290.1	293.1	291.4	286.1	278.0	267.1	253.9	237.7	217.3	191.5	160.2	122.1
$Id =$	47.2	92.3	136.3	179.2	220.6	260.2	297.4	331.2	360.0	380.67	385.2	337.1
$\overline{Id} =$	470.8	490.1	478.6	454.3	422.7	386.5	347.1	305.3	261.7	216.7	170.5	123.3
$Di =$	63.9	118.6	167.3	210.7	250.8	287.9	321.9	352.4	378.1	395.4	395.4	338.0
$\overline{Di} =$	339.2	370.5	375.8	370.4	359.0	343.7	324.6	300.0	269.1	231.8	186.4	130.6
$Tr =$	52.1	98.1	140.3	180.6	219.2	255.9	290.1	321.1	347.3	365.5	367.8	319.1
$\overline{Tr} =$	358.4	387.9	390.6	382.4	368.2	349.6	327.1	298.4	263.7	223.6	178.5	127.2
$v =$	50.4	90.6	124.1	151.7	173.5	189.9	198.8	199.8	193.9	180.4	159.6	125.4

TABLE 2. Bounds ($n = 124$)

greatest element. We define the *spectral partition bound*

$$\phi(\mathcal{N}) := \sum_{l=1}^{s} \ln \Lambda_l(\mathcal{N}) . \tag{3.1}$$

Proposition 3.1 $z \leq \phi(\mathcal{N})$.

Proof: Let S be any subset of N having cardinality s.

$$\ln \det C[S, S] \leq \sum_{k=1}^{n} \ln \det C[S \cap N_k, S \cap N_k] \tag{3.2}$$

$$= \sum_{k=1}^{n} \sum_{\lambda \in \Lambda(S \cap N_k)} \ln \lambda \tag{3.3}$$

$$\leq \sum_{k=1}^{n} \sum_{l=1}^{|S \cap N_k|} \ln \Lambda_l(N_k) \tag{3.4}$$

$$\leq \sum_{l=1}^{s} \ln \Lambda_l(\mathcal{N}) . \tag{3.5}$$

Note that (3.2) holds by (2.1), (3.3) holds since the product of all eigenvalues of a matrix is its determinant, (3.4) holds by the eigenvalue interlacing inequalities (see, e.g., Horn and Johnson, 1985, pp. 185-6, Theorem 4.3.8), and (3.5) holds by allowing S to range over the s-subsets of N.   □

We note that

- $\phi(\{N, \varnothing, \varnothing, \ldots, \varnothing\}) = v$ (thus subsuming the bound of Ko *et al.*, 1995)
- $\phi(\{\{1\}, \{2\}, \ldots, \{n\}\}) = \psi_1$.

Next, we demonstrate a revealing situation in which neither of these two partitions is the best possible.

Example 4 *Let* $s := n/2$ *and* $S := \{1, 2, \ldots, n/2\}$; *so* $N \setminus S = \{n/2 + 1, n/2+2, \ldots, n\}$. *Let* $C[S, S] := nI_s + 1_{s \times s}$, $C[N-S, N-S] := (3n/4)I_s + 1_{s \times s}$ *and* $C[S, N \setminus S] := C[N \setminus S, S] := 0_{s \times s}$. *We have* $\Lambda(S) = \{3n/2, n, n, \ldots, n\}$, $\Lambda(N-S) = \{5n/4, 3n/4, 3n/4, \ldots, 3n/4\}$, *and obviously* $\Lambda(N) = \Lambda(S) \cup \Lambda(N-S)$. *Therefore* $v = \phi(\{N, \varnothing, \varnothing, \ldots, \varnothing\}) = \phi(\{S, N \setminus S, \varnothing, \varnothing, \ldots, \varnothing\}) = \ln(3n/2)(5n/4)n^{n/2-2}$; *note that we picked up the* $5n/4$ *from* $\Lambda(N-S)$. *We also have* $\phi(\{\{1\}, \{2\}, \ldots, \{n\}\}) = \psi_1 = \ln(n+1)^{n/2}$; *note that we did not pick up any of* $\Lambda(N-S)$, *but the bound has deteriorated by chopping up* S *and using Fischer's inequality. Finally, we observe that* $\phi(\{S, \{n/2+1\}, \{n/2+2\}, \ldots, \{n\}\}) = \ln \det C[S, S]$, *since all of the diagonal entries of* $C[N-S, N-S]$ *are* $3n/4+1$, *which is less than all eigenvalues of* $C[S, S]$. *So this last partition establishes that* S *has maximum entropy, while the other partitions mentioned above do not.*

Example 4 suggests the following sufficient optimality criterion.

Proposition 3.2 *Let* S *be a feasible subset of* N. *If*

$$\Lambda_s(S) \geq \max\{C_{jj} \ : \ j \in N \setminus S\},$$

then $z = \ln \det C[S, S]$.

Proof: We simply observe that under the hypothesis we have $\phi(\{S, \{s+1\}, \{s+2\}, \ldots, \{n\}\}) = \ln \det C[S, S]$.   □

Next, we turn to the problem of finding the best partition $\mathcal{N}$. That is,

$$\text{MIN}\Phi: \quad \min\{\phi(\mathcal{N}) \ : \ \mathcal{N} \text{ is a partition of } N\}.$$

We suggest a heuristic, outlined in Fig. 1, for MINΦ. We experimented with the heuristic on an example from Anstreicher *et al.*, 1999 having $n = 63, s = 31$. For the "local-search" of Step 2, we repeatedly evaluated the spectral bound for $\mathcal{O}(n^2)$ "nearby" partitions and selected the move that gave the best improvement. We used the moves described in Fig. 2. At a first pass, we used the moves 2a until no further improvement was possible. Then we proceeded further using 2a–d. We worked with both the original and complementary problems. The results are displayed in Table 3. We subtracted the optimal value (obtained from Anstreicher *et al.* (1999)) from the bounds to obtain the gaps. The first row consists of the gaps after Step 1. The second (resp., third) row consists of the gaps after all 2a (resp., 2a–d) moves were completed. Also, for Step 1a, since we wish to see how the bounding idea performs when we have an extremely good heuristic for finding an S with high entropy, we actually used an optimal S. The results are exceptionally good. The best bound obtained (2.521062) is much better than the ordinary spectral bound ($v = 5.707025$) and also significantly better than the best bound obtained ($\bar{\psi}_1 = 3.252440$) without applying the local-search of Step 2. In addition, our results suggests that our particular local-search moves are rather robust, since the final bound obtained does not depend very much on the initial partition selected in Step 1. Finally, we mention that for the best bound obtained (i.e., $\bar{\phi}(\mathcal{N})$ starting with 1b), the final partition had block sizes of: 3, 4, 4, 5, 5, 7, 9, 11, 15.

1a. Use heuristic methods to find an $S \subset N$, $|S| = s$ with a high value of $\ln \det C[S, S]$ (e.g., greedy and exchange heuristics for the problem are discussed in Ko *et al.*, 1995); then take the initial partition $\mathcal{N} = \{S, \{j_1\}, \{j_2\}, ..., \{j_{n-s}\}\}$, where $N \setminus S = \{j_1, j_2, ..., j_{n-s}\}$ (and attempt to establish the optimality of S with Proposition 3.2).

1b. Alternatively, use the initial partition $\mathcal{N} = \{N, \varnothing, \varnothing, ..., \varnothing\}$ which guarantees a bound no worse than the spectral bound v.

1c. Alternatively, use the initial partition $\mathcal{N} = \{\{1\}, \{2\}, ..., \{n\}\}$ which guarantees a bound no worse than the diagonal bound ψ_1.

2. Use a local-search on the space of partitions, to decrease $\phi(\mathcal{N})$.

FIGURE 1. MINΦ Heuristic

	original			complementary		
	1a	1b	1c	1a	1b	1c
1	5.5121	$v = 5.7070$	$\psi_1 = 7.9250$	3.3524	$v = 5.7070$	$\bar{\psi}_1 = 3.2524$
2a	4.5767	4.5793	5.0606	2.6555	2.6077	2.6294
2a–d	4.5767	4.5793	4.5774	2.6302	2.5211	2.6273

TABLE 3. Gaps using local search ($n = 63$, $s = 31$)

2a. *(single-element move)* $j \in N_k$, $l \neq k$: $N_k \leftarrow N_k - j$, $N_l \leftarrow N_l + j$.

2b. *(simple exchange)* $j \in N_k$, $i \in N_l$, $l \neq k$: $N_k \leftarrow N_k - j + i$, $N_l \leftarrow N_l - i + j$.

2c. *(one new two-block or two new one-blocks)* $j \in N_k$, $i \in N_l$, $i \neq j$, $N_h = \varnothing$, $N_g = \varnothing$: $N_k \leftarrow N_k - j$, $N_l \leftarrow N_l - i$, $N_h \leftarrow N_h + i$, $N_g \leftarrow N_g + j$.

2d. *(merge two blocks)* $k \neq l$: $N_k \leftarrow N_k \cup N_l$, $N_l \leftarrow \varnothing$.

FIGURE 2. Local-Search Moves

The spectral partition bound does not take advantage of the side-constraints. We can improve the spectral partition bound by adapting the Lagrangian methodology employed in Lee (1998). As before, we consider a partition $\mathcal{N} = \{N_1, N_2, ..., N_n\}$ of N and define the diagonal matrix D_w by (1.6). Let $\Lambda(N_k, A, w)$ denote the multiset of $|N_k|$ eigenvalues of $(D_w C D_w)[N_k, N_k]$ and $\Lambda(\mathcal{N}, A, w)$ denote the multiset union of $|N| = n$ elements from the sets $\Lambda(N_k, A, w)$. We introduce the *Lagrangian spectral partition bound*

$$\phi(\mathcal{N}, A, b) := \min_{w \in \Re_+^M} \phi(\mathcal{N}, A, b, w),$$

where

$$\phi(\mathcal{N}, A, b, w) := \sum_{l=1}^{s} \ln \Lambda_l(\mathcal{N}, A, w) + \sum_{i \in M} w_i b_i.$$

Following the ideas in Lee (1998), we have

Proposition 3.3 $z \leq \phi(\mathcal{N}, A, b)$;

Proposition 3.4 *The function $\phi(\mathcal{N}, A, b, w)$ is convex in w,*

and we can give an expression for subgradients of $\phi(\mathcal{N}, A, b, w)$.

Finally, although we concentrated on spectral bounds, other bounds can also be strengthened using our ideas. We will discuss experiments with some of these possibilities in a forthcoming paper in which we report on results of incorporating our new bounds in a branch-and-bound code.

References

Anstreicher, K.M., Fampa, M., Lee, J., and Williams, J. (1999). Using continuous nonlinear relaxations to solve constrained maximum-entropy sampling problems. *Math. Prog., Ser. A*, **85**, 221-240.

Fischer, E.S. (1908). Über den Hadamardschen Determinantensatz. *Archiv für Mathematik und Physik*, **13**, 32-40.

Horn, R.A. and Johnson, C.R. (1985). *Matrix Analysis*. Cambridge University Press, Cambridge.

Ko, C.-W., Lee, J., and Queyranne, M. (1995). An exact algorithm for maximum entropy sampling. *Operations Research*, **43**, 684-691.

Lee, J. (1998). Constrained maximum-entropy sampling. *Operations Research*, **46**, 655-664.

Lee, J. (2000). Semidefinite-programming in experimental design. In Wolkowicz, H., Saigal, R., and Vandenberghe, L. (eds.), *Handbook of Semidefinite Programming*, Kluwer, Dordrecht.

Nemhauser, G.L. and Wolsey, L.A. (1988). *Integer and Combinatorial Optimization*. Wiley, New York.

Sebastiani, P. and Wynn, H.P. (2000). Maximum entropy sampling and optimal Bayesian experimental design. *Journal of the Royal Statistical Society*, **B62**, 145-157.

Shewry, P. and Wynn, H.P. (1987). Maximum entropy sampling. *Journal of Applied Statistics*, **14**, 165-170.

Zidek, J.V., Sun, W., and Le, N.D. (2000). Designing and integrating composite networks for monitoring multivariate Gaussian pollution fields. *Journal of Applied Statistics*, **49**, 63-79.

Residuals

H. Läuter

ABSTRACT: We consider different definitions for residuals and their distributions with a view to their use in bootstrapping. It turns out that order
statistics play an important role in finding good residuals.

KEYWORDS: bootstrap; estimation of distributions; order statistics; residuals

1 Introduction

Consider the regression model

$$Y_i = x_i(\beta) + \varepsilon_i, \quad i = 1, ..., n$$

where the ε_i's are i.i.d. random variables with $E\varepsilon_i = 0$, $\beta \in \mathbb{R}^k$ and x_i's are
known functions. In several problems one needs the distribution of the ε_i
or one is interested in the estimation of the distribution. Here the power of
regularizations of 'ill-posed equations for estimates is to be mentioned. In
such problems the solutions are unstable and one knows that the optimal
regularizations depend on the type of distributions.

Another problem is bootstrapping of statistics. Here sufficiently good estimates for the distribution of the ε_i are necessary. At least the estimates
have to be consistent and quite efficient. The finite sample properties of the
estimates determine the effectiveness of bootstrap methods. So we have in
mind moderate sample sizes, e.g. $n \leq 10k$.

The usual approach is based on the determination of the residuals $\hat{\varepsilon}_j :=
Y_j - x_j(\hat{\beta})$ where

$$\hat{\beta} = \arg\min_{\beta} \sum_{i=1}^{n} (Y_i - x_i(\beta))^2 \ .$$

The desired estimator for the distribution of ε_i is given by the empirical

$mODa6$, A.C.Atkinson, P.Hackl and W.G.Müller, eds., Physica, Heidelberg, 2001.

distribution of the centralized residuals

$$\varepsilon_j^* := \hat{\varepsilon}_j - \bar{\hat{\varepsilon}}$$

where $\bar{\hat{\varepsilon}} := \frac{1}{n}\sum_{i=1}^n \hat{\varepsilon}_i$.

We discuss in detail the most simple case $x_i(\beta) = \beta \in \mathbb{R}^1$, so the variables $Y_1, ..., Y_n$ are i.i.d. We denote by f the Lebesgue density of ε_i. The density of the first residual $Y_1 - \bar{Y}_n$ will be denoted by $\hat{f}$. We demonstrate that

$$\frac{\hat{f}(t)}{f(t)} \to 0 \quad \text{as } t \to \infty .$$

If $\hat{\hat{f}}$ is the density of the standardized residual $\sqrt{\frac{n}{n-1}}(Y_1 - \bar{Y}_n)$ then $\hat{\hat{f}}(t) := \sqrt{\frac{n-1}{n}}\hat{f}(\sqrt{\frac{n-1}{n}}t)$ and we have for many interesting distributions

$$\frac{\hat{\hat{f}}(t)}{f(t)} \to 0, \quad t \to \infty .$$

These expressions say that the distribution to be estimated is more 'long-tailed than the estimated distribution of $Y_1 - \bar{Y}_n$ or $\sqrt{\frac{n}{n-1}}(Y_1 - \bar{Y}_n)$. This is also a consequence of robustness theory (Hampel *et al.*, 1986). So this procedure is not appropriate for estimating the distribution of the errors ε_i. Normal distributions take a special place in this context. For those we have

$$\frac{\hat{\hat{f}}(t)}{f(t)} \to 1 \quad t \to \infty.$$

As an alternative method we propose residuals coming from a robust estimate for β. For that we use the median and prove that, at least in important classes of distributions, the distribution of $Y_1 - \text{med}(Y_1, ..., Y_n)$ has the same tail behavior as the distribution to be estimated. The classes we consider are the normal and exponential distributions as well as mixtures of these.

2   Joint distribution of Y_1 and medY

In this section we will show that the joint distribution of Y_1 and medY has no density with respect to Lebesgue–measure.

Theorem 2.1 *Let $Y := (Y_1, ..., Y_n)$ be a vector of i.i.d. random variables and assume that*

1. n is odd,

2. *the distribution of Y_i is absolutely continuous with respect to Lebesgue–measure,*

3. *the density f of Y_i is symmetric.*

Then the conditional joint distribution of Y_1 and $\mathrm{med}Y$ under $Y_1 \neq \mathrm{med}Y$ is continuous. The probability for $Y_1 = \mathrm{med}Y$ is $\frac{1}{n}$.

Proof. We consider the probability

$$P(Y_1 \in \Delta_{t_1}, \mathrm{med}Y \in \Delta_{t_2})$$

$$= P(Y_1 < \mathrm{med}Y, Y_1 \in \Delta_{t_1}, \mathrm{med}Y \in \Delta_{t_2}) \tag{2.1}$$

$$+ P(Y_1 = \mathrm{med}Y, Y_1 \in \Delta_{t_1}, \mathrm{med}Y \in \Delta_{t_2}) \tag{2.2}$$

$$+ P(Y_1 > \mathrm{med}Y, Y_1 \in \Delta_{t_1}, \mathrm{med}Y \in \Delta_{t_2}) \tag{2.3}$$

with $\Delta_t = [t, t + \delta)$. Here (2.1) and (2.3) will lead to the continuous part and (2.2) to the discontinuity.

First we determine the conditional density of $(Y_1, \mathrm{med}Y)$ under the condition $Y_1 \neq \mathrm{med}Y$. Beginning with (2.1) and with

$$P(Y_{i:n} \in \Delta_{t_1}, \mathrm{med}Y \in \Delta_{t_2})$$

$$= \sum_{j=1}^{n} P(Y_j = Y_{i:n}, Y_{i:n} \in \Delta_{t_1}, \mathrm{med}Y \in \Delta_{t_2})$$

$$= nP(Y_1 = Y_{i:n}, Y_{i:n} \in \Delta_{t_1}, \mathrm{med}Y \in \Delta_{t_2}),$$

we get

$$P(Y_1 < \mathrm{med}Y, Y_1 \in \Delta_{t_1}, \mathrm{med}Y \in \Delta_{t_2})$$

$$= \sum_{i=1}^{\frac{n-1}{2}} P(Y_1 = Y_{i:n}, Y_{i:n} \in \Delta_{t_1}, \mathrm{med}Y \in \Delta_{t_2})$$

$$= \frac{1}{n} \sum_{i=1}^{\frac{n-1}{2}} P(Y_{i:n} \in \Delta_{t_1}, \mathrm{med}Y \in \Delta_{t_2}) . \tag{2.4}$$

If $t_1 > t_2$ and δ is sufficiently small the probability (2.4) is zero because we always can choose a δ with $t_2 + \delta < t_1$. If $t_1 = t_2$ then (2.4) converges to 0 with $\delta \to 0$. The joint distribution of $Y_{i:n}$ and $Y_{\frac{n+1}{2}:n}$ exists and we will denote the density by $f_{i,n}(t_1, t_2)$ (Arnold, 1992). For

$$\tilde{f}_{<,n}(t_1, t_2) := \lim_{\delta \to 0} \frac{1}{\delta^2} 1_{\{t_1 < t_2\}}(t_1, t_2) P(Y_1 < \mathrm{med}Y, Y_1 \in \Delta_{t_1}, \mathrm{med}Y \in \Delta_{t_2})$$

we get from (2.4)

$$\tilde{f}_{<,n}(t_1,t_2) = \frac{1}{n}\sum_{i=1}^{\frac{n-1}{2}} f_{i,n}(t_1,t_2)\mathbf{1}_{\{t_1<t_2\}}(t_1,t_2)$$

$$= \frac{1}{n}\sum_{i=1}^{\frac{n-1}{2}} \frac{n!}{(i-1)!\left(\frac{n-1}{2}-i\right)!\left(\frac{n-1}{2}\right)!}F(t_1)^{i-1}\times$$

$$\times[F(t_2)-F(t_1)]^{\frac{n-1}{2}-i}[1-F(t_2)]^{\frac{n-1}{2}}\times$$

$$\times f(t_1)f(t_2)\mathbf{1}_{\{t_1<t_2\}}(t_1,t_2)$$

$$= \frac{(n-1)!}{\left(\frac{n-1}{2}\right)!\left(\frac{n-3}{2}\right)!}f(t_1)f(t_2)[1-F(t_2)]^{\frac{n-1}{2}}F(t_2)^{\frac{n-3}{2}}\times$$

$$\times\mathbf{1}_{\{t_1<t_2\}}(t_1,t_2). \tag{2.5}$$

Now the term (2.3) is considered. For

$$\tilde{f}_{>,n}(t_1,t_2) := \lim_{\delta\to 0}\frac{1}{\delta^2}\mathbf{1}_{\{t_1>t_2\}}(t_1,t_2)P(Y_1 > \text{med}Y, Y_1\in\Delta_{t_1}, \text{med}Y\in\Delta_{t_2})$$

we get with the same arguments

$$\tilde{f}_{>,n}(t_1,t_2) := \frac{1}{n}\sum_{i=\frac{n+3}{2}}^{n} \frac{n!}{\left(\frac{n-1}{2}\right)!\left(i-\frac{n+3}{2}\right)!(n-i)!}F(t_2)^{\frac{n-1}{2}}\times$$

$$\times[F(t_1)-F(t_2)]^{i-\frac{n+3}{2}}[1-F(t_1)]^{i-\frac{n+3}{2}}\times$$

$$\times f(t_1)f(t_2)\mathbf{1}_{\{t_1>t_2\}}(t_1,t_2)$$

$$= \frac{(n-1)!}{\left(\frac{n-1}{2}\right)!\left(\frac{n-3}{2}\right)!}f(t_1)f(t_2)F(t_2)^{\frac{n-1}{2}}[1-F(t_2)]^{\frac{n-3}{2}}\times$$

$$\times\mathbf{1}_{\{t_1>t_2\}}(t_1,t_2). \tag{2.6}$$

Then we obtain

$$\tilde{f}_n(t_1,t_2) := \tilde{f}_{<,n}(t_1,t_2) + \tilde{f}_{>,n}(t_1,t_2)$$

$$= \frac{(n-1)!}{\left(\frac{n-1}{2}\right)!\left(\frac{n-3}{2}\right)!}f(t_1)f(t_2)F(t_2)^{\frac{n-3}{2}}[1-F(t_2)]^{\frac{n-1}{2}}\times$$

$$\times\mathbf{1}_{\{t_1<t_2\}}(t_1,t_2)$$

$$+\frac{(n-1)!}{\left(\frac{n-1}{2}\right)!\left(\frac{n-3}{2}\right)!}f(t_1)f(t_2)F(t_2)^{\frac{n-1}{2}}[1-F(t_2)]^{\frac{n-3}{2}}\times$$

$$\times\mathbf{1}_{\{t_1>t_2\}}(t_1,t_2).$$

This function is obviously continuous for $t_1\neq t_2$. Because

$$P\left(Y_1 = Y_{\frac{n+1}{2}:n}\right) = \frac{1}{n}$$

the manifold $\{t_1 = t_2\}$ of smaller dimension has a positive probability measure. ∎

REMARK 1: The function $\tilde{f}_n$ is not a density function, we have

$$\int \tilde{f}_n(t_1, t_2) dt_1 dt_2 = \frac{n-1}{n}.$$

This is not surprising because there is a positive weight on the line $t_1 = t_2$.

One finds that

$$\tilde{\tilde{f}}_n(t_1, t_2) := \frac{n}{n-1} \tilde{f}_n(t_1, t_2)$$

is the conditional density of $(Y_1, \text{med}Y)$ under $Y_1 \neq \text{med}Y$. From Theorem 2.1 we get immediately

REMARK 2: Under the conditions of Theorem 2.1 the joint distribution of Y_1 and $\text{med}Y$ has no density with respect to Lebesgue–measure.

3 Distribution of $Y_1 - \text{med}Y$ and $Y_1 - \bar{Y}$

In the previous section it was shown that the usual approach for getting the density of $\tilde{\varepsilon}_1 := Y_1 - \text{med}Y$ by convoluting the joint density fails. So we use a straightforward way to get it. We get for any odd number n

$$P(Y_1 - \text{med}Y \in \Delta_t) = P\left(Y_1 - Y_{\frac{n+1}{2}:n} \in \Delta_t\right)$$

$$= \begin{cases} \frac{1}{n} \sum_{i=1}^{\frac{n-1}{2}} P(Y_{i:n} - Y_{\frac{n+1}{2}:n} \in \Delta_t) & \text{if } t < 0 \\ \frac{1}{n} \sum_{i=\frac{n+3}{2}}^{n} P(Y_{i:n} - Y_{\frac{n+1}{2}:n} \in \Delta_t) & \text{if } t > 0 \end{cases}$$

and

$$\lim_{\delta \to 0} P(Y_1 - \text{med}Y \in \Delta_0) = \frac{1}{n}.$$

Let $\tilde{f}_{i,n}$ be the density of $Y_{i:n} - Y_{\frac{n+1}{2}:n}$ with $i \neq \frac{n+1}{2}$. We define for $t \neq 0$

$$\tilde{f}_n(t) := \lim_{\delta \to 0} \frac{1}{\delta} P(Y_1 - \text{med}Y \in \Delta_t),$$

then we have for $t < 0$

$$\tilde{f}_n(t) = \frac{1}{n} \sum_{i=1}^{\frac{n-1}{2}} \tilde{f}_{i,n}(t) = \frac{1}{n} \sum_{i=1}^{\frac{n-1}{2}} \int_{-\infty}^{+\infty} f_{i,\frac{n+1}{2}}(s, s-t) ds$$

$$= \frac{1}{n} \int\limits_{-\infty}^{+\infty} \sum_{i=1}^{\frac{n-1}{2}} \frac{n!}{(i-1)!\left(\frac{n+1}{2}-i-1\right)!\left(n-\frac{n+1}{2}\right)!} F(s)^{i-1} \times$$

$$\times [F(s-t)-F(s)]^{\frac{n+1}{2}-i-1}[1-F(s-t)]^{n-\frac{n+1}{2}} f(s)f(s-t)ds$$

$$= \int\limits_{-\infty}^{+\infty} \frac{(n-1)!}{\left(\frac{n-1}{2}\right)!\left(\frac{n-3}{2}\right)!} f(s)f(s-t)[1-F(s-t)]^{\frac{n-1}{2}} F(s-t)^{\frac{n-3}{2}} ds.$$

Similarly we get for $t > 0$

$$\tilde{f}_n(t) = \int\limits_{-\infty}^{+\infty} \frac{(n-1)!}{\left(\frac{n-1}{2}\right)!\left(\frac{n-3}{2}\right)!} f(s)f(s-t)[1-F(s-t)]^{\frac{n-3}{2}} F(s-t)^{\frac{n-1}{2}} ds$$

and then we have

$$\tilde{f}_n(t) = \begin{cases} \int\limits_{-\infty}^{+\infty} \frac{(n-1)!}{\left(\frac{n-1}{2}\right)!\left(\frac{n-3}{2}\right)!} f(s)f(s-t)[1-F(s-t)]^{\frac{n-1}{2}} F(s-t)^{\frac{n-3}{2}} ds, \\ \hspace{8cm} t < 0 \\ \int\limits_{-\infty}^{+\infty} \frac{(n-1)!}{\left(\frac{n-1}{2}\right)!\left(\frac{n-3}{2}\right)!} f(s)f(s-t)[1-F(s-t)]^{\frac{n-3}{2}} F(s-t)^{\frac{n-1}{2}} ds, \\ \hspace{8cm} t > 0. \end{cases}$$

$$(3.1)$$

Since

$$\int\limits_{-\infty}^{\infty} \tilde{f}_n(t)dt = \frac{n-1}{n},$$

$\tilde{f}_n$ is not a density function. The conditional density of $\tilde{\varepsilon}_1$ under the condition $\tilde{\varepsilon}_1 \neq 0$ is given by

$$\bar{\tilde{f}}_n(t) = \begin{cases} \frac{n}{n-1} \int\limits_{-\infty}^{+\infty} \frac{(n-1)!}{\left(\frac{n-1}{2}\right)!\left(\frac{n-3}{2}\right)!} f(s)f(s-t)[1-F(s-t)]^{\frac{n-1}{2}} F(s-t)^{\frac{n-3}{2}} ds, \\ \hspace{8cm} t < 0 \\ \frac{n}{n-1} \int\limits_{-\infty}^{+\infty} \frac{(n-1)!}{\left(\frac{n-1}{2}\right)!\left(\frac{n-3}{2}\right)!} f(s)f(s-t)[1-F(s-t)]^{\frac{n-3}{2}} F(s-t)^{\frac{n-1}{2}} ds, \\ \hspace{8cm} t > 0 \end{cases}$$

$$(3.2)$$

This is a representation of the conditional density of $\tilde{\varepsilon}_1$ in a general form. In the next section we use these representations for several distributions. To complete this section we will describe the way to compute the density of $\hat{\varepsilon}_1 := Y_1 - \bar{Y}$. Because of

$$\hat{\varepsilon}_1 = Y_1 - \bar{Y} = Y_1 - \frac{1}{n}\sum_{i=1}^{n} Y_i = \frac{n-1}{n} Y_1 - \frac{1}{n}\sum_{i=2}^{n} Y_i$$

we get the density of $\hat{\varepsilon}_1$ ($*$ means convolution)

$$\hat{f}_n(t) := f_{Y_1 - \bar{Y}}(t) = f_{\frac{n-1}{n} Y_1} * f_{\frac{1}{n} Y_2} * \ldots * f_{\frac{1}{n} Y_n}(t) \ .$$

Here we denote by f_Z the density of a variable Z.

Because of $\mathrm{Var}(Y_1 - \bar{Y}) = \frac{n-1}{n} \mathrm{Var} Y_1$ we use the convenient standardization

$$\hat{\bar{\varepsilon}}_1 = \sqrt{\frac{n}{n-1}}(Y_1 - \bar{Y}) \tag{3.3}$$

with its density

$$\hat{\bar{f}}_n(t) = \sqrt{\frac{n-1}{n}} \hat{f}_n \left(\sqrt{\frac{n-1}{n}} t \right). \tag{3.4}$$

4 Consequences and recommendations

The correspondence between the distribution of ε_i and those of $\hat{\varepsilon}_1 = Y_1 - \bar{Y}$ or with $\tilde{\varepsilon}_1 = Y_1 - \mathrm{med} Y$ depends strongly on the distribution of ε_i. We consider three types of distributions.

- Normal distributions

- Exponential distributions

- Mixing of two normal distributions (multimodal distributions)

The formula (3.1), (3.2), (3.4) are used and computed for the considered distributions. As a summary we formulate the results which were obtained partly by numerical investigations (Aßmus and Läuter, 2000):

1. Under the normal distribution the distribution of ε_i coincides with the distribution of $\sqrt{\frac{n}{n-1}}(Y_1 - \bar{Y})$.

2. Under exponential distributions the tails of the distribution of ε_i are detected by $Y_1 - \mathrm{med} Y$ but not by $Y_1 - \bar{Y}$. This remains true if the distibution is long-tailed.

3. If the distribution of ε_i is a mixture of two normal distributions then this distribution is better approximated by $Y_1 - \bar{Y}$.

4. If the distibution has less mass around 0, i.e. under a mixture of two unimodal distributions, then the approximation is not good around 0 with $Y_1 - \mathrm{med} Y$.

5. If there is no information about the rough form of the distribution then, from a statistical point of view, the use of the traditional residuals $Y_1 - \bar{Y}, \ldots, Y_n - \bar{Y}$ is not well justified.

Recommendation:

The estimates $\bar{Y}$ and med$Y$ are extreme cases of the general winsorized estimates for EY_1. The computations show that the distribution of ε_i should be approximated by those of $Z := Y_1 - \bar{Y}$ or of $\rho Z, \rho$ a real number, and for a winsorized estimate $\tilde{Y}$.

Acknowledgments: I am indebted to the referees for important hints in the representation of this paper.

References

Aßmus, J. and Läuter, H. (2000). Distributions of residuals. Preprint 14, University of Potsdam.

Arnold, B.C., Balakrishnan, N., and Nagaraja, H.N. (1992). *A First Course in Order Statistics*. Wiley, New York.

Hampel, F.R., Ronchetti, E.M., Rousseeuw, P.J., and Stahel, W.A. (1986). *Robust Statistics*. Wiley, New York.

Asymptotically Optimal Sequential Discrimination between Markov Chains

M.B. Malyutov
I.I. Tsitovich

ABSTRACT: An asymptotic lower bound is derived involving a second additive term of order $\sqrt{|\ln \alpha|}$ as $\alpha \to 0$ for the mean length of a sequential strategy s for discrimination between two statistical models for Markov chains. The parameter α is the maximal error probability of s. A sequential strategy is outlined attaining (or almost attaining) this asymptotic bound uniformly over the distributions of models including those from the indifference zone.

KEYWORDS: controlled Markov chain; sequential test; second order optimality; mean length of a strategy

1 Introduction and setting of the problem

The aim of the present paper is to extend the results of Malyutov and Tsitovich (2000a) , derived for independent observations, to sequential discrimination between Markov chains. Among feasible applications: detection of CpG islands in a genome sequence (Durbin *et al.*, 1998) and dynamic speech recognition by a phoneme sequence.

Our outline follows the pattern of Malyutov and Tsitovich (2000a).

No control case. Let X be a finite set with m_X elements and let $\mathcal{P}$ be a Borel set of transition probability matrices for Markov chains on state space X satisfying the conditions formulated in the next paragraph. We denote by $p(x, y), x \in X, y \in X$, the elements of the matrix $P \in \mathcal{P}$.

Under the convention $0/0 := 1$ we assume that for some $C > 0$

$$\sup_{P,Q \in \mathcal{P}} \max_{x \in X, y \in X} \frac{p(x, y)}{q(x, y)} \leq C < \infty. \tag{1.1}$$

Suppose also that every $P \in \mathcal{P}$ determines an aperiodic and irreducible

---

mODA6, A.C.Atkinson, P.Hackl and W.G.Müller, eds., Physica, Heidelberg, 2001.

Markov chain (see e.g. Ross, 1997). This implies the existence and unique-
ness of a stationary distribution $\mu = \mu_P$ with $\mu_P(x) > 0$ for every $x \in X$.
It follows from (1.1) that $p(x, y) = 0$ for any $P \in \mathcal{P}$ entails $q(x, y) = 0$ for
all $Q \in \mathcal{P}$.

Denote by $I(x, P, Q) := \sum_{y \in X} p(x, y) \ln \frac{p(x,y)}{q(x,y)}$ the (Kullback-Leibler) divergence (or the relative entropy).

The set $\mathcal{P}$ is partitioned into Borel sets $\mathcal{P}_0$, $\mathcal{P}_1$ and the indifference zone
$\mathcal{P}_+ = \mathcal{P} \setminus (\mathcal{P}_1 \cup \mathcal{P}_0)$. We test $H_0 : P \in \mathcal{P}_0$ versus $H_1 : P \in \mathcal{P}_1$, with every
decision good for $P \in \mathcal{P}_+$.

Suppose that the divergence between the hypotheses is positive, i.e.

$$\min_{i=0,1} \inf_{P \in \mathcal{P}_i, Q \in \mathcal{P}_{1-i}} \max_{x \in X} I(x, P, Q) \geq \delta_0 > 0. \tag{1.2}$$

Let $\mathcal{F}_n$ be the σ-algebra generated by a Markov chain $X_i, i = 0, 1, \ldots, n$
with a transition matrix P. Denote by p_0 a generally unknown marginal
distribution of X_0. The probability law of $X_i, i = 0, 1, \ldots,$ is denoted by
$\mathbf{P}_P$ and the expectation is denoted by $\mathbf{E}_P$.

A *strategy* s consists of a stopping time N and a measurable binary decision
δ; $\delta = r$ means that H_r , $r = 0, 1$, is accepted.

For an $\alpha > 0$ introduce α-*strategies* s satisfying

Condition $G(\alpha)$: $\max_{r=0,1} \sup_{P \in \mathcal{P}_r} \mathbf{P}_P(\delta = 1 - r) \leq \alpha$.

Let $\mathbf{E}_P^s N$ be the *mean length* (MEAL) of a strategy s. Our first aim is to
find a lower bound for the MEAL of α-strategies as $\alpha \to 0$.

Define $I(\mu, P, Q) := \sum_{x \in X} \mu(x) I(x, P, Q)$, where μ is a probability distribution on X, $I(P, Q) := I(\mu_P, P, Q)$ and $I(P, \mathcal{R}) := \inf_{Q \in \mathcal{R}} I(P, Q)$ for
$\mathcal{R} \subset \mathcal{P}$; $A(P) := \mathcal{P}_{1-r}$ for $P \in \mathcal{P}_r$ as the alternative set in $\mathcal{P}$ for P. For
$P \in \mathcal{P}_+$, if $I(P, \mathcal{P}_0) \leq I(P, \mathcal{P}_1)$, then $A(P) := \mathcal{P}_1$, otherwise, $A(P) := \mathcal{P}_0$.
Finally $k(P) = I(P, A(P))$. It follows from (1.2) that $k(P) > 0$ since
$\mu_P(x) > 0$ for all $x \in X$ and $P \in \mathcal{P}$.

We prove in our last section that for every $P \in \mathcal{P}$ and α-strategy s

$$\mathbf{E}_P^s N \geq \frac{|\ln \alpha|}{k(P)} + O(\sqrt{|\ln \alpha|}). \tag{1.3}$$

In Theorem 2 we construct an α-strategy s^* attaining equality in (1.3).

Controlled experiments. Let sets $\mathcal{P}^u$ of matrices $P^u \in \mathcal{P}$ labeled by
controls $u \in U = \{1, \ldots, m\}$ be given. Chernoff (1972) shows the relevance
of mixed controls for controlled sequential testing. Introduce a **mixed con-
trol** $\mathbf{u} := (\mathbf{u}(x), x \in X), \mathbf{u}(x) = (\kappa_1(x), \ldots, \kappa_m(x))$, where $\kappa_u(x) \geq 0$ and
$\sum_{u=1}^{m} \kappa_u(x) = 1$; U^* is the set of all mixed controls, $P^{\mathbf{u}} = (p^{\mathbf{u}}(x, y))$,
where $p^{\mathbf{u}}(x, y) := \sum_{u=1}^{m} \kappa_u(x) p^u(x, y)$. Introduce the set $\bar{\mathcal{P}}$ of all m-tuples
$\mathbf{P} := (P^u, u \in U)$, and $\mathcal{P}^{\mathbf{u}} := \{P^{\mathbf{u}}, \mathbf{P} \in \bar{\mathcal{P}}\}$.

As before, the set $\bar{\mathcal{P}}$ is partitioned into sets $\bar{\mathcal{P}}_0$, $\bar{\mathcal{P}}_1$ and the indifference
zone $\bar{\mathcal{P}}_+ := \bar{\mathcal{P}} \setminus (\bar{\mathcal{P}}_1 \cup \bar{\mathcal{P}}_0)$, and we test $H_0 : \mathbf{P} \in \bar{\mathcal{P}}_0$ versus $H_1 : \mathbf{P} \in \bar{\mathcal{P}}_1$,
with every decision good for $\mathbf{P} \in \bar{\mathcal{P}}_+$.

We assume that (1.1) holds for every set of matrices $\mathcal{P}^u$, $u \in U$, and the following refinement of (1.2) holds

$$\min_{i=0,1} \max_{u \in U} \inf_{P \in \mathcal{P}_i, Q \in \mathcal{P}_{1-i}} \max_{x \in X} I(x, P^u, Q^u) \geq \delta_0 > 0. \tag{1.4}$$

After obtaining the n-th observation, the experimenter either decides to stop or chooses a mixed control for the $(n+1)$st experiment. Let $\mathcal{F}_n$ be the σ-algebra generated by the observations and controls up to time n. We suppose that the $(n+1)$st experiment is predictable, i.e. the corresponding distribution on U is $\mathcal{F}_n$-measurable, the strategy length N is a stopping time under the flow $\mathbf{F} := \{\mathcal{F}_n, n = 0, 1, \ldots\}$, and a decision δ is $\mathcal{F}_N$-measurable. A strategy S consists now of a predictable mixed control chain $\mathbf{u}(\cdot)$, a stopping time N, and a decision δ. For more details about a relevant probability space and controlled strategies, see, for instance, Malyutov (1983).

Define for $\mathbf{P}, \mathbf{Q} \in \bar{\mathcal{P}}$

$$I_{\mathbf{u}}(\mathbf{P}, \mathbf{Q}) := \sum_{x \in X} \mu_{P^u}(x) \sum_{i=1}^{m} \kappa_i(x) I(x, P^i, Q^i), \tag{1.5}$$

for a mixed control $\mathbf{u}$ and introduce

$$k^*(\mathbf{P}) := \max_{\mathbf{u} \in U^*} \max_{i=0,1} \inf_{\mathbf{Q} \in \bar{\mathcal{P}}_i} I_{\mathbf{u}}(\mathbf{P}, \mathbf{Q}) > 0. \tag{1.6}$$

The positivity of $k^*(\mathbf{P})$ for $\mathbf{P} \in \bar{\mathcal{P}}_0 \cup \bar{\mathcal{P}}_1$ is implied by ((1.4): we assume it for $\mathbf{P} \in \bar{\mathcal{P}}_+$. Let $\mathbf{u}^* = \mathbf{u}^*(\mathbf{P})$ be a control such that

$$k^*(\mathbf{P}) = \max_{i=0,1} \inf_{\mathbf{Q} \in \bar{\mathcal{P}}_i} I_{\mathbf{u}}(\mathbf{P}, \mathbf{Q}),$$

where $\mathbf{u}^*$ exists by (1.1).

We introduce the alternative set $A(\mathbf{P})$ as before: $A(\mathbf{P}) := \bar{\mathcal{P}}_{1-r}$ for $\mathbf{P} \in \bar{\mathcal{P}}_r$, for $\mathbf{P} \in \bar{\mathcal{P}}_+$, if $I_{\mathbf{u}^*}(\mathbf{P}, \bar{\mathcal{P}}_0) \leq I_{\mathbf{u}^*}(\mathbf{P}, \bar{\mathcal{P}}_1)$, then $A(\mathbf{P}) := \bar{\mathcal{P}}_1$, otherwise, $A(\mathbf{P}) := \bar{\mathcal{P}}_0$.

Due to space limitations, we only formulate a lower bound (2.1) of first order for controlled discrimination, and outline an asymptotically optimal α- strategy S satisfying (2.1).

2 Results

The lower bounds are given by the following:

Theorem 2.1 *i.Every α-strategy s for the no-control problem satisfies (1.3) for every $P \in \mathcal{P}$.*

ii. For controlled experiments and every $\mathbf{P} \in \bar{\mathcal{P}}$ the following inequality holds as $\alpha \to 0$

$$\mathbf{E_P^S} N \geq \frac{|\ln \alpha|}{k^*(\mathbf{P})} (1 + o(1)).$$

(2.1)

The strategies s^*, S^* discussed in Theorem 2.2 below, are introduced in our next section.

Theorem 2.2 *i. For every $P \in \mathcal{P}$ under appropriate parameters, s^* is an α-strategy and*

$$\mathbf{E_P^{s^*}} N \leq \frac{|\ln \alpha|}{k(P)} + O\left(\sqrt{|\ln \alpha|}\right).$$

(2.2)

ii. For every $\mathbf{P} \in \hat{\mathcal{P}}$ under appropriate parameters S^* is an α-strategy and

$$\mathbf{E_P^{S^*}} N \leq \frac{|\ln \alpha|}{k^*(P)} (1 + o(1)).$$

3 Asymptotically optimal strategy

Here we introduce our strategy $s^* = s^*(n)$ for the no-control case depending on the parameter n. Strategy s^* consists of conditionally i.i.d. loops. The loop terminating by the event (3.2) is the final loop of s^*. Every loop contains two phases.

Based on the first $L = K_1[\sqrt{|\ln \alpha|}] + 1, K_1 > 0$, observations of a loop, we estimate the matrix P using a standard estimator $\hat{P}$ via conditional frequencies. Since the chain is ergodic and finite, $\mu_P(x) > 0$ for all $x \in X$ and we can get

$$I(P, \hat{P}) \leq K_3 \sqrt{|\ln \alpha|},$$

(3.1)

where K_3 is the same for all α.

Let us enumerate the measurements of the second phase anew and introduce $L_k(\hat{P}, Q) = \sum_{i=1}^{k} z(\hat{P}, Q, X_{i-1}, X_i)$, where $\hat{P}$ is the estimate of P in the first phase. We stop observations at the first moment M such that

$$\inf_{Q \in A_n(\hat{P})} L_M(\hat{P}, Q) > 1 + |\ln \alpha|$$

(3.2)

or

$$M > 2k(\hat{P})^{-1} |\ln \alpha|$$

(3.3)

and accept the hypothesis H_r (i.e. $\delta = r$), if (3.2) holds and $1 - r$ is the index of the set $A(\hat{P})$. In all other cases we begin a new loop.

For controlled experiments our procedure $S^* = S^*(n)$, as before, has conditionally i.i.d. loops until the final success almost similar to (3.2). If an analogue of (3.3) holds, a new loop begins. Every loop contains two phases. In the first phase we obtain estimates $\hat{P}^u$ of P^u for every $u \in U$ based on the $L = K_1[\sqrt{|\ln \alpha|}] + 1$ observations with this control.

We use the control $\mathbf{u}(i) = \mathbf{u}_n^*(\hat{P})$, where $\mathbf{u}_n(\mathbf{P})$ is a measurable approximation of the function $\mathbf{u}(\mathbf{P})$, for the i-th measurement of the second phase, stop observations and take the decision δ as in the strategy s^*.

4   Proof

We prove here only the first part of Theorem 2.1, other statements will be proved in our subsequent publications. Two cases are considered separately:
A1. $P \in \mathcal{P}_+$,   and   A2. $P \in \mathcal{P}_r, r = 0, 1$.
In A1 we assume for definiteness that $A(P) = \mathcal{P}_1$, and consider arbitrary $Q^{\{0\}} \in \mathcal{P}_0$ and $Q^{\{1\}} \in \mathcal{P}_1$. Define

$$z(P,Q,x,y) := \ln \frac{p(x,y)}{q(x,y)}, \ L_k(P,Q) := \sum_{i=1}^{k} z(P,Q,X_{i-1},X_i)$$

where $(\ldots, X_{i-1}, X_i, \ldots)$ is one realization of the Markov chain and

$$M_j(\alpha) := \begin{cases} \inf\{k : L_k(P, Q^{\{j\}}) \geq -\ln\alpha\}, \\ \infty \ \text{if} \ \sup_k L_k(P, Q^{\{j\}}) < -\ln\alpha, \end{cases}$$

$j = 0, 1$.
Since the random variables $z_i^j := z(P, Q^{\{j\}}, X_{i-1}, X_i)$ are bounded by (1.1),

$$\mathbf{E}_P^s \left(z(P, Q^{\{j\}}, X_{i-1}, X_i) | \mathcal{F}_{i-1} \right) = I(X_{i-1}, P, Q^{\{j\}}) := I_i^j$$

are also bounded, and $z_i^j - I_I^j$ are bounded $\mathcal{F}_i$- martingale-differences. It follows from the Wald identity that

$$\mathbf{E}_P^s \sum_{i=1}^{M_j(\alpha)} I(X_{i-1}, P, Q^{\{j\}}) \leq |\ln\alpha| + C_1, j = 0, 1, \tag{4.1}$$

where the constant $C_1 = \ln C$ and C is introduced in (1.1).
Define the *static projection* of a sequential design with stopping time $M_i(\alpha)$ as

$$\mu_\alpha^{(j)}(x) := (\mathbf{E}_P^s M_j(\alpha))^{-1} \sum_{i=0}^{\infty} \mathbf{P}_P^s(X_i = x, i < M_j(\alpha))$$

and $\mu^{(i)} := (\mu^{(i)}(x), x \in X), i = 0, 1$. Theorem 2.2.1, a., Malyutov (1983) extends the Wald identity to controlled experiments implying

$$\mathbf{E}_P^s \sum_{i=1}^{M_j(\alpha)} I(X_{i-1}, P, Q^{\{j\}}) = \mathbf{E}_P^s M_j(\alpha) I(\mu^{(j)}, P, Q^{\{j\}}), j = 0, 1. \tag{4.2}$$

Since our Markov chain is finite and ergodic, there exists a constant C_2 such that

$$\left|I(\mu^{(j)},P,Q^{\{j\}}) - I(P,Q^{\{j\}})\right| \le C_2 |\ln\alpha|^{-1}, j = 0,1. \qquad (4.3)$$

Using (4.2), and (4.3) the formula (4.1) may be rewritten as

$$\mathbf{E}_P^S M_j(\alpha) \le \frac{|\ln\alpha| + C_3}{I(P,Q^{\{j\}})}, \qquad (4.4)$$

with the same constant C_3 for all α and $Q^{\{j\}}$.

Introduce $N_j := \min(M_j(\alpha), N\{\delta = 1 - j\})$, where $N\{\delta = 1 - j\} := N$, if H_j is rejected and $N\{\delta = 1 - j\} := \infty$, if H_j is accepted. It follows from these definitions that

$$M - N \le \sum_{j=0}^{1}(M_j(\alpha) - N_j), \qquad (4.5)$$

where $M := \min(M_0(\alpha), M_1(\alpha))$. From Wald (1947, p.197) we get

$$\mathbf{E}_P^s L_{N_j}(P,Q^{\{j\}}) \ge |\ln\mathbf{P}_{Q^{\{j\}}}^s(N_j < \infty)|.$$

Therefore it follows from the Wald identity that

$$\mathbf{E}_P^s \sum_{i=1}^{N_j} I(X_{i-1}, P, Q^{\{j\}}) \ge |\ln\mathbf{P}_{Q^{\{j\}}}^s(N_j < \infty)|.$$

If we transform the last expressions in the same way as we transformed (4.1) into (4.4), then

$$\mathbf{E}_P^s N_j \ge \frac{|\ln\mathbf{P}_{Q^{\{j\}}}^s(N_j < \infty)| + C_4}{I(P,Q^{\{j\}})}, j = 0,1, \qquad (4.6)$$

with the same constant C_4 for all α and $Q^{\{j\}}$.

The error probability of testing simple hypotheses $H_0' : P = Q^{\{0\}}$ versus $H_1' : P = Q^{\{1\}}$ does not exceed α for any α-strategy. Besides, the definition of stopping times $M_j(\alpha)$ implies that $\mathbf{P}_{Q^{\{j\}}}^s(M_j(\alpha) < \infty) \le \alpha, j = 0,1$. Hence

$$\mathbf{P}_{Q^{\{j\}}}^s(N_j < \infty) \le 2\alpha, j = 0,1, \qquad (4.7)$$

and we get from (4.4), (4.5), (4.6), and (4.7),

$$\mathbf{E}_P^s(M - N) \le \frac{1 + C_3 + C_4}{I(P,Q^{\{0\}})} + \frac{1 + C_3 + C_4}{I(P,Q^{\{1\}})}. \qquad (4.8)$$

Thus (4.8) entails:

$$\mathbf{E}_P^s N \ge \mathbf{E}_P^s M - C_5, \qquad (4.9)$$

where the constant C_5 is the same for all α and $Q^{\{j\}}$.

Introduce

$$I := I(Q_0, Q_1) := \max\left(\sum_{i=1}^{M} I(X_{i-1}, P, Q^{\{0\}}), \sum_{i=1}^{M} I(X_{i-1}, P, Q^{\{1\}})\right).$$

It follows now from the definition of I and the stopping times $M_j(\alpha)$ that

$$|\ln\alpha| \le \max(L_M(P, Q^{\{0\}}), L_M(P, Q^{\{1\}})) \le I + \max_{j=0,1} \zeta_M^{\{j\}}, \qquad (4.10)$$

where $\zeta_k^{\{j\}} := \sum_{i=1}^{k}(z_i^j - I_i^j))$ are sums of bounded martingale-differences under the flow $\mathbf{F} := (\mathcal{F}_i, i \ge 0)$. The following inequalities are proved for such martingales in the Appendix of Malyutov and Tsitovich (2000b):

$$\mathbf{E}_P^s|\zeta_M^{\{j\}}| \le C_6(\mathbf{E}_P^s M)^{\frac{1}{2}}, j = 0, 1, \qquad (4.11)$$

where the constant C_6 is the same for all α, $Q^{\{j\}}$, and $\mathbf{E}_P^s M$.
Therefore, from (4.10) and (4.11) we get

$$|\ln\alpha| \le \mathbf{E}_P^s I + \mathbf{E}_P^s\left(|\zeta_M^{\{0\}}| + |\zeta_M^{\{1\}}|\right) \le \mathbf{E}_P^s I + -C_7(\mathbf{E}_P^s M)^{\frac{1}{2}}, \qquad (4.12)$$

where the constant C_7 is the same for all α and $Q^{\{j\}}$.
Let

$$\mu^*(x) := M^{-1}\sum_{j=0}^{\infty} I(X_j = x, j < M),$$

where $I(A)$ is the indicator function for the event A, and $\mu^* := (\mu^*(x), x \in X)$. The chain is finite and ergodic, thus $|\mathbf{E}_P^s\mu^*(x) - \mu_P(x)| \le C_8 (\mathbf{E}_P^s M)^{-1}$ for the same constant C_8 for all $x \in X$. Thus

$$\mathbf{E}_P^s I = \mathbf{E}_P^s(M \max(I(\mu^*, P, Q), I(\mu^*, P, Q^{\{1\}}))) \le$$

$$\max(I(P, Q), I(P, Q^{\{1\}}))\mathbf{E}_P^s M + 2C_1 C_8 m_X.$$

It follows from (4.12) that

$$\mathbf{E}_P^s M \ge \left(\max(I(P, Q^{\{0\}}), I(P, Q^{\{1\}}))\right)^{-1}|\ln\alpha| - C_9 - C_{10}(\mathbf{E}_P^s M)^{-\frac{1}{2}}, \qquad (4.13)$$

with the same constants C_9 and C_{10} for all α and $Q^{\{j\}}$.
If we replace $\mathbf{E}_P^s M$ with its upper bound from (4.4), then from (4.13) we get a lower bound

$$\mathbf{E}_P^s M \ge |\ln\alpha|\left(\max(I(P, Q^{\{0\}}), I(P, Q^{\{1\}}))\right)^{-1} - C_{11}(|\ln\alpha|)^{\frac{1}{2}}, \qquad (4.14)$$

with the same constant C_{11} for all α and $Q^{\{j\}}$. From the definition of $k(P)$ it follows that there exist a sequence $Q_n^{\{0\}}$ and $Q_n^{\{1\}}$ such that $\max(I(P, Q^{\{0\}}), I(P, Q^{\{1\}})) \le k(P) + n^{-1}$. It follows from (4.9) and (4.14) that

$$\mathbf{E}_P^s N \geq |\ln\alpha|k(P)^{-1} - C_{11}(|\ln\alpha|)^{\frac{1}{2}} - C_5 - 2k(P)^{-2}|\ln\alpha|n^{-1}.$$

If a sequence $n = n(\alpha)$ is chosen such that $|\ln\alpha|^{\frac{1}{2}}n^{-1} \to 0$ as $\alpha \to 0$, then the theorem follows from the last inequality.

For the easy case A2 the theorem follows from Wald's lower bound for the MEAL of the sequential ratio likelihood probability test after transformation of likelihood ratios as in the previous case. The proof is complete.

Remark. Our proof of Theorem 1.ii uses a solution of the following control problem: choose a predictable control function $u(\cdot)$ to minimize the MEAL $\mathbf{E}_P^S M$, until

$$\sum_{i=1}^{M} I(X_i, P^{u(i)}, Q^{u(i)}) \geq |\ln\alpha|$$

for all $\mathbf{Q}$ in $A(\mathbf{P})$ for the *controlled Markov chain* X_i. We now indicate why

$$\mathbf{E}_P^S M \geq |\ln\alpha|k^*(\mathbf{P})^{-1}(1 + o(1)),$$

as $\alpha \to 0$: Derman (1962) proved that the optimal strategy is stationary Markov, generating a Markov chain with regeneration (renewal) points as transition times to the initial state with mean $\mu = k^*(P)$ equal to the mean reward per unit time (if we choose Q in a proper way) and $EM = (|\ln\alpha|/\mu)(1 + o(1))$ (Ross (1997), sect. 4.10 and Proposition 7.1).

References

Chernoff, H. (1972). *Sequential Analysis and Optimal Design*. Philadelphia: SIAM.

Derman, C. (1962). On Sequential Decisions and Markov Chains, *Management Science*, **9**, 16-24.

Durbin, R., Eddy, S., Krogh, A., and Mitchison, G. (1998). *Biological Sequence Analysis*, Cambridge: University Press.

Malyutov, M.B. (1983). Lower Bounds for the Mean Length of Sequentially Designed Experiments. *Soviet Math. (Izv. VUZ.)* **27**, 21-47.

Malyutov, M.B., and Tsitovich, I.I. (2000a). Second Order Optimal Sequential Tests. *"Optimum Design 2000"*,. Eds. Atkinson, A.C., Bogacka, B., and Zhigljavsky, A., Kluwer, Netherlands. 67-78.

Malyutov, M.B., and Tsitovich, I.I. (2000b), Asymptotically Optimal Sequential Testing Hypotheses. *Problems of Information Transmission* **30**, 4, 98-112.

Ross, S (1997). *Introduction to Probability Models*, 6th edition. New York: Academic Press.

Wald, A. (1947). *Sequential Analysis*. New York: Wiley.

Optimum Experimental Designs for a Modified Inverse Linear Model

I. Martínez
I. Ortiz
C. Rodríguez

ABSTRACT: In this paper some optimal designs for a modified inverse linear model are studied. D-optimal designs are obtained from analytical results. In order to study some of the characteristics of this model c-, L- and D_L-optimal designs are examined empirically. Finally, the behavior of the resulting designs is compared.

KEYWORDS: efficiency; inverse linear model; optimal designs

1 Introduction

The modified inverse linear model is a specific case of the rational model as expressed by:

$$Y = \frac{\theta_1 + \theta_2 x + \theta_3 x^2}{1 + \theta_4 x} + \varepsilon, \qquad x \in [a, b], \tag{1.1}$$

where the controlled variable x is chosen from the interval $[a, b]$ and $\theta_1, \theta_2, \theta_3$ and θ_4 are unknown parameters. In addition, Y is the response variable and ε is the experimental error with mean zero and variance σ^2, which is assumed to be 1 without loss of generality. This model has been used in several previous papers to investigate the relationship between crop-yield and fertilizer applied (Sparrow, 1979a and 1979b) and in plant density experiments studying yield per unit area (see McCullagh and Nelder, 1989 p. 291, where this model is used when the response distribution is gamma).

Optimal designs for rational models are currently the focus of great interest. Many significant results may be found in Haines (1992), where explicit algebraic expressions for the support point of D-optimal designs are shown for a particular inverse quadratic model; in He, Studden and Sun (1996),

mODa6, A.C.Atkinson, P.Hackl and W.G.Müller, eds., Physica, Heidelberg, 2001.

three-point Bayesian D-optimal designs for an inverse linear model are obtained; in Song and Wong (1998), D-optimal designs for the Michaelis-Menten model are built up when the variance of the response depends on the independent variable and in Dette, Haines and Imhof (1999) D-optimal designs for a weighted polynomial regression model with a determinate efficiency function are presented.

Let

$$Y = f(x, \theta) + \varepsilon, \qquad x \in \mathcal{X},$$

be a general regression model with $f(x, \theta)$ a nonlinear function of θ, where θ is a $k \times 1$ vector of parameters and $\mathcal{X}$ denotes the design space. If $f(x, \theta)$ is differentiable with respect to θ, the $k \times 1$ gradient vector of $f(x, \theta)$ is defined as $\nabla f(x, \theta)$ with j-th entry

$$\nabla f(x, \theta)_j = \frac{\partial f(x, \theta)}{\partial \theta_j}, \quad j = 1, \dots, k.$$

For an approximate design ξ with n support points $x_1, \dots, x_n$ and weight $\xi(x_i)$ for each, the $k \times k$ information matrix is

$$M(\xi, \theta) = \sum_{i=1}^{n} \xi(x_i) \nabla f(x_i, \theta) \nabla f(x_i, \theta)^T,$$

and the variance function $d(x, \xi, \theta)$ is defined as

$$d(x, \xi, \theta) = \nabla f(x, \theta)^T M^{-1}(\xi, \theta) \nabla f(x, \theta). \tag{1.2}$$

The goal of an optimum design is the optimization of a function of the information matrix. In D-optimality, the determinant of the information matrix is maximized. This criterion minimizes the generalized variance of the parameter estimates. L- and D_L-optimality criteria are used when r linear combinations, $L^T \theta$, are of interest and L is a known $(k \times r)$-matrix. The L-optimality criterion minimizes $tr(L^T M^{-1}(\xi, \theta) L)$. The D_L-optimality criterion minimizes $|L^T M^{-1}(\xi, \theta) L|$. When only one linear combination, $c^T \theta$, is of interest, the criterion is known as c-optimality. For a nonlinear combination, $g(\theta)$, linearization by a Taylor expansion is used, in the form of $c^T(\theta)\theta$, with $c(\theta)_j = \frac{\partial g(\theta)}{\partial \theta_j}$ $j = 1, ..., k$.

For nonlinear models, the information matrix depends on some unknown parameters and additional information about these parameters is necessary to calculate the optimal design. If this additional information is an initial value of the parameters, the optimal design is said to be locally optimal design. A generalization of this method uses a prior distribution of the parameters to obtain Bayesian optimal designs, considering the expectation of the design criterion over this distribution.

In this paper, some results for a modified inverse linear model which yield optimal designs for the criteria mentioned above are presented. It is organized as follows. In Section 2, several lemmas and a theorem are presented that obtain the support points of the D-optimal designs as a coefficient vector corresponding to the greatest eigenvalue of a particular matrix. In Section 3, L-, D_L- and c-optimality criteria are studied. Optimal designs for these criteria are obtained using the Gauss program. Finally, in Section 4, some numerical examples of the criteria studied are shown, and these optimal designs are compared for efficiency .

2 D-optimal designs

The following lemma provides a first result concerning the support of the D-optimal design.

Lemma 2.1 *The D-optimal design for model* (1.1) *in the design space* $[a, b]$ *with* $0 \le a < b$ *is equally supported on four points* $\{x_1, x_2, x_3, x_4\}$, *two of which are the extremes of the interval, i.e.* $x_1 = a$ *and* $x_4 = b$.

Proof: Study of the variance function (1.2), shows that the D-optimal design is supported at four points, the number of unknown parameters in the model, so the D-optimal design is equally supported at them (Fedorov, 1972). The variance function for a four-point D-optimal design is:

$$d(x, \xi^*, \theta) = 4 \sum_{i=1}^{4} \left(\frac{1 + \theta_4 x_i}{1 + \theta_4 x} \right)^4 \prod_{\substack{j=1 \\ j \ne i}}^{4} \frac{(x - x_j)^2}{(x_i - x_j)^2}. \qquad (2.1)$$

By the General Equivalence Theorem, a design ξ^* is a locally D-optimal design, for a best guess θ^0, if $d(x, \xi^*, \theta^0) \le k$, $x \in \mathcal{X}$. Equality is reached at the support points. From study of the behavior of the function (2.1) at x_1 and x_4 it may be concluded that these points cannot be interior points of the design space, $d'(x_1, \xi^*, \theta) < 0$ and $d'(x_4, \xi^*, \theta) > 0$ and $d(x_1, \xi^*, \theta) = 4 = d(x_4, \xi^*, \theta)$, so they are therefore the extremes of this interval. $\square$

The determinant of the information matrix for an equally supported four-point design is:

$$|M(\xi, \theta)| = \frac{[\theta_3 + \theta_4(\theta_1\theta_4 - \theta_2)]^2}{4^4 \prod_{i=1}^{4}(1 + \theta_4 x_i)^4} \prod_{i=1}^{4} \prod_{j>i}^{4} (x_i - x_j)^2. \qquad (2.2)$$

For independent observations, $M^{-1}(\xi, \theta)$ is proportional to the asymptotic covariance matrix for the maximum likelihood estimate for θ.

It is important to note that the D-optimal design for model (1.1) is also the D-optimal design for the linear model

$$Y = \beta_0 + \beta_1 x + \beta_2 x^2 + \beta_3 x^3 + \varepsilon \qquad (2.3)$$

with an efficiency function $\lambda(x) = (1 + \theta_4 x)^{-4}$.

Let $Q(x) = (x-a)(x-x_2)(x-x_3)(x-b)$ be a 4th-degree monic polynomial, in which zeros are the D-optimal design support points. The lemmas below are obtained from the equivalent model (2.3) and allow D-optimal designs to be found analytically in the following Theorem.

Lemma 2.2 *A number α exists such that $Q(x)$ satisfies the differential equation (DE)*

$$[(1 + \theta_4 x)Q''(x) - 4\theta_4 Q'(x)](x - a)(x - b) = -4\theta_4 (x - \alpha)Q(x). \quad (2.4)$$

Proof: The variance function can be expressed as:

$$d(x, \xi^*, \theta) = \frac{1}{(1 + \theta_4 x)^4} \sum_{i=1}^{4} L_i^2(x)(1 + \theta_4 x_i)^4,$$

where $L_i(x)$, $i = 1, \ldots, 4$ are the fundamental Lagrange interpolation polynomials induced by the support points. Since $d'(x, \xi^*, \theta)$ is zero at x_2 and x_3 then

$$d'(x_j, \xi^*, \theta) = 2L_j'(x_j) - \frac{4\theta_4}{1 + \theta_4 x_j} = 0 \quad j = 2, 3,$$

as $L_j'(x_j) = \frac{Q''(x_j)}{2Q'(x_j)}$ then the following equation is obtained:

$$(1 + \theta_4 x_j)Q''(x_j) - 4\theta_4 Q'(x_j) = 0.$$

$(1 + \theta_4 x)Q''(x) - 4\theta_4 Q'(x)$ is a 3rd-degree polynomial. Its roots are the same as those of the polynomial $Q(x)$. But, as points $x_1 = a$ and $x_4 = b$ are support points then

$$((1 + \theta_4 x_j)Q''(x_j) - 4\theta_4 Q'(x_j))(x - a)(x - b) = c(x - \alpha)Q(x),$$

then $c = -4\theta_4$ and DE (2.4) is obtained. $\square$

Lemma 2.3 *If DE (2.4) has a 4th-degree polynomial solution then α is a real value.*

Proof: The previous DE can be represented as a Sturm-Liouville system as follows

$$(p(x)Q'(x))' + q(x)Q(x) = \alpha r(x)Q(x),$$

with

$$p(x) = \frac{1}{4\theta_4(1 + \theta_4 x)^4}, \quad q(x) = \frac{x}{(x - a)(x - b)(1 + \theta_4 x)^5}$$

and

$$r(x) = \frac{1}{(x - a)(x - b)(1 + \theta_4 x)^5}.$$

The operator defined in the Sturm-Liouville system is a self-adjoint operator, so all the eigenvalues are real values, (see Tolstov, 1976 p 251). $\square$

Let $Q(x) = \sum_{i=0}^{4} q_i x^i$ ($q_4 = 1$) with coefficient vector denoted by q and $q = (q_0, q_1, q_2, q_3, q_4)^T$.

Lemma 2.4 *The coefficient vector q of the polynomial $Q(x)$ satisfies the equation $Aq = \alpha q$ with*

$$A = \begin{pmatrix} \gamma_0 & \mu_0 & \tau_0 & 0 & 0 \\ \delta_1 & \gamma_1 & \mu_1 & \tau_1 & 0 \\ 0 & \delta_2 & \gamma_2 & \mu_2 & \tau_2 \\ 0 & 0 & \delta_3 & \gamma_3 & \mu_3 \\ 0 & 0 & 0 & \delta_4 & \gamma_4 \end{pmatrix} \tag{2.5}$$

where

$$\delta_i = 1 + \frac{(i-1)(i-6)}{4} \quad i = 1, \ldots, 4;$$

$$\tau_i = \frac{(i+1)(i+2)ab}{4\theta_4} \quad i = 0, \ldots, 2;$$

$$\gamma_i = \frac{i}{4}\left[(5-i)(a+b) + \frac{i-1}{\theta_4}\right] \quad i = 0, \ldots, 4;$$

$$\mu_i = -\frac{i+1}{4}\left[(4-i)ab + \frac{i(a+b)}{\theta_4}\right] \quad i = 0, \ldots, 3,$$

and q is the unique eigenvector with $q_4 = 1$ corresponding to eigenvalue α of matrix A.

Proof: DE (2.4) can be written as

$$\left[(1 + \theta_4 x)\sum_{i=2}^{4} i(i-1)q_i x^{i-2} - 4\theta_4 \sum_{i=1}^{4} i q_i x^{i-1}\right](x-a)(x-b)$$

$$= -4\theta_4(x-\alpha)\sum_{i=0}^{4} q_i x^i$$

and identifying the coefficients

$$\left[\frac{(i+1)(i+2)ab}{4\theta_4}\right]q_{i+2} + \left[-\frac{i+1}{4}\left((4-i)ab + \frac{i(a+b)}{\theta_4}\right)\right]q_{i+1}$$

$$+ \left[\frac{i}{4}\left((5-i)(a+b) + \frac{i-1}{\theta_4}\right)\right]q_i + \left[1 + \frac{(i-1)(i-6)}{4}\right]q_{i-1} = \alpha q_i,$$

then $\tau_i q_{i+2} + \mu_i q_{i+1} + \gamma_i q_i + \delta_i q_{i-1} = \alpha q_i$ $\quad i = 0, \ldots, 4$ with $q_{-1} = q_5 = q_6 = 0$, and it may be concluded that $Aq = \alpha q$ and α is an eigenvalue of the matrix A. $\square$

Theorem 2.1 *The locally D-optimal design ξ^* for model (1.1) and $\theta_4 = \theta_4^0$, in the design space $[a, b]$, with $0 \le a < b$, is equally supported at four points that can be calculated as:*

    a) *The zeros of the polynomial $Q(x)$, where its coefficient vector q is the only eigenvector with $q_4 = 1$ corresponding to the greatest eigenvalue of the matrix A.*

or

    b) *The solutions of the equations*

$$\frac{2}{x_2 - a} - \frac{2}{x_3 - x_2} - \frac{2}{b - x_2} - \frac{4\theta_4^0}{1 + \theta_4^0 x_2} = 0$$

$$\frac{2}{x_3 - a} + \frac{2}{x_3 - x_2} - \frac{2}{b - x_3} - \frac{4\theta_4^0}{1 + \theta_4^0 x_3} = 0.$$

and points a and b.

Proof: To see a), recall that from Lemma 2.4 the coefficient vector of $Q(x)$ is known to be an eigenvector of the matrix (2.5). It may now be proven that this corresponds to the maximum eigenvalue of the matrix. Let n_i and n_j be the number of zeros of the polynomial solutions u_i and u_j corresponding to the eigenvalues α_i and α_j, respectively. The first step is to prove that if $\alpha_i < \alpha_j$ then $n_i < n_j$. Let $k_1 < k_2$ be two consecutive zeros of u_i. Assume $u_i(x) > 0$ and $u_j(x) > 0$ for $x \in (k_1, k_2)$ and let $w(x) = p(x) \left(u_i(x) u_j'(x) - u_i'(x) u_j(x) \right)$, then as $w(k_1) < 0$ and $w(k_2) > 0$, some value $x \in (k_1, k_2)$ exists with $w'(x) > 0$, but $w'(x)$ can be expressed as $w'(x) = u_i(x) u_j(x) r(x)(\alpha_j - \alpha_i) < 0$, which is a contradiction. As a and b are zeros of u_i and u_j, for two consecutive zeros of u_i there is some zero of u_j, then $n_i < n_j$. Thus the solution with four zeros in $[a, b]$ must correspond to the greatest eigenvalue of the matrix A.

For b), maximizing the determinant of the information matrix (2.2) is equivalent to maximizing the function:

$$\Phi = \sum_{\substack{i,j=1 \\ i<j}}^{4} 2 \log(x_j - x_i) - 4 \sum_{i=1}^{4} \log(1 + \theta_4 x_i).$$

From the partial derivatives of Φ with respect to x_2 and x_3, equations in which $x_1 = a$, $x_4 = b$ and $\theta_4 = \theta_4^0$ are obtained. $\square$

Considering prior distributions over the parameters, $\pi(\theta)$, a Bayesian D-optimal design maximizes the expectation of the logarithm of the determinant of the information matrix

$$\phi(\xi) = \mathbb{E}_\theta[\log |M(\xi, \theta)|] = \int_\Theta \log |M(\xi, \theta)| \pi(d\theta).$$

For model (1.1), it is sufficient to specify a prior distribution on θ_4. However, in this case, it is not certain whether the Bayesian D-optimal designs are supported at four points since this depends on the prior distribution.

For model (1.1), if the Bayesian four-point D-optimal design exists, this design is:

$$\xi^* = \left\{ \begin{matrix} a & x_2 & x_3 & b \\ \frac{1}{4} & \frac{1}{4} & \frac{1}{4} & \frac{1}{4} \end{matrix} \right\},$$

where x_2 and x_3 are the solutions of:

$$\frac{2}{x_2 - a} - \frac{2}{x_3 - x_2} - \frac{2}{b - x_2} - E_{\theta_4}\left[\frac{4\theta_4}{1 + \theta_4 x_2} \right] = 0$$

$$\frac{2}{x_3 - a} + \frac{2}{x_3 - x_2} - \frac{2}{b - x_3} - E_{\theta_4}\left[\frac{4\theta_4}{1 + \theta_4 x_3} \right] = 0,$$

where $E_{\theta_4}[.]$ denotes the expectation for the prior distribution $\pi(\theta_4)$.

3   Other optimal designs

Two important characteristics of model (1.1) are the point of maximum value of the response, x_{max}, and the maximum response, y_{max}. These characteristics are nonlinear functions of the parameters of the model. c_θ-Optimal designs to estimate each characteristic and L_θ- or D_L-optimal designs to estimate both of them are sought. The Equivalence Theorem for the L_θ-optimality criterion indicates that a design ξ^* is locally L_θ-optimal, for $\theta = \theta^0$, if

$$\nabla f(x, \theta^0)^T M^{-1}(\xi^*, \theta^0) L L^T M^{-1}(\xi^*, \theta^0) \nabla f(x, \theta^0) - tr(L^T M^{-1}(\xi^*, \theta^0) L) \leq 0,$$

and equivalently, a design ξ^* is locally D_L-optimal, for $\theta = \theta^0$, if

$$\nabla f(x, \theta^0)^T M^{-1}(\xi^*, \theta^0) L (L^T M^{-1}(\xi^*, \theta^0) L)^{-1} L^T M^{-1}(\xi^*, \theta^0) \nabla f(x, \theta^0) \leq r.$$

The point x_{max} is given by

$$x_{max} = g_1(\theta) = \frac{-\sqrt{\theta_3} - \sqrt{\theta_3 - \theta_2 \theta_4 + \theta_1 \theta_4^2}}{\theta_4 \sqrt{\theta_3}},$$

when $c_1(\theta)$, by Taylor series expansion, is $c_1(\theta)^T = \frac{1}{\kappa}\left(-\theta_4, 1, \frac{\theta_1 \theta_4 - \theta_2}{\theta_3}, \right.$ $\left. \frac{2\theta_3 - \theta_2 \theta_4 + \kappa}{\theta_4^2} \right)$ with $\kappa = 2\theta_3 \sqrt{\frac{\theta_3 + \theta_4(\theta_1 \theta_4 - \theta_2)}{\theta_3}}$ and, for applicability of the model, $\theta_3 < 0, \theta_4 > 0, \theta_2^2 > 4\theta_1 \theta_3, \theta_2 > \theta_1 \theta_4 > 0$.

The maximum value, y_{max}, is obtained by inserting x_{max} in $f(x, \theta)$

$$y_{max} = g_2(\theta) = f(x_{max}, \theta).$$

The expression and linearization of $c_2(\theta)$ are omitted for brevity. The two vectors $c_i(\theta)$ are used to build up the matrix L for the study of L_θ- and D_L-optimal designs for x_{max} and y_{max}.

Gauss programs can be used to obtain numerically c_θ-, L_θ- and D_L-optimal designs for some initial values of the parameters. The following remarks are derived from these designs.

Remark 1: Locally optimal designs are supported at four points having different weights, including the extremes of the support interval.

Remark 2: The c_θ-optimal designs for both characteristics are supported at the same points.

Remark 3: The support points of the c_θ-optimal designs depend only on θ_4 and their weights are a function of all the parameters. In L_θ- and D_L-optimality the support points and their weights depend on all the parameters.

Remark 4: When θ_4 increases, the interior support points approach the lower extreme of the support interval and the weight at the extremes decreases.

Efficiency is used as a means of comparison of the behaviour of the various optimal designs. The efficiencies of the D-, c_θ-, L_θ- and D_L-optimal designs make it possible to find out whether one optimal design behaves as well as another with regard to a criterion.

Remark 5: Nevertheless, c_θ- and L_θ-optimal designs are robust with regard to changes in the initial values of θ_1, θ_2 and θ_3 because their efficiencies are high.

These studies demonstrate that the efficiencies decrease when the initial value of θ_4 increases. D-optimal designs have c_θ- and L_θ-efficiencies of less than 85%. They reach 95% when θ_4 is a small value. In general, c_θ-optimal designs for x_{max} (y_{max}) are highly efficient with regard to the criteria for y_{max} (x_{max}) and more efficient than L_θ-optimality and less than D_L-optimality. c_θ-optimal designs have low D-efficiencies and the behaviour of c_θ-optimal designs for x_{max} is worse than for y_{max}. L_θ- and D_L-optimal designs are very efficient ($> 75\%$ and $> 85\%$ respectively) for the different properties. D-efficiency of L_θ-optimal designs and c_θ-optimal designs is very similar.

4   Some numerical examples

The examples below illustrate the results obtained. Using Sparrow's results, the following design and estimates for the parameters are obtained for the design space $[0, 200]$:

$$\xi_s = \left\{ \begin{array}{ccccccccc} 0 & 25 & 50 & 75 & 100 & 125 & 150 & 175 & 200 \\ \frac{1}{9} & \frac{1}{9} & \frac{1}{9} & \frac{1}{9} & \frac{1}{9} & \frac{1}{9} & \frac{1}{9} & \frac{1}{9} & \frac{1}{9} \end{array} \right\}.$$

$$\hat{\theta}_1 = 3.94 \quad \hat{\theta}_2 = 0.074 \quad \hat{\theta}_3 = -0.00017 \quad \hat{\theta}_4 = 0.0062.$$

The locally D-optimal design for the modified inverse linear model (1.1) is

$$\xi_D^* = \left\{ \begin{matrix} 0 & 35.8727 & 121.786 & 200 \\ \frac{1}{4} & \frac{1}{4} & \frac{1}{4} & \frac{1}{4} \end{matrix} \right\}.$$

To study the point of maximum value of the response, the locally c_θ-optimal design is

$$\xi_{c_1}^* = \left\{ \begin{matrix} 0 & 35.0856 & 128.128 & 200 \\ 0.1884 & 0.3662 & 0.3116 & 0.1338 \end{matrix} \right\},$$

and for the maximum response, the locally c_θ-optimal design is

$$\xi_{c_2}^* = \left\{ \begin{matrix} 0 & 32.0856 & 128.128 & 200 \\ 0.1515 & 0.3220 & 0.3496 & 0.1769 \end{matrix} \right\}.$$

If the two properties are studied simultaneously, the locally L_θ-optimal design is

$$\xi_L^* = \left\{ \begin{matrix} 0 & 35.0917 & 128.123 & 200 \\ 0.1877 & 0.3653 & 0.3123 & 0.1347 \end{matrix} \right\},$$

and the locally D_L-optimal design is

$$\xi_{D_L}^* = \left\{ \begin{matrix} 0 & 39.3596 & 117.8070 & 200 \\ 0.1785 & 0.3058 & 0.3007 & 0.2150 \end{matrix} \right\}.$$

The efficiency of the designs obtained is used to compare their usefulness. The behaviour of the Sparrow, D-optimal, c-optimal, L- and D_L-optimal designs, for x_{max} and y_{max}, is studied for the criteria considered.

		D-efficiency	c_1-efficiency	c_2-efficiency
Sparrow design	ξ_S^*	0.80	0.60	0.61
D-optimal design	ξ_D^*	1	0.83	0.85
c-optimal design (x_{max})	$\xi_{c_1}^*$	0.92	1	0.97
c-optimal design (y_{max})	$\xi_{c_2}^*$	0.93	0.97	1
L-optimal design	ξ_L^*	0.92	1	0.97
D_L-optimal design	$\xi_{D_L}^*$	0.97	0.87	0.89
		L-efficiency	D_L-efficiency	
Sparrow design	ξ_S^*	0.60	0.75	
D-optimal design	ξ_D^*	0.83	0.95	
c-optimal design (x_{max})	$\xi_{c_1}^*$	1	0.90	
c-optimal design (y_{max})	$\xi_{c_2}^*$	0.97	0.92	
L-optimal design	ξ_L^*	1	1.90	
D_L-optimal design	$\xi_{D_L}^*$	0.87	1	

These tables show the high efficiencies of the c_θ- and L_θ-optimal designs, however, as noted earlier, to calculate these designs best guesses are required for all parameters. For D-optimality, only a guess is needed for θ_4. To obtain Bayesian designs, a prior distribution on θ_4 is considered. In the first case, this prior distribution is a continuous uniform distribution centred on Sparrow's estimates and with width four asymptotic standard errors, $\pi_c(\theta_4) = \mathcal{U}[0.0062 \pm 0.004]$, whereby the Bayesian D-optimal design is

$$\xi_{D_c}^* = \left\{ \begin{matrix} 0 & 36.3102 & 122.7883 & 200 \\ \frac{1}{4} & \frac{1}{4} & \frac{1}{4} & \frac{1}{4} \end{matrix} \right\}.$$

If the prior distribution on θ_4 is a discrete distribution

$$\pi_d(\theta_4) = \left\{ \begin{matrix} 0.004 & 0.0062 & 0.008 \\ \frac{1}{3} & \frac{1}{3} & \frac{1}{3} \end{matrix} \right\}$$

the Bayesian D-optimal design is:

$$\xi_{D_d}^* = \left\{ \begin{matrix} 0 & 36.1061 & 122.3008 & 200 \\ \frac{1}{4} & \frac{1}{4} & \frac{1}{4} & \frac{1}{4} \end{matrix} \right\}.$$

In conclusion, for these prior continuous and discrete uniform distributions, the Bayesian D-optimal designs are supported at four points having the same weight.

Thus the optimal designs for model (1.1) depend on the criteria and the properties of the model chosen. An advantage of L_θ-optimal designs is that they are highly efficient for the study of several different properties. D-optimal designs can be obtained from the analytical results in Section 2. However these designs may be very different from c_θ- and L_θ-optimal designs.

References

Atkinson, A.C., Chaloner, K., Herberg, A.M., and Juritz, J. (1993). Optimum experimental designs for properties of a compartmental model. *Biometrics*, **49**, 325-337.

Dette, H., Haines, L.M., and Imhof, L. (1999). Optimal designs for rational models and weighted polynomial regression. *The Annals of Statistics*, **27**, 1272-1293.

Fedorov, V.V. (1972). *Theory of Optimal Experiments*. Academic Press, New York.

Haines, L.M. (1992). Optimal design for inverse quadratic polynomials. *South African Statistical Journal*, **26**, 25-41.

He, Z., Studden, J., and Sun, D. (1996). Optimal designs for rational models. *The Annals of Statistics*, **24**, 2128-2147.

Huang, M.-N.L., Chang, F.C., and Wong, W.K. (1995). *D*-optimal designs for polynomial regression without an intercept. *Statistica Sinica*, **5**, 441-458.

McCullagh, P. and Nelder, J.A. (1989). *Generalized Linear Models*. Chapman and Hall, London.

Song, D. and Wong, W.K. (1998). Optimal two-point designs for the Michaelis-Menten model with heteroscedastic errors. *Communications in Statistics*, **27**, 1503-1516.

Sparrow, P.E. (1979a). The comparison of five response curves for representing the relation between the annual dry-matter yield of grass herbage and fertilizer nitrogen. *Journal of Agricultural Science, Cambridge*, **93**, 513-520.

Sparrow, P.E. (1979b). Nitrogen response curves of spring barley. *Journal of Agricultural Science, Cambridge*, **92**, 307-317.

Tolstov, G.P. (1976). *Fourier Series*. Dover Publications, New York.

Permutation Tests for Effects in Unbalanced Repeated Measures Factorial Designs

D. Mazzaro
F. Pesarin
L. Salmaso

ABSTRACT: We deal with permutation testing for balanced and unbalanced repeated measures designs and we consider a replicated homoscedastic (balanced or unbalanced) factorial design with fixed effects (Milliken, 1984) as the basic experimental plan. It is worth noting that the new permutation approach, presented here, is exact and, being conditional to the sufficient statistic represented by the data matrix it does not require normality of error terms in the linear model for responses. A comparative simulation study has been performed in order to evaluate the power of such exact separate tests.

KEYWORDS: conditional inference; exact tests; growth curves; synchronized permutations; two-way ANOVA

1 Introduction

In this paper, starting from the derivation of exact separate permutation tests for effects in balanced and unbalanced designs, we extend such solutions to repeated measures designs. For a review on such designs and related testing of hypotheses we refer the reader to Milliken and Johnson (1984). Such extension is easily obtained by using the nonparametric combination methodology (Pesarin, 1992, 1999).

mODa6, A.C.Atkinson, P.Hackl and W.G.Müller, eds., Physica, Heidelberg, 2001.

2   Synchronized permutations in 2^2 balanced and unbalanced designs

Let us remember that permutation tests are conditional inferential procedures, where conditioning is with respect to a set of joint sufficient statistics under the null hypothesis. Hence, the permutation approach for a two-way layout should be based on such a set of, possibly minimal, joint sufficient statistics. To this end, let us first consider the balanced, fixed effects and homoscedastic case, where two factors have only two levels each, hence the model matrix is a 2^2 complete factorial design. A replicated 2^2 complete factorial design is a $(M \times N)$ matrix, $M = n \cdot N$, $n > 1$, $N = 2^2$, where for each distinct combination of levels n replicates, which identify a *block*, are considered.

In this framework, let us suppose the model matrix which is a Hadamard matrix, is in normal form and let us refer to the usual linear model for data responses:

$$y_{ijr} = \mu + a_i + b_j + (ab)_{ij} + \varepsilon_{ijr},$$

where: y_{ijr} are the experimental responses, μ is the general mean, a_i, b_j are the main effects, $(ab)_{ij}$ is the interaction effect, $i = 1, 2$, $j = 1, 2$, $r = 1, ..., M$, $M = n \cdot 2^2$, n is the number of times each combination of factor's levels is replicated, and ε_{ijr} are exchangeable experimental errors, with zero mean and unknown continuous distribution $\mathcal{P}$. For the sake of notation, let us consider: $a_1 = -a_2 = a$, $b_1 = -b_2 = b$, $(ab)_1 = -(ab)_2 = (ab)$. Thus the overall null hypothesis is written as:

$$H_0 : \{(a = 0) \cap (b = 0) \cap (ab = 0)\},$$

and the overall alternative as: $H_1 : \{H_0 \text{ is not true}\}$.

Let us briefly discuss the hypotheses being tested. Usually the major experimenter's interest is in testing separately for two main effects and for interaction. Hence, there are three separate null hypotheses which are of interest: $H_{0A} : \{a = 0\}$, $H_{0B} : \{b = 0\}$ and $H_{0AB} : \{ab = 0\}$ and the emphasis is to find three separate tests. Thus, what is needed is, for instance, to test for $H_{0A} : \{a = 0\}$ against $H_{1A} : \{a \neq 0\}$, irrespective of whether H_{0B} or H_{0AB} are true or not. In order to reach this aim within a permutation framework, we must find a proper set of jointly sufficient statistics for all three testing sub-problems: H_{0A} irrespective of $H_{0B} \cup H_{0AB}$, H_{0B} irrespective of $H_{0A} \cup H_{0AB}$ and H_{0AB} irrespective of $H_{0A} \cup H_{0B}$. In other terms, we should find a set of sufficient statistics under $\{H_{0A}, H_{0B}, H_{0AB}\}$. According to this setting, such a set of jointly sufficient statistics is: $\mathbf{y} = {}^{\mathsf{T}}(\mathbf{y}_{11}, \mathbf{y}_{12}, \mathbf{y}_{21}, \mathbf{y}_{22})$, that is the vector of four data blocks (Pesarin, 1999). The following figure represents the effects of all combinations of factor's levels for are replicated 2^2 factorial.

FIGURE 1. Effects of level combinations in a 2^2 factorial.

	A_1	A_2
B_1	$a,\ b,\ ab$	$-a,\ b,-ab$
B_2	$a,-b,-ab$	$-a,-b,\ ab$

Let us first consider two partial statistics for comparing factor A separately at levels 1 and 2 of factor B: $T_{A|1} = \sum_r y_{11r} - \sum_r y_{21r}$ and $T_{A|2} = \sum_r y_{12r} - \sum_r y_{22r}$. Let us suppose that ν_1^* data from block A_1B_1 are randomly selected and exchanged with other ν_1^*, randomly selected from block A_2B_1. Moreover, let us consider ν_2^* data are randomly exchanged from blocks A_1B_2 and A_2B_2. After elementary calculations, permutation structures of two partial statistics are:

$$T_{A|1}^* = 2(n - 2\nu_1^*)(a + ab) + n(\bar{\varepsilon}_{11}^* - \bar{\varepsilon}_{21}^*)$$

and

$$T_{A|2}^* = 2(n - 2\nu_2^*)(a - ab) + n(\bar{\varepsilon}_{21}^* - \bar{\varepsilon}_{22}^*),$$

where $\bar{\varepsilon}_{ij}^* = \sum_r \varepsilon_{ijr}^*/n$ are means of permutation errors in the ij-th block. It is worth noting that effects a and ab are confounded on both $T_{A|1}^*$ and $T_{A|2}^*$. However, if $\nu_1^* = \nu_2^* = \nu^*$, that is if we synchronize permutations on two partial statistics, $T_A^* = T_{A|1}^* + T_{A|2}^*$ has the permutation structure given by: $4(n - 2\nu^*) \cdot a + n(\bar{\varepsilon}_{11}^* + \bar{\varepsilon}_{12}^* - \bar{\varepsilon}_{21}^* - \bar{\varepsilon}_{22}^*)$. Thus T_A^*, being dependent only on effect a and on a linear combination of exchangeable errors, gives a separate exact permutation test for H_{0A}, independently of whether H_{0B} or H_{0AB} are true or not.

Moreover, ${}^aT_{AB}^* = T_{A|1}^* - T_{A|2}^*$ has the permutation structure given by: $4(n - 2\nu^*) \cdot (ab) + n(\bar{\varepsilon}_{11}^* - \bar{\varepsilon}_{12}^* - \bar{\varepsilon}_{21}^* + \bar{\varepsilon}_{22}^*)$. Thus ${}^aT_{AB}^*$, being dependent only on the interaction effect (ab) and on a linear combination of exchangeable errors, gives a separate exact permutation test for H_{0AB}, independently of whether H_{0A} or H_{0B} are true or not.

In order to complete the testing analysis, we must also take into consideration partial statistics for contrasting factor B separately for levels 1 and 2 of factor A: $T_{B|1} = \sum_r y_{11r} - \sum_r y_{12r}$ and $T_{B|2} = \sum_r y_{21} - \sum_r y_{22}$ and act accordingly. Hence, let us again consider synchronized permutations between paired blocks (A_1B_1, A_1B_2) and (A_2B_1, A_2B_2) by supposing to randomly exchange ν^* data. Thus: $T_B^* = T_{B|1}^* + T_{B|2}^*$, whose permutation structure is: $4(n - 2\nu^*) \cdot b + n(\bar{\varepsilon}_{11}^* - \bar{\varepsilon}_{12}^* + \bar{\varepsilon}_{21}^* - \bar{\varepsilon}_{22}^*)$, being dependent only on effect b and on a linear combination of exchangeable errors, is a separate exact permutation test for H_{0B} independently of whether H_{0A} or H_{0AB} are true or not.

Moreover, ${}^bT_{AB}^* = T_{B|1}^* - T_{B|2}^*$, whose permutation structure is: $4(n - 2\nu^*) \cdot (ab) + n(\bar{\varepsilon}_{11}^* - \bar{\varepsilon}_{12}^* - \bar{\varepsilon}_{21}^* + \bar{\varepsilon}_{22}^*)$, being dependent only on effect (ab) and on a linear combination of exchangeable errors, is a separate exact permutation test for H_{0AB} independently of whether H_{0A} or H_{0B} are true or not.

For testing the interaction effect we have two tests $^aT^*_{AB}$ and $^bT^*_{AB}$ with different permutation distributions, hence we need to consider only one test, and this is $T^*_{AB} = {^aT^*_{AB}}$ or $T^*_{AB} = {^bT^*_{AB}}$ indifferently.

It is worth noting that the permutation distribution of T^*_q, $q = A, B, AB$, under H_{0q} depends only on permutation combinations of errors, so that each test is permutationally exact (Pesarin, 1999) because errors are exchangeable by assumption. Tests T^*_q are proper tests for one-sided alternatives, whereas $(T^*_q)^2$ or $|T^*_q|$ are proper tests for two-sided alternatives.

Further, let us note that: (i) if the permutation in the first pair of blocks exchanges elements in the same positions as the permutation applied to the remaining pair of blocks, that is the same permutation is applied to each couple of blocks (constrained synchronized permutations, CSP), then the cardinality of the permutation support of each test statistic is $\binom{2n}{n}$, where n is the number of replicates in each block, for one-sided alternatives; otherwise (ii) if permutations applied to each couple of blocks exchange elements in different positions (unconstrained synchronized permutations, USP), then cardinality of the permutation support of each test statistic becomes $\sum_{\nu^*} \binom{n}{\nu^*}^4$, for one-sided alternatives. From now on, we mainly refer to USP as the principal synchronized permutation strategy which can be applied both to balanced and unbalanced factorial designs. Unbiasedness and consistency of the three tests are straightforward.

Let us consider an unbalanced 2^2 full factorial design, where $n_{ij} \geq 1$ is the number of runs (sample sizes) for the ij-th block, and it may vary from block to block. In order to obtain separate exact permutation tests for main effects and the interaction, we must consider weighted intermediate statistics. Weights are derived by considering that, within the synchronized permutation approach, no effect must be confounded with others in the testing process. This may require further restrictions in the available permutations. If we consider the intermediate statistics for comparing factor A separately at levels 1 and 2 of B, we randomly exchange $\nu^* \leq \min_{ij} n_{ij}$ elements between each pair of blocks. Thus, the intermediate statistics become:

$$^wT^*_{A|1} = \sum_{r=1}^{n_{11}} y^*_{11r} \cdot w^*_{11} - \sum_{r=1}^{n_{21}} y^*_{21r} \cdot w^*_{21} \text{ and } {^wT^*_{A|2}} = \sum_{r=1}^{n_{12}} y^*_{12r} \cdot w^*_{12} - \sum_{r=1}^{n_{22}} y^*_{22r} \cdot w^*_{22},$$

where permutation weights w^*_{ij} associated with blocks are introduced in order to obtain a separate exact permutation test for a. From analysis of permutation structures of the test statistic for a: $^wT^*_A = {^wT^*_{A|1}} + {^wT^*_{A|2}}$, we are able to determine the weights w^*_{ij}:

$$^wT^*_A = w^*_{11} \sum_{r=1}^{n_{11}} y^*_{11r} - w^*_{21} \sum_{r=1}^{n_{21}} y^*_{21r} + w^*_{12} \sum_{r=1}^{n_{12}} y^*_{12r} - w^*_{22} \sum_{r=1}^{n_{22}} y^*_{22r} =$$

$$= w^*_{11}[(n_{11}-2\nu^*)(\mu+a+b+(ab))+\widetilde{\varepsilon}^*_{11}] + w^*_{12}[(n_{12}-2\nu^*)(\mu+a-b-(ab))+\widetilde{\varepsilon}^*_{12}] -$$

$$-w_{21}^*[(n_{21}-2\nu^*)(\mu-a+b-(ab))+\widetilde{\varepsilon}_{21}^*]-w_{22}^*[(n_{22}-2\nu^*)(\mu-a-b+(ab))+\widetilde{\varepsilon}_{22}^*],$$

where $\widetilde{\varepsilon}_{ij}^* = \sum_r^{n_{ij}} \varepsilon_{ijr}^*$ are sum of permutation errors relative to the ij-th block.

Now, in order to get a separate test for the effect of factor a, we introduce the following system of equations to determine the weights w_{ij}^*:

$$\begin{cases} \mu[w_{11}^*(n_{11}-2\nu^*) + w_{12}^*(n_{12}-2\nu^*) - w_{21}^*(n_{21}-2\nu^*) - w_{22}^*(n_{22}-2\nu^*)] = 0 \\ b[w_{11}^*(n_{11}-2\nu^*) - w_{12}^*(n_{12}-2\nu^*) + w_{21}^*(n_{21}-2\nu^*) - w_{22}^*(n_{22}-2\nu^*)] = 0 \\ (ab)[w_{11}^*(n_{11}-2\nu^*) - w_{12}^*(n_{12}-2\nu^*) - w_{21}^*(n_{21}-2\nu^*) + w_{22}^*(n_{22}-2\nu^*)] = 0 \end{cases}$$

hence we obtain: $w_{11}^* = \dfrac{(n_{22} - 2\nu^*)}{(n_{11} - 2\nu^*)}, w_{12}^* = \dfrac{(n_{22} - 2\nu^*)}{(n_{12} - 2\nu^*)}, w_{21}^* = \dfrac{(n_{22} - 2\nu^*)}{(n_{21} - 2\nu^*)},$
where, without loss of generality, $w_{22}^* = 1$, and $n_{22} = \min_{ij} n_{ij}$. Of course, in order for these coefficients to be all well-defined (by non-null denominators), we must remove from our analysis all permutations in which $\nu^* = n_{ij}/2, \forall i, j$, unless $i = j = 2$. Thus, the permutation structure of the separate test for the effect of factor A becomes:

$$^wT_A^* =$$

$$= \frac{(n_{22} - 2\nu^*)}{(n_{11} - 2\nu^*)} \sum_{r=1}^{n_{11}} y_{11r}^* + \frac{(n_{22} - 2\nu^*)}{(n_{12} - 2\nu^*)} \sum_{r=1}^{n_{12}} y_{12r}^* - \frac{(n_{22} - 2\nu^*)}{(n_{21} - 2\nu^*)} \sum_{r=1}^{n_{21}} y_{21r}^* - \sum_{r=1}^{n_{22}} y_{22r}^*$$

$$= 4(n_{22}-2\nu^*)a+(n_{22}-2\nu^*)\widetilde{\varepsilon}_{11}^*+(n_{22}-2\nu^*)\widetilde{\varepsilon}_{12}^*-(n_{22}-2\nu^*)\widetilde{\varepsilon}_{21}^*-(n_{22}-2\nu^*)\widetilde{\varepsilon}_{22}^*.$$

Of course, the observed value of the test statistic is:

$$^wT_A^{ob} = \frac{n_{22}}{n_{11}} \sum_{r=1}^{n_{11}} y_{11r} + \frac{n_{22}}{n_{12}} \sum_{r=1}^{n_{12}} y_{12r} - \frac{n_{22}}{n_{21}} \sum_{r=1}^{n_{21}} y_{21r} - \sum_{r=1}^{n_{22}} y_{22r}.$$

In order to prove the unbiasedness of such a test, we will show that:

$$\Pr\{^wT_A^* \geq {}^wT_A^{ob}|\mathbf{y}(a,b,ab)\} \leq \Pr\{^wT_A^* \geq {}^wT_A^{ob}|\mathbf{y}(0,b,ab)\}.$$

Let us consider $H_{1A} : a > 0$, we get:

$$\Pr\{4(n_{22} - 2\nu^*)a + \widetilde{\overline{\varepsilon}}^* \geq 4n_{22}a + \overline{\varepsilon}|\mathbf{y}(a,b,ab)\} =$$
$$= \Pr\{-8\nu^*a + \widetilde{\overline{\varepsilon}}^* \geq \overline{\varepsilon}|\mathbf{y}(a,b,ab)\} \leq \Pr\{\widetilde{\overline{\varepsilon}}^* \geq \overline{\varepsilon}|\mathbf{y}(0,b,ab)\},$$

where $\widetilde{\overline{\varepsilon}}^*$ is the linear combination of $\widetilde{\varepsilon}_{11}^*$, $\widetilde{\varepsilon}_{12}^*$, $\widetilde{\varepsilon}_{21}^*$ and $\widetilde{\varepsilon}_{22}^*$, and $\overline{\varepsilon}$ is the linear combination of $\widetilde{\varepsilon}_{11}$, $\widetilde{\varepsilon}_{12}$, $\widetilde{\varepsilon}_{21}$ and $\overline{\varepsilon}_{22}$. Furthermore, such test is consistent (see Hoeffding, 1952; Pesarin, 1999).

It is easy to prove that weights w_{ij}^* are also suitable for obtaining exact, unbiased and consistent test statistics for testing separately the effects b and (ab):

$$^wT_B^* = w_{11}^* \sum_r^{n_{11}} y_{11r}^* - w_{12}^* \sum_r^{n_{12}} y_{12r}^* + w_{21}^* \sum_r^{n_{21}} y_{21r}^* - w_{22}^* \sum_r^{n_{22}} y_{22r}^*,$$
$$^wT_{AB}^* = w_{11}^* \sum_r^{n_{11}} y_{11r}^* - w_{12}^* \sum_r^{n_{12}} y_{12r}^* - w_{21}^* \sum_r^{n_{21}} y_{21r}^* + w_{22}^* \sum_r^{n_{22}} y_{22r}^*.$$

It is worth noting that in unbalanced designs, only USP can be adopted because shuffling the data within blocks at each permutation is needed. The permutation support of each test statistic is: $\sum_{\nu^*=0}^{\min n_{ij}} \binom{n_{11}}{\nu^*} \times \binom{n_{12}}{\nu^*} \times \binom{n_{21}}{\nu^*} \times \binom{n_{22}}{\nu^*}$, with $0 \le \nu^* \le \min n_{ij}$, $\nu^* \ne n_{ij}/2$, $\forall i,j$, unless $i = j = 2$.

3   Permutation tests for balanced and unbalanced repeated measures designs

In this section we extend the exact synchronized permutation solution for 2^2 factorials to balanced and unbalanced repeated measures designs and we consider a replicated unbalanced factorial design with fixed effects (Milliken and Johnson, 1984) as the basic experimental plan. In order to deal with repeated measures designs from the point of view of nonparametric permutation testing we need the introduction of the nonparametric combination methodology and due to the lack of space we refer the reader to Pesarin (1999) for a complete review on this method. In order to produce the nonparametric combination of dependent permutation tests, as in the case of repeated measures designs, it is possible to use a Monte Carlo procedure called Conditional Simulation Technique. We also refer to Pesarin (1999) for details on such technique. The experimental responses are measured in L time occasions. The usual linear model for single responses is: $y = \{y_{ij(\ell)r} = \mu_{(\ell)} + a_{ij(\ell)} + b_{ij(\ell)} + (ab)_{ij(\ell)} + \varepsilon_{ij(\ell)r}, j = 1,2, i = 1,2, r = 1,...,n_{ij} \cdot 2^2, i = 1,2, j = 1,2, l = 1,...,L \}, \sum_{i,j} n_{ij} = N$, where $y_{ij(\ell)r}$ are the experimental responses, $\mu_{(\ell)}$ is the general mean for the ℓ-th measure, $a_{(\ell)}$ is the effect of factor A in the ℓ-th measure, $b_{(\ell)}$ is the effect of factor B in the ℓ-th measure; $(ab)_{(\ell)}$ is the interaction effect A and B in the ℓ-th measure, $\varepsilon_{ij(\ell)r}$ are exchangeable experimental errors in the ℓ-th measure from an unknown distribution $\mathcal{P}$ with zero mean and variance σ_ℓ^2, finally n_{ij} is the number of observations for each factor's levels combination. Thus, the total sample size is $L \cdot N$.

The overall system of hypotheses is H_0: $\cap_{\ell=1,...,L} \{[H_{0A(\ell)}] \cap [H_{0B(\ell)}] \cap [H_{0AB(\ell)}]\}$, against the alternative H_1: {at least one is false}, where the three separate hypotheses for the ℓ-th measure are $H_{0A(\ell)}$: $\{a_{(\ell)} = 0\}$ vs $H_{1A(\ell)}$: $\{a_{(\ell)} \ne 0\}$, $H_{0B(\ell)}$: $\{b_{(\ell)} = 0\}$ vs $H_{1B(\ell)}$: $\{b_{(\ell)} \ne 0\}$, $H_{0AB(\ell)}$: $\{(ab)_{(\ell)} = 0\}$ vs $H_{1AB(\ell)}$: $\{H_{0AB(\ell)} \ne 0\}$, so that, the null hypothesis H_0 is true if all three separate sub hypotheses are true.

Let us consider the three partial tests for effects in every measure. For example, the ℓ-th permutation test for the effect of factor A is: $^wT_{A(\ell)}^* = ^wT_{A|1,(\ell)}^* + ^wT_{A|2,(\ell)}^*$, where $^wT_{A|1,(\ell)}^* = w_{11}^*(\sum_{r=1}^{n_{11}} y_{11(\ell)r}^*)/n_{11} - w_{21}^*(\sum_{r=1}^{n_{21}} y_{21(\ell)r}^*)/n_3$, $^wT_{A|2,(\ell)}^* = w_{12}^*(\sum_{r=1}^{n_{12}} y_{12(\ell)r}^*)/n_{12} - w_{22}^*(\sum_{i=1}^{n_{22}} y_{22(\ell)r}^*)/n_{22}$, weights w_{ij}^*

have been previously defined, and $y_{ij(\ell)r}^*$ is the permutation value of $y_{ij(\ell)r}$ according to the synchronized permutation approach. Conditions on the weights w_{ij}^* and the cardinality of the permutation support of the test statistics are defined as for the simple unbalanced case. It is easy to note that in case of balanced designs weights $w_{ij}^* = 1$. Now, in order to test the overall null hypothesis, we jointly consider the L measures, then the permutation solution is based on the nonparametric combination method which allows us to combine partial tests on single measures.

It should be noted that, as it is usual in multidimensional problems, random numbers ν^* of elements to exchange between blocks are invariant with respect to $\ell = 1, ..., L$. Hence, individual response vectors are exchanged. In presence of multiresponses again an easy application of the nonparametric combination method allows us to combine overall tests on single responses.

4 Simulation study

In the following we present a comparative simulation study to evaluate the power behaviour of the above presented synchronized permutation solution for balanced and unbalanced repeated measures designs. At first we make power comparisons between the parametric F test and the synchronized permutation test for the simple 2^2 factorial and for the corresponding repeated measures design with $L = 2$ (i.e. two repeated measures). In the case of repeated measures designs, the error terms, in the usual linear model for responses, i.e. $\mathbf{y} = \{y_{ij(\ell)r} = \mu_{(\ell)} + a_{ij(\ell)} + b_{ij(\ell)} + (ab)_{ij(\ell)} + \varepsilon_{ij(\ell)r},$ $j = 1, 2, i = 1, 2, r = 1, ..., n_{ij} \cdot 2^2, i = 1, 2, j = 1, 2, l = 1, ..., L \}$, are taken accordingly to the following autoregressive model: $\varepsilon_{ij(\ell)r} = \frac{\varepsilon_{ij(\ell-1)r}}{2} + \frac{\varepsilon_{ij(\ell)r}}{2}$, and the number of replicates is $n_{ij} = 3$, $\forall i, j$, in balanced designs. In each power study one effect has been taken as a non-active effect in order to show the behaviour of such tests also under the null hypothesis.

For unbalanced designs, the simulation study examines the power behaviour of the synchronized permutation approach, noting that in such a case a general parametric solution does not exists because the F-statistic depends on weights chosen for defining the usual side conditions on main factors (see e.g. Salmaso, 2000). For the balanced case, a Cauchy distribution with zero location parameter and unit scale parameter is considered for the error terms. For the unbalanced case the simulations presented below are evaluated only in the case of a standard normal distribution for the error components. The number of replicates, considered for the unbalanced design, are: $n_{11} = 3$, $n_{12} = 1$, $n_{21} = 3$, $n_{22} = 1$. In tables for separate tests, the location shift, δ, for each effect, is reported; of course if $\delta = 0$, then the corresponding effect is not active.

The parametric F test is pointed out by the abbreviation *param* and the permutation test is pointed out by the abbreviation *perm*. For the sim-

TABLE 1. Cauchy distribution (two-sided tests):

Separate Tests

A₁: δ = 0.5			B₁: δ= 0.7			AB₁: δ= 0.0		
α	param	perm	α	param	perm	α	param	perm
0.0122	0.0660	0.1807	0.0122	0.1340	0.2880	0.0122	0.0018	0.0227
0.0244	0.1050	0.2070	0.0244	0.1838	0.3137	0.0244	0.0065	0.0300
0.0366	0.1388	0.2345	0.0366	0.2290	0.3395	0.0366	0.0092	0.0410
0.0488	0.1703	0.2552	0.0488	0.2612	0.3638	0.0488	0.0170	0.0462
0.0610	0.1915	0.2758	0.0610	0.2908	0.3885	0.0610	0.0235	0.0590
0.0976	0.2587	0.3292	0.0976	0.3593	0.4385	0.0976	0.0530	0.0948

A₂: δ= 0.9			B₂: δ= 1.1			AB₂: δ= 0.6		
α	param	perm	α	param	perm	α	param	perm
0.0122	0.0560	0.1505	0.0122	0.0838	0.2080	0.0122	0.0230	0.0887
0.0244	0.0922	0.1775	0.0244	0.1290	0.2315	0.0244	0.0460	0.1090
0.0366	0.1213	0.2057	0.0366	0.1645	0.2625	0.0366	0.0680	0.1303
0.0488	0.1492	0.2213	0.0488	0.1900	0.2823	0.0488	0.0815	0.1442
0.0610	0.1638	0.2375	0.0610	0.2127	0.3053	0.0610	0.1015	0.1603
0.0976	0.2225	0.2898	0.0976	0.2863	0.3548	0.0976	0.1462	0.2060

Combined Tests

A				B			
α	param	perm (F)	perm (T)	α	param	perm (F)	perm (T)
0.0122	0.0673	0.1608	0.0742	0.0122	0.1288	0.2318	0.1373
0.0244	0.1095	0.2040	0.2570	0.0244	0.1805	0.2878	0.3593
0.0366	0.1417	0.2468	0.2675	0.0366	0.2193	0.3367	0.3680
0.0488	0.1708	0.2778	0.2988	0.0488	0.2517	0.3658	0.3950
0.0610	0.1940	0.3050	0.3155	0.0610	0.2845	0.3955	0.4083
0.0976	0.2595	0.3675	0.3658	0.0976	0.3475	0.4587	0.4633

AB			
α	param	perm (F)	perm (T)
0.0122	0.0145	0.0310	0.0050
0.0244	0.0253	0.0570	0.1075
0.0366	0.0413	0.0882	0.1138
0.0488	0.0522	0.1070	0.1335
0.0610	0.0680	0.1303	0.1400
0.0976	0.978	0.1752	0.1895

ulations regarding combined tests two different combining functions are considered for the synchronized permutation solution: Fisher (*perm (F)*) and Tippett (*perm (T)*). For the synchronized permutation solution the exact permutation distribution is computed at each Monte Carlo iteration; the number of Monte Carlo iterations is 4000; α is the type I error rate and the nominal attainable significance levels are reported in each table. The simulations in Table 1 and 2 show a general good behaviour of the synchronized permutation solution both in the normal case and using an heavy tailed distribution.

5 Conclusions

By using the above presented procedure, if we consider the general case with L times and K variables of interest, we can obtain the following different analyses: i) partial tests on only one response (ℓ) for the analysis of single variables (k) with respect to the treatments carried out and their interaction; ii) combined tests on all L responses of each variable (k) with respect to the treatments carried out and their interaction; iii) combined tests on all K variables for each response (ℓ) with respect to the treatments

TABLE 2. Standard normal distribution (two-sided tests for unbalanced designs):

Separate Tests

A_1: $\delta = 0.4$		B_1: $\delta = 0.5$		AB_1: $\delta = 0.0$	
α	perm	α	perm	α	perm
0.0501	0.0780	0.0501	0.0921	0.0501	0.0455
0.1875	0.3265	0.1875	0.3683	0.1875	0.1775
0.3895	0.6948	0.3895	0.6782	0.3895	0.3780
0.5700	0.8501	0.5700	0.8501	0.5700	0.5675
0.7134	0.9002	0.7134	0.8995	0.7134	0.7060
0.8433	0.9431	0.8433	0.9486	0.8433	0.8347

A_2: $\delta = 0.5$		B_2: $\delta = 0.9$		AB_2: $\delta = 0.2$	
α	perm	α	perm	α	perm
0.0501	0.0884	0.0501	0.0989	0.0501	0.0621
0.1875	0.3779	0.1875	0.3955	0.1875	0.2581
0.3895	0.7053	0.3895	0.7470	0.3895	0.5124
0.5700	0.8711	0.5700	0.8970	0.5700	0.7137
0.7134	0.9186	0.7134	0.9328	0.7134	0.8099
0.8433	0.9556	0.8433	0.9689	0.8433	0.9011

carried out and their interaction; iv) final combined test on all K variables for all L responses with respect to the treatments carried out and their interaction.

References

Folks, J.L. (1984). Combinations of independent tests. In Krishnaiah, P.R. and Sen, P. K. (eds.), *Handbook of Statistics, vol. 4*, Elsevier Science Publishers, North Holland, 113-121.

Hoeffding, W. (1952). The large-sample power based on permutation of observations, *Annals Math. Statistics*, **23**, 168-192.

Milliken, G.A. and Johnson, D.E. (1984). *Analysis of Messy Data, Designed Experiments, vol. 1*. Van Nostrand Reinhold Company, New York.

Pesarin, F. (1992). A resampling procedure for nonparametric combination of several dependent tests, *J. Ital. Statist. Soc.*, **1**, 87-101.

Pesarin, F. (1999). *Permutation testing of multidimensional hypotheses by nonparametric combination of dependent tests*. CLEUP, Padua.

Salmaso, L. (2000). Synchronized permutation tests in 2^k factorial designs, *International Journal of Applied Science and Computations* (to appear).

The Influence of the Design on the Breakdown Point of ℓ_1-type M-estimators

I. Mizera
Ch.H. Müller

ABSTRACT: Mizera and Müller (1999) showed that the breakdown of M-estimators with bounded score function in the general linear model depends via a quantity $\mathcal{M}(X,r)$ only on the design X and the variation exponent r of the objective function. In the case $r = 0$, e.g. for the Cauchy M-estimator, the quantity $\mathcal{M}(X,r)$ is the maximum number of design points at which the unknown parameter vector is not identifiable. We study the case $r = 1$, the case of the so-called ℓ_1-type M-estimators—including the ℓ_1-estimator itself, Huber's M-estimator, and others—where the quantity $\mathcal{M}(X) = \mathcal{M}(X,1)$ has the potential of detecting the presence of leverage points in the design. We give a numerical algorithm for computing $\mathcal{M}(X)$ and calculate it for several examples, showing how $\mathcal{M}(X)$ can be used as a diagnostic tool for detecting leverage points and how the design influences the breakdown point of ℓ_1-type M-estimators.

KEYWORDS: Breakdown point; Optimum design; Leverage point; ℓ_1-estimator

1 Introduction

Consider the general linear model with N-dimensional univariate response vector $y = (y_1, y_2, \ldots, y_N)^\top$ and $N \times p$ matrix of design points $X = (x_1, x_2, \ldots, x_N)^\top$:

$$y = X\beta + \varepsilon$$

—where, as usual, β is the vector of unknown parameters, and ε the vector of random errors. Let

$$m(\beta, E, X) = \frac{\sum_{n \in E} |x_n^\top \beta|}{\sum_{n=1}^{N} |x_n^\top \beta|},$$

mODa6, A.C.Atkinson, P.Hackl and W.G.Müller, eds., Physica, Heidelberg, 2001.

and let

$$m(E, X) = \max_{\beta} m(\beta, E, X). \tag{1.1}$$

Finally, let

$$\mathcal{M}(X) = \min\{\text{card } E : m(E, X) \geq 1/2\}. \tag{1.2}$$

The relationship of $\mathcal{M}(X)$ to the breakdown point of the ℓ_1 and ℓ_1-type estimators was pointed out by He et al. (1990) and also by Ellis and Morgenthaler (1992), who added more insights and facts about the exact fit degree; finally it was firmly established by Mizera and Müller (1999) in the framework of their general theory for M-estimators in the linear regression model. In this theory, (1.2) is a special case of a certain general quantity, depending on the regular variation exponent r of the objective function of the M-estimator. The value $r = 1$ corresponds to the ℓ_1-estimator (and other ℓ_1-type estimators—Huber's estimator, for instance); the breakdown point is equal to $\mathcal{M}(X)/N$. If $r = 0$, then the breakdown point depends on the design only via the *identification index*

$$\mathcal{N}(X) = \max_{\beta \neq 0} \text{card}\{n : x_n^\top \beta = 0\}$$

and is equal to $\lfloor (N - \mathcal{N}(X) + 1)/2 \rfloor / N$. This value is the maximal possible for equivariant estimators: the inequality

$$\mathcal{M}(X) \leq \left\lfloor \frac{N - \mathcal{N}(X) + 1}{2} \right\rfloor \tag{1.3}$$

holds. For more details, we refer the reader to Mizera and Müller (1999). It is important to note that the version of the breakdown theory under discussion keeps the design points fixed and admits outliers only in the response. Although fractional values of r are possible, only integer values of r seem of practical interest; the values $r = 0$ and $r = 1$ correspond to robust M-estimators, $r = 2$ to the classical least squares estimator.

In this note, we further examine the case $r = 1$. We study some consequences of the breakdown theory in optimum design considerations and diagnostics of leverage/influential points. We illustrate, with several examples, how $\mathcal{M}(X)$ may serve for evaluation of designs and how $\mathcal{M}(X)$ and related quantities may be used as a regression diagnostic tool. To this end, we need a numerical algorithm for their computation.

2   Computation of $\mathcal{M}(X)$

For computing $\mathcal{M}(X)$ numerically, the following observations are useful. First, if $\mathcal{N}(X) < N$, then $\mathcal{M}(X) - 1$ is equal to the maximal cardinality of a set $E \subset \{1, 2, \ldots, N\}$ such that $m(E, X) < 1/2$. For $c \neq 0, m(c\beta, E, X) =$

$m(\beta, E, X)$; thus, since we may restrict the domain of maximization to a compact set

$$A = \left\{ \beta : \sum_{n=1}^{N} |x_n^\mathsf{T}\beta| = 1 \right\},$$

the maximum is attained for every E. Let $\mathcal{X}$ be *the set* of all design points x_n (that is, multiple design points enter $\mathcal{X}$ only once). Let $\mathcal{D}$ be the set of all subsets D of $\mathcal{X}$ with the property that the linear space spanned by the elements of D is $(p-1)$-dimensional. Let $D^\perp$ be the orthogonal complement of this space, the set of all β such that $x_n^\mathsf{T}\beta = 0$ for all $x_n \in D$. Let

$$\mathcal{B} = \bigcup_{D \in \mathcal{D}} (A \cap D^\perp).$$

Let $\mathcal{S} = \{-1, 1\}^N$. For $s \in \mathcal{S}$ we define

$$A_s = \{ \beta \in A : |x_n^\mathsf{T}\beta| = s_n x_n^\mathsf{T}\beta \text{ for } n = 1, 2, \ldots, N \}$$

—if $\beta \in A_s$, then $\operatorname{sign}(x_n^\mathsf{T}\beta) = s_n$ or 0.

Proposition 1 *For every $s \in \mathcal{S}$, the set A_s is compact and convex (possibly empty). All extremal points of A_s lie in $\mathcal{B}$; $m(\cdot, E, X)$ is linear on A_s for every E.*

Proof. The set A_s is a subset of A, hence it is bounded. The properties that determine A_s are preserved under a passage to the limit; hence A_s is closed. Convex combinations also remain in A_s; since

$$\sum_{n=1}^{N} |x_n^\mathsf{T}\beta| = \sum_{n=1}^{N} s_n x_n^\mathsf{T}\beta,$$

all convex combinations of elements of A_s belong to A. Also linearity of $m(\cdot, E, X)$ on A_s follows in the analogous way.

Fix $s \in \mathcal{S}$. Let β be an extremal point of A_s and let C be the set of all x_n such that $x_n^\mathsf{T}\beta = 0$. Suppose that the dimension of the linear space spanned by C does not exceed $p - 2$. Then the linear space spanned by $C \cup \{\sum_{n=1}^{N} s_n x_n\}$ has dimension at most $p - 1$; consequently, there is a $\tilde{\beta} \neq 0$ such that $x_n^\mathsf{T}\tilde{\beta} = 0$ for all $x_n \in C$ and $\sum_{n=1}^{N} s_n x_n^\mathsf{T}\tilde{\beta} = 0$. For all $x_n \in \mathcal{X} \setminus C$, $|x_n^\mathsf{T}\beta| \geq \eta$ for some $\eta > 0$; we may suppose that $|x_n^\mathsf{T}\tilde{\beta}| \leq \eta/2$ for all $n = 1, 2, \ldots, N$, taking a suitable nonzero multiple of $\tilde{\beta}$ otherwise. As a consequence,

$$\operatorname{sign}(x_n^\mathsf{T}\beta + x_n^\mathsf{T}\tilde{\beta}) = \operatorname{sign}(x_n^\mathsf{T}\beta) = \operatorname{sign}(x_n^\mathsf{T}\beta - x_n^\mathsf{T}\tilde{\beta}) \qquad (2.1)$$

for all $x_n \in \mathcal{X} \setminus C$, and also for all $x_n \in C$. Therefore,

$$\sum_{n=1}^{N} \left| x_n^\mathsf{T}(\beta + \tilde{\beta}) \right| = \sum_{n=1}^{N} s_n x_n^\mathsf{T}(\beta + \tilde{\beta}) = \sum_{n=1}^{N} s_n x_n^\mathsf{T}\beta + \sum_{n=1}^{N} s_n x_n^\mathsf{T}\tilde{\beta} = 1,$$

and analogously $\sum_{n=1}^{N} |x_n^{\top}(\beta - \tilde{\beta})| = 1$. We obtain, from (2.1), that β is a convex combination of $\beta + \tilde{\beta}$ and $\beta - \tilde{\beta}$, both elements of A_s. This contradicts the extremality of β; hence, the space spanned by C has dimension at least $p - 1$. It cannot have dimension p since then $x_n^{\top}\beta = 0$ for all $x_n \in C$ would imply $\beta = 0$, a contradiction of the fact that $\beta \in A$. Thus, the dimension is $p - 1$. $\square$

Proposition 2 *For every $E \subseteq \{1, 2, \ldots, N\}$, the maximum $m(E, X)$ of $m(\beta, E, X)$ over β is attained on some point of $\mathcal{B}$ satisfying*

$$\frac{1}{2} \operatorname{card} \mathcal{B} \leq \binom{\operatorname{card} \mathcal{X}}{p - 1}. \tag{2.2}$$

Proof. Since $D^{\perp}$ is a one-dimensional linear subspace, the set $A \cap D^{\perp}$ contains, for any $D \in \mathcal{D}$, at most two elements; this proves (2.2). The remaining part of the statement follows from Proposition 1 and the fact that $A = \bigcup_{s \in S} A_s$—since the maximum of a linear function over a compact convex set is attained in one of its extremal points. $\square$

Proposition 2 says that for computing $\mathcal{M}(X)$, we have only to inspect $\beta \in \mathcal{B}$. Note that, in fact, we have to consider only one β from each $A \cap D^{\perp}$, the second one is $-\beta$, with the same $|x_n^{\top}\beta|$'s for all $n = 1, 2, \ldots, N$. For simplicity, we used $\{x_1, \ldots, x_N\}$ instead of $\mathcal{X}$ in our computer program—despite the fact that the speed of the algorithm would improve if $\mathcal{X}$ were used for designs with many repetitions. The computer program written in S (the R implementation, version 1.0.0) consists of a main routine called M.X and a recursive function M.X.rek that determines all subsets of $\{x_1, \ldots, x_N\}$ with $p - 1$ elements. The main function M.X is a function of X and has the form

```
N <- number of rows of X
p <- rank of X
cardE.opt <- N
E.opt <- 1:N
beta.opt <- vector of p zeros
for (n in 1:N) {
    subset <- n
    result <- result of M.X.rek at X, subset, E.opt,
              cardE.opt, beta.opt, N, p }
```

The recursive function M.X.rek is a function of X, N, p, the subset E and the candidates cardE.opt, E.opt and beta.opt for $\mathcal{M}(X)$, the minimal cardinality of E such that $m(E, X) \geq 1/2$, and $\arg\max_{\beta} m(\beta, E.opt, X)$, respectively. Its form is

```
sall <- 1:n
if (length of subset < p-1)
    for (snew in sall[-subset])
        if (snew > max(subset)) {
```

```
              subset.new <- cbind(subset,new)
              result <- result of M.X.rek at X, subset.new, E.opt,
                         cardE.opt, beta.opt, N, p }
       else{
          beta <- an element of the orthogonal complement of X[subset,]
          Xbeta <- abs(X %*% beta) # vector of |x_n^\top \beta|
          Xorder <- order(Xbeta)
          Xbeta.order <- Xbeta[Xorder] # Xbeta ordered
          n <- p-1
          while(n<N & 2*sum(Xbeta.order[1:n])-sum(Xbeta)<0.000001)
                {cardE <- N-n
                 n <- n+1}
          if (cardE <- cardE.opt) {
              cardE.opt <- cardE
              E.opt <- Xorder[n:N]
              beta.opt <- beta }}
```

3 $\mathcal{M}(X)$ as a design criterion

A usual way to choose an optimal design is to maximize the information matrix $\mathcal{I}(X) = X^\top X/N$ or a function of it. This ensures the smallest variance of the least squares estimator—but also of the ℓ_1-estimator, if no outliers are present. As soon as outliers can appear, an additional criterion on the design should be adopted, ensuring maximal possible robustness of the given estimator. For ℓ_1 estimators this calls for $\mathcal{I}(X)$ and $\mathcal{M}(X)$ as large as possible. The following examples indicate that the D-optimal designs, the designs maximizing $\det \mathcal{I}(X)$, also maximize $\mathcal{M}(X)$ (however, we have no rigorous proof of this assertion so far).

Example 1. Simple regression
In Mizera and Müller (1999), an explicit formula for $\mathcal{M}(X)$ for simple regression was given, when $x_n = (1, t_n)^\top$. This implies that any equispaced design—and thus the D-optimal design—maximizes $\mathcal{M}(X)$.

Example 2. Quadratic regression
For the quadratic regression with $x_n = (1, t_n, t_n^2)^\top$, we calculated $\mathcal{M}(X)$ for four different designs. For a design with 40 repetitions at $-1, 0, 1$ and for that with 30 repetitions at $-1, -0.3, 0.3, 1$, we obtained $\mathcal{M}(X) = 20$. For equispaced design with the t_n's from -1 up to 1, the result was smaller: $\mathcal{M}(X) = 18$. Putting more repetitions at $t_n = 0$ reduces $\mathcal{M}(X)$ as well: for instance, we obtained $\mathcal{M}(X) = 15$ for a design with 30 repetitions at -1 and 1 and 60 repetitions at 0. Apparently, 20 is the maximal value for $\mathcal{M}(X)$—and it is reached by the D-optimal design.

Note that $20/120 = 1/6$ is the maximal possible breakdown point for any equivariant estimator at the D-optimal design; this follows from (1.3) since $\mathcal{N}(X) = 80$. Thus, at the D-optimal design, the highest breakdown point is attained by the ℓ_1-estimator—this is not true for other designs, however.

For example, for the equispaced design (design points from -1 up to 1), the maximal breakdown point is $59/120$ (according to (1.3)); this is much larger than $18/120$ provided by the ℓ_1-estimator.

Example 3. Two-factor model

For a model with two factors each with three levels 1, 2, 3 and without interactions we compared the following two designs (each level combination with one observation):

Factor 1	1	1	1	2	2	2	3	3	3
Factor 2	1	1	2	2	2	3	1	3	3
Factor 1	1	1	1	2	2	2	3	3	3
Factor 2	1	2	3	1	2	3	1	2	3

The result was $\mathcal{M}(X) = 1$ for the first design and $\mathcal{M}(X) = 2$ for the second design. The D-optimal design again gives the maximal breakdown point. Since $\mathcal{N}(X) = 1$ for the first design and $\mathcal{N}(X) = 2$ for the second design, the ℓ_1-estimator has at both designs the maximal breakdown point—according to (1.3).

4   Applications for leverage/influence diagnostics

Ellis and Morgenthaler (1992) introduced $m(E, X)$ as an ℓ_1 analog of the quantity

$$\max_{\beta} \frac{\sum_{n \in E} (x_n^\top \beta)^2}{\sum_{n=1}^{N} (x_n^\top \beta)^2}, \tag{4.1}$$

proposed as a diagnostic for the classical least squares method quantifying the influence/leverage of the group of observations represented by E. They interpreted (4.1) as the squared cosine of the minimal angle between two directions, one belonging to the linear space generated by the design points, the other to the linear space generated by the unit vectors e_i, $i \in E$. Particularly, if $E = \{i\}$, then (4.1) is equal to h_{ii}, the i-th diagonal element of the hat (projection) matrix $X(X^\top X)^{-1} X^\top$, whose diagnostic value is well-known; see Atkinson (1985), or Rousseeuw and Leroy (1987).

As already mentioned, in the theory of Mizera and Müller (1999) both (1.1) and (4.1) are special cases of the same general quantity, corresponding to values $r = 1$ and $r = 2$, respectively, of the regular variation exponent of the objective function of the M-estimator. In view of this breakdown point theory, $m(E, X)$ can be interpreted as a "breaking power", a capability to break the ℓ_1-estimator down, if the response is altered at the points with indices in E. This suggests the interpretation of $m(E, X)$ as the measure of leverage of the observations corresponding to the set E.

Many of regression diagnostics appearing in the literature are designed to evaluate the potential of observations to influence the least squares fit.

Their rationale is often based on the influence function of the least squares estimator, or some of its finite-sample substitutes. As a result, they usually depend on h_{ii} as well as on the residual $y_i - x_i^\top \widehat{\beta}$. The influence function of the ℓ_1-estimator—unlike that of the least squares estimator—involves only the sign of the residual, suggesting that potential influence measures designed for the ℓ_1 estimator—like (1.1)—may depend only on the design matrix.

To evaluate the leverage of individual points, it is natural to look at values of $m(\{i\}, X)$ (we use the shorthand $m(i, X)$). As with other measures of influence, it is important to know what values are "large". To this end, $\mathcal{M}(X)$ provides a valid comparison, evaluating how "stable" is the design: any E with $m(E, X) \geq 1/2$ is a "strategic set of points", capable of breaking down the estimator completely. Moreover, the set $E.opt$ on which the minimal cardinality appearing in $\mathcal{M}(X)$ is attained, is often the set of "principal outliers"; for instance, if there is one extremely outlying covariate point, then its index is contained in $E.opt$.

Example 4. Simple regression with one outlier

We considered again the simple regression, as in Example 1. The 30 points t_i were generated from the uniform distribution on $[-1, 1]$. Then we added an outlying design point $(1, T)$, where T took 10 equispaced values from 0 to 20. We plotted the dependence of $\mathcal{M}(X)$ on T; for comparison, we include also the analogous plot for one minus the diagonal element of the hat matrix corresponding to the outlying design point.

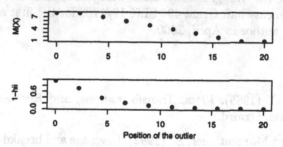

Example 5. The Hertzsprung-Rusell data

These data were introduced and analyzed by Rousseeuw and Leroy (1987). Using again simple regression for the data, $E.opt$ coincides with the 5 marked points, four of them corresponding to so-called "giants".

Example 6. Quadratic regression again

In certain situations, the influence may be distributed more uniformly: for instance, in polynomial regression with equispaced design, the points near the boundary are more influential than those in the middle and the influence is symmetric. However, $E.opt$ returned by M.X may reflect just one side of the data points. To appreciate the full picture, we recommend looking at the individual influence of the points expressed by $m(i, X)$, and then trying various possible competitors of $E.opt$. We plotted $m(i, X)$ for the

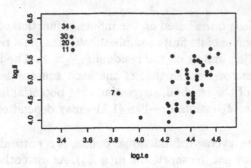

equispaced design in quadratic regression like in Example 2 (20 points from
−1 to 1); for comparison, we plot again also the corresponding diagonal
elements of the hat matrix. Apparently, points with indices 1, 2 and 3

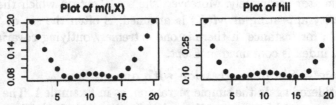

possess the same leverage as those from $E.opt$: 18, 19 and 20. Looking at
$m(E, X)$ for both sets, we realize that they are indeed the same.

Acknowledgments: We gratefully acknowledge support of the Deutsche
Forschungsgemeinschaft Grant 436 SLK 17/1/00 for the first author's visit
of the second author in April, 2000.

References

Atkinson, A. C. (1985). *Plots, Transformations, and Regression.* Clarendon Press, Oxford.

Ellis, S. P. and Morgenthaler, S. (1992). Leverage and breakdown in L_1 regression. *Journal of the American Statistical Association*, **87**, 143-
148.

He, X., Jurečková, J., Koenker, R. and Portnoy, S. (1990). Tail behavior
of regression estimators and their breakdown points. *Econometrica*
58, 1195–1214.

Mizera, I. and Müller, Ch. H. (1999). Breakdown properties and variation
exponents of robust M-estimators in linear models. *Annals of Statistics* **27**, 1164–1177.

Rousseeuw, P. J. and Leroy A. M. (1987). *Robust Regression and Outlier
Detection.* Wiley, New York.

Analytical Properties of Locally D-optimal Designs for Rational Models

V.B. Melas

ABSTRACT: The paper is devoted to studying locally D-optimal designs for rational regression models. He *et al.* (1997) have shown that the problem considered is equivalent to that of D-optimal designs for polynomial regression with polynomial variance function. Both problems were studied in a number of papers with the help of two usual approaches: numerical construction of optimal designs and constructing the designs in a closed analytical form. It appears that the latter form is available only in some special cases. Here another approach is developed. Points of optimal designs are studied as functions of the parameters. It turns out that these functions can be expanded in Taylor series. An example of such expansion for a model with six parameters is given.

KEYWORDS: optimal experimental designs, nonlinear regression, rational models, locally D-optimal designs.

1 Introduction

The present paper is devoted to studying locally D-optimal designs for rational models on a finite or infinite interval. Rational functions generate an important class of functions and can be used, particularly, for approximating an arbitrary continuous function.

The problem has been studied in a number of papers. He *et al.* (1996) have shown that the problem of constructing locally D-optimal designs is equivalent to that of finding D-optimal designs for polynomial regression models with variance functions which are also polynomials. The connection between the problems is discussed in Dette *et al.* (1999). Futher references can be found in those papers. Two basic approaches to constructing locally D-optimal designs were used: the numerical construction of optimal de-

mODa6, A.C.Atkinson, P.Hackl and W.G.Müller, eds., Physica, Heidelberg, 2001.

signs and constructing designs in a closed analytical form. The numerical construction is simple but the analytical solution appears to be available only in some particular cases.

In this paper another approach is developed. It is based on studying locally D-optimal design points as functions of values of parameters. We call such functions optimal design-functions. These functions are implicitly given by a differential equation system which is introduced on the basis of known necessary conditions for an extremum point of a function of a general form. It is shown that the Jacobi matrix of this system is nonsingular under mild restrictions on the parameter values.

This allows us to establish, with the help of the known Implicit Function Theorem, that optimal design functions are real analytical ones. With the help of recurrence formulae (see Melas, 2000) these functions can be expanded in a Taylor series.

Note that the approach applied here was introduced in Melas (1978) for another type of nonlinear regression model.

The paper is organized as follows. Section 2 contains the description of the model and a formal outline of the problem. Section 3 is devoted to a number of preliminary results. In Section 4 optimal design functions are introduced and studied. In Section 5 an example of Taylor expansion of these functions is given. Section 6 contains concluding remarks.

2   Outline of the problem

Let experimental results $y_1, \ldots, y_n \in \mathbf{R}$ be described by the equation

$$y_j = \eta(x_j, \Theta) + \varepsilon_j, \quad j = 1, \ldots, n, \tag{2.1}$$

where
$\{\varepsilon_j\}$ are i. i. d. random values such that

$$E\varepsilon_j = 0, \quad E\varepsilon_j^2 = \sigma^2 > 0, \tag{2.2}$$

$$\eta(x, \Theta) = P(x, \Theta_1)/Q(x, \Theta_2), \tag{2.3}$$

$$\Theta^T = (\Theta_1^T, \Theta_2^T), \quad \Theta_1^T = (\vartheta_1, \ldots, \vartheta_{m-k}), \quad \Theta_2^T = (\vartheta_{m-k+1}, \ldots, \vartheta_m),$$

$$P(x, \Theta_1) = \sum_{i=1}^{m-k} \vartheta_i x^{i-1}, \quad Q(x, \Theta_2) = x^k + \sum_{i=1}^{k} \vartheta_{i+m-k} x^{i-1},$$

Θ is the vector of unknown parameters, $x_j \in \mathfrak{X}$, $\mathfrak{X}$ is an interval in $\mathbf{R}$. Let us consider the case $\mathfrak{X} = [0, d]$, $d < \infty$. The case of an arbitrary interval $\mathfrak{X}$ can be studied in a similar way. As usual, a discrete probability measure given by the table

$$\xi = \{x_1, \ldots, x_n; \mu_1, \ldots, \mu_n\},$$

where $x_i \in \mathfrak{X}$, $x_i \neq x_j$ $(i \neq j)$, $\mu_i > 0$, $\sum \mu_i = 1$, will be called an (experimental) design.

Let $\Theta \in \Omega$, where Ω is a given bounded set in $\mathbf{R}^m$.

Let us introduce the following assumptions:

(a) the ratio in the right side of (2.3) is irreducible with $\Theta \in \Omega$,

(b) $Q(x, \Theta_2) \neq 0$ for $x \in \mathfrak{X}$, $\Theta \in \Omega$,

(c) $m \geq 2k$.

It can be verified (see the following section) that under these assumptions $\det M(\xi, \Theta) \neq 0$ with $n \geq m$, $\Theta \in \Omega$, where

$$M(\xi, \Theta) = \left(\sum_{l=1}^{n} \frac{\partial \eta(x_l, \Theta)}{\partial \vartheta_i} \frac{\partial \eta(x_l, \Theta)}{\partial \vartheta_j} \mu_l \right)_{i,j=1}^{m}$$

is the information matrix.

Let $\mu_j N$ observations be performed at points x_j, $j = 1, \dots, n$, where the magnitudes $\mu_j N$ are rounded in some way to integers. Denote by $\hat{\Theta}_N$ the least squares estimator of Θ_{tr}, the true value of the parameter vector. From the known results (Jennrich, 1969) $\hat{\Theta}_N \to \Theta_{\mathrm{tr}}$ with $N \to \infty$ almost surely, and the magnitude $\frac{1}{\sqrt{N}}(\hat{\Theta}_N - \Theta_{\mathrm{tr}})$ is asymptotically normally distributed with covariance matrix

$$V = \frac{1}{N} M^{-1}(\xi, \Theta), \quad \Theta = \Theta_{\mathrm{tr}}.$$

The design which maximizes $\det M(\xi, \Theta)$ under some fixed value $\Theta = \Theta^{(0)}$ will be called, as usual, a locally D-optimal design.

It can easily be checked that the design does not depend on Θ_1. The present paper considers the dependence of the design points on Θ_2.

3 Preliminary results

Consider also the condition

(d) the polynomial $Q(x, \Theta_2)$ with $\Theta \in \Omega$ has k distinct and negative roots.

It is easy to verify that under conditions (a), (b), (c) and (d) the right hand side of equality (2.3) can be written in the form

$$\eta_1(x, \tilde{\Theta}) = \sum_{i=1}^{l} \tilde{\vartheta}_i x^{i-1} + \sum_{i=1}^{k} \tilde{\vartheta}_{l+i}/(x + \gamma_i), \qquad (3.1)$$

where $l = m - 2k$, $\tilde{\Theta} = (\tilde{\vartheta}_1, \ldots, \tilde{\vartheta}_{m-k}, \gamma_1, \ldots, \gamma_k)^T$ while $\tilde{\vartheta}_i \neq 0$, $i = l+1, \ldots, l+k$ and with $l = 0$ the first term should be deleted.

Further we will show that under these conditions, locally D-optimal designs for the regression function (2.3) are also such designs for the regression function (3.1) with the corresponding choice of $\tilde{\Theta}$ and vice versa.

Let ξ be an arbitrary design with m points,

$$\xi = \{x_1, \ldots, x_m; \mu_1, \ldots, \mu_m\}. \tag{3.2}$$

Without loss of generality we will assume that

$$0 \leq x_1 < x_2 < \ldots < x_m \leq d.$$

Consider the regression function (2.3). The following lemma holds.

Lemma 3.1 *Let conditions (a), (b) and (c) be fulfilled. Then the determinant of the information matrix for design (3.2) and regression function (2.3) has the form*

$$\det M(\xi, \Theta) = \prod_{i=1}^{m} \mu_i \left(C \prod_{1 \leq i < j \leq m} (x_j - x_i) / \prod_{i=1}^{m} Q^2(x_i) \right)^2, \tag{3.3}$$

where $Q(x) = Q(x, \Theta_2)$, $C = C(\Theta)$ does not depend on the design and $C(\Theta) \neq 0$.

The proof of this lemma and of the next one is based on elementary operations on rows of the matrix

$$F = \left(\frac{\partial \eta(x_i, \Theta)}{\partial \vartheta_j} \right)_{i,j=1}^{m}$$

and the application of the formula for Vandermonde determinants since $M(\xi, \Theta) = F^T \Lambda F$, where $\Lambda = \text{diag}\{\mu_1, \ldots, \mu_m\}$.

Lemma 3.2 *Let $\gamma_1 > \gamma_2 > \ldots > \gamma_k > 0$, $\tilde{\vartheta}_i \neq 0$, $i = l+1, \ldots, l+k$. Then the determinant of the information matrix for design (3.2) and regression function (3.1) has the form (3.3) with the replacement of Θ by $\tilde{\Theta}$, while $Q(x) = \prod_{j=1}^{k}(x + \gamma_i)$ and $C = C(\tilde{\Theta}) \neq 0$.*

From Lemmas 3.1 and 3.2 and the Binet–Cauchy formula it follows that the problem of finding locally D-optimal design for model (2.3) is equivalent to the similar problem for model (3.1) in the above mentioned sense.

Consider now the problem of the number of points in a locally D-optimal design.

Theorem 3.1 *If conditions (a), (b) and (c) are fulfilled then a locally D-optimal design for the regression function (2.3) at an arbitrary point $\Theta^{(0)} \in \Omega$ exists, is unique and consists of m points with weights equal to $1/m$. If $m > 2k$ then two points of the design coincide with the boundary of $\mathfrak{X}$. If $m = 2k$ and also condition (d) is fulfilled, then the minimal point of the design coincides with 0 and the maximal one is less than d for sufficiently large d.*

The theorem is a slight modification of Theorem 5 in He *et al.* (1996).

4 Definition and properties of optimal design-functions

Let us restrict our attention to the case $m = 2k$ and regression function (3.1). The general case can be considered in a similar way.

Assume that $\tilde{\vartheta}_1, \ldots, \tilde{\vartheta}_k \neq 0$ and fixed, $\gamma_1 > \gamma_2 > \ldots > \gamma_k > 0$. Due to Theorem 3.1 and the equivalence of models (2.3) and (3.1) there exists a unique locally D-optimal design which has the form (3.2) with $\mu_1 = \mu_2 = \ldots = \mu_m = 1/m$ and does not depend on $\tilde{\vartheta}_1, \ldots, \tilde{\vartheta}_k$. Denote the points of this design by $x_1^*(\gamma), \ldots, x_m^*(\gamma)$, where $\gamma = (\gamma_1, \ldots, \gamma_k)$,

$$0 \leq x_1^*(\gamma) < x_2^*(\gamma) < \ldots < x_m^*(\gamma) \leq d.$$

Due to Theorem 3.1 $x_1^*(\gamma) \equiv 0$, $x_m^*(\gamma) < d$, if d is a sufficiently large number. Let

$$\Gamma = \{\gamma; \gamma_1 > \gamma_2 > \ldots > \gamma_k \geq \varepsilon\},$$
$$\bar{\Gamma} = \{\gamma; \gamma_1 \geq \gamma_2 \geq \ldots \geq \gamma_k \geq \varepsilon\},$$

where ε is an arbitrarily small positive number.

Assign to any vector $\tau = (\tau_1, \ldots, \tau_{m-1})$, $0 < \tau_1 < \ldots < \tau_{m-1}$ the design $\xi = \{x_1, \ldots, x_m; \frac{1}{m}, \ldots, \frac{1}{m}\}$, where $x_1 = 0$, $x_{i+1} = \tau_i$, $i = 1, \ldots, m-1$. Introduce the function

$$\varphi(\tau, \gamma) = \left[\prod_{1 \leq i < j \leq m} (x_j - x_i)^2 \Big/ \prod_{i=1}^{m} \prod_{j=1}^{k} (x_i + \gamma_j)^4 \right]^{1/m}.$$

Due to Lemma 3.2 we have for $\gamma \in \Gamma$

$$\varphi(\tau, \gamma) = C \left(\det M(\xi, \tilde{\Theta}) \right)^{1/m},$$

where C does not depend on τ.

Consider the equations

$$\frac{\partial \varphi(\tau, \gamma)}{\partial \tau_i} = 0, \quad i = 1, \ldots, m-1. \tag{4.1}$$

Note that for any fixed $\gamma \in \Gamma$ the unique solution of these equations is the vector

$$\tau = \tau(\gamma) = (\tau_1(\gamma), \ldots, \tau_{m-1}(\gamma)) = (x_2^*(\gamma), \ldots, x_m^*(\gamma)).$$

With the help of the Kiefer–Wolfowitz equivalence theorem (Kiefer and Wolfowitz, 1960) it can easily be verified that the conditions (4.1) are equivalent to local D-optimality of the design ξ corresponding to a vector τ. Due to continuity arguments equation system (4.1) possesses a unique solution for any fixed $\gamma \in \bar{\Gamma}$.

Thus a vector function $\tau(\gamma)$ is uniquely determined by equation system (4.1).

Definition 4.1 *The vector function $\tau(\gamma)$ which for any fixed γ satisfies equation system (4.1) will be called the optimal design function.*

Consider the Jacobi matrix of equations (4.1)

$$J(\tau, \gamma) = \left(\frac{\partial^2}{\partial \tau_i \partial \tau_j} \varphi(\tau, \gamma) \right)_{i,j=1}^{m-1}.$$

Denote $J^* = J^*(\gamma) = J(\tau(\gamma), \gamma)$.

Using the known representation (see Karlin, Studden, 1966, Ch. X)

$$m(\det M)^{1/m} = \inf_A \operatorname{tr} AM/(\det A)^{1/m},$$

where the infinum is taken over all positive definite matrices A, it can be verified (see Melas (2001) for details) that the condition

$$\left(f^T(x) M^{-1}(\xi, \Theta) f(x) \right)''_{x^2} < 0, \quad x = x_i, \quad i = 2, \ldots, m-1,$$

where $f(x) = \left(\frac{\partial \eta(x, \Theta)}{\partial \vartheta_i} \right)_{i=1}^m$ and $\xi = \{x_1, \ldots, x_m; \frac{1}{m}, \ldots, \frac{1}{m}\}$ is a locally D-optimal design is a sufficient condition for nonsingularity of the matrix J^*.

The verification of this condition in our case is straightforward.

Thus the matrix $J^*(\gamma)$ is a nonsingular matrix for any $\gamma \in \bar{\Gamma}$. Now the following theorem is an immediate corollary of the well known Implicit Function Theorem (Gunning and Rossi, 1965).

Theorem 4.1 *Under the above assumptions the optimal design-function is uniquely determined with $\gamma \in \bar{\Gamma}$ and is a real analytical vector function satisfying the differential equations*

$$J(\tau(\gamma), \gamma) \frac{\partial \tau(\gamma)}{\partial \gamma_j} = L_j(\tau(\gamma), \gamma), \quad j = 1, \ldots, k,$$

where

$$L_j(\tau, \gamma) = \left(\frac{\partial^2}{\partial \gamma_j \partial \tau_i} \varphi(\tau, \gamma) \right)_{i=1}^{m-1}.$$

Because of Theorem 4.1 the optimal design function can be expanded in a Taylor series in the vicinity of every point $\gamma \in \bar{\Gamma}$.

In the following section we will present an example of such an expansion basing on recurrence formulae introduced in Melas (2001).

5 Examples

Consider the regression function (3.1) with $m = 2k$. For convenience rewrite this function in the form

$$\eta(x, \Theta) = \sum_{i=1}^{k} \frac{\vartheta_i}{x + \vartheta_{i+k}}.$$

Assume that $\vartheta_i \neq 0$, $\vartheta_{k+i} > 0$, $i = 1, \ldots, k$, $x \in \mathfrak{X} = [0, d]$, where d is sufficiently large.

For $k = 1$ it can be verified by an immediate calculation that the locally optimal design is $\{0, 1/\vartheta_2; 1/2, 1/2\}$.

For $k = 2$ the optimal design-function can be found in an explicit form as well. By some algebra we can verify that

$$x_3^* = \sqrt{\vartheta_3 \vartheta_4}, \quad x_{2,4}^* = \frac{\sqrt{\vartheta_3 \vartheta_4}}{2} \left(-\Delta/2 - 1 \mp \sqrt{(\Delta/2 + 1)^2 - 4} \right),$$

$$\Delta = -(\vartheta_3 + \vartheta_4) - 3 - \sqrt{(\vartheta_3 + \vartheta_4 + 3)^2 + 24}.$$

For $k = 3$ a closed analytical solution seems not to be available. Let us construct the expansion of the optimal design-function in a Taylor series. It can easily be checked that, after multiplying all parameters $\vartheta_{k+1}, \ldots, \vartheta_{2k}$ by one and the same quantity, all design points are multiplied by this quantity. Therefore without loss of generality assume that $(\vartheta_{k+1} + \ldots + \vartheta_{2k})/k = 1$.

Set $u = (1 - \vartheta_4)(1 - \vartheta_5)$, $v = (2 - \vartheta_4 - \vartheta_5)$. From Theorem 4.1 the points $x_{i+1}^*(\vartheta_4, \vartheta_5, \vartheta_6)$, $i = 1, \ldots, 5$ can be given as

$$\tau_i(u, v) = \sum_{s_1=0}^{\infty} \sum_{s_2=0}^{\infty} \tau_{i(s_1, s_2)} u^{s_1} v^{s_2}. \tag{5.1}$$

Several first coefficients of this expansion are represented in Table 1.1.

TABLE 1. Coefficients for the rational model, $k = 3$

$\tau \backslash s$	0,0	0,1	1,0	0,2	1,1	2,0
τ_1	0.0928	0.0449	0	-0.0155	0.0540	-0.0449
τ_2	0.3616	0.1514	0	-0.0481	0.1577	-0.1514
τ_3	1.0000	0.3333	0	-0.0955	0.2864	-0.3333
τ_4	2.7654	0.6861	0	-0.1803	0.4991	-0.6861
τ_5	10.7802	1.9661	0	-0.5087	1.3892	-1.9661

The optimal design function taking account of the first twenty Taylor co-efficients with fixed value $u = 0$ (i.e. $\vartheta_4 = 1$ or $\vartheta_5 = 1$) is presented in Figure 1. This figure allows us to find the locally D-optimal design points for a given γ.

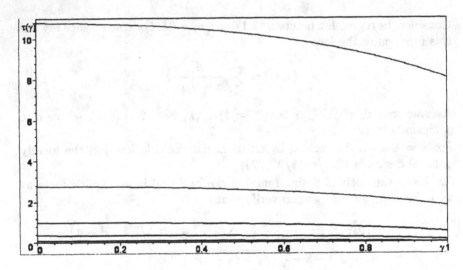

FIGURE 1. Optimal design function.

The number of coefficients to be taken into account for particular values of u and v can be estimated from the known inequality for the ratio of determinants of information matrices of arbitrary and optimal designs.

6 Concluding remarks

The approach developed here enables us to study analytically the dependence of locally *D*-optimal design points on the values of the parameters. The points considered as the functions of the values are proved to be real analytical functions. These functions can be presented by their Taylor series, whose coefficients can be evaluated by recurrence formulae.

Acknowledgments: This work was partly supported by RFBR grant No. 00-01-00495.

References

Dette, H., Haines, L.M., Imhof, L. (1999) Optimal designs for rational models and weighted polynomial regression. *Ann. Statist.*, **27**, 1272–1293.

Gunning R.C., Rossi H. (1965). *Analytical functions of several complex variables.* Prentice-Hall, Inc., NewYork.

He, Z., Studden, W.J., Sun, D. (1996). Optimal designs for rational models. *Ann. Statist.*, **24**, 2128–2147.

Jennrich, R. J. (1969). Asymptotic properties of non-linear least squares estimators. *Ann. Math. Stat.*, **40**, 633–643.

Kiefer, J. and J. Wolfowitz. (1960). The equivalence of two extremum problems. *Canadian Journal of Mathematics*, **14**, p. 363–366.

Melas, V.B. (1978). Optimal designs for exponential regression. *Math. Operationsforsch. Statist.* Ser. Statistics, **9**, 45-59.

Melas, V.B. (2001). Optimal designs for nonlinear models. Springer, Heidelberg (to appear).

Understanding Aliasing Using Gröbner Bases

G. Pistone
E. Riccomagno
H.P. Wynn

ABSTRACT: The now well-established Gröbner basis method in experimental design (see the authors' monograph "Algebraic Statistics") had the understanding of aliasing as a key motivation. The basic method asks: given an experimental design, what is estimable, or more generally what is the alias structure? The paper addresses the following related question: given a set of conditions which the design is known to satisfy, what can we say about the alias structure? Some classical and non-classical construction methods are included.

KEYWORDS: aliasing; Gröbner basis; polynomial conditions

1 The Gröbner basis method

We summarise the method briefly in a number of steps.

Step 1 Define a design $D \subset \mathbf{R}^d$ as a set of n distinct points: $D = \left[a^{(i)}\right]_{i=1}^{n}$.

Step 2 Set up a series of polynomial equations whose solutions give precisely D. This, mathematically, amounts to representing the design as a zero dimensional algebraic variety. The design ideal, Ideal(D), is the set of all polynomials whose zeros include the design points.

Step 3 Select a so-called monomial ordering τ. This is a total well-ordering on the monomials such that $x^\alpha \prec_\tau x^\beta$ implies $x^\alpha x^\gamma \prec_\tau x^\beta x^\gamma$ for all $\gamma \neq 0$.

Step 4 Generate a Gröbner basis for Ideal(D), given τ, namely a special representation of D as the solutions of polynomial equations

$$\{g_j(x) = 0 : j = 1, \ldots, k\}.$$

mODa6, A.C.Atkinson, P.Hackl and W.G.Müller, eds., Physica, Heidelberg, 2001.

The full details are omitted.

Step 5 List the leading terms $l_j(x) = \mathrm{LT}\,(g_j(x))$, $j = 1, \ldots, k$ with respect to the monomial ordering τ.

Step 6 List all monomials not divisible by any leading term $l_j(x)$ ($j = 1, \ldots, k$). Call this list $\mathrm{Est}_\tau(D)$ and note that

(i) $\#\mathrm{Est}_\tau(D) = n$, that is the sample size of the design.

(ii) If $\mathrm{Est}_\tau(D) = \{x^\alpha : \alpha \in L\}$ then the monomial terms are the basis of a saturated and estimable regression model

$$\sum_{\alpha \in L} \vartheta_\alpha x^\alpha$$

with non singular X-matrix

$$X = \{x^\alpha\}_{x \in D, \alpha \in L}\,.$$

(iii) $\mathrm{Est}_\tau(D)$ is an order ideal, that is if $x^\alpha \in \mathrm{Est}_\tau(D)$ then $x^\beta \in \mathrm{Est}_\tau(D)$ for any β component wise smaller than α.

A lot can be said about this process with regard to appropriate computer algebra. For example methods are available for directly computing the Gröbner basis and $\mathrm{Est}_\tau(D)$ from D. See Pistone *et al.* (2000) for details. For the present paper we note simply that the equations $\{g_j(x) = 0 : j = 1, \ldots, k\}$ defining the design essentially also give some alias structure. For example each leading term can be written

$$l_j(x) = \sum_{\alpha \in L} \vartheta_\alpha^{(j)} x^\alpha$$

so that $g_j(x) = l_j(x) - \sum_{\alpha \in L} \vartheta_\alpha^{(j)} x^\alpha$. That is to say "higher order" terms with respect to τ can be written in terms of polynomials constructed from monomials in $\mathrm{Est}_\tau(D)$. Apart from the fact that the equations are dependent on τ, all the alias structure can be captured from such equations. A generic member of the ideal $\mathrm{Ideal}(D)$ is written $\sum_{j=1}^{k} s_j(x)g_j(x)$ where the $s_j(x)$'s are generic polynomials. Setting this to zero for arbitrary $s_j(x)$ gives all possible alias relations.

2   A theorem on aliasing

Historically there have been many combinatorial constructions of experimental design of which the standard Abelian group construction of symmetric and asymmetric factorial design is perhaps the most celebrated. In such constructions one exhibits a set of conditions which the designs must satisfy. For example to construct a 2^{3-1} the equations are

$$x_1^2 = 1, \qquad x_2^2 = 1, \qquad x_3^2 = 1, \qquad x_1 x_2 x_3 = 1.$$

In the previous section this could be considered as Step 2.

In this section we discuss some simple properties that can be predicted for $\text{Est}_\tau(D)$ given the construction equations but in advance of computing $\text{Est}_\tau(D)$ itself.

The equation $x_1 x_2 x_3 = 1$ for the 2^{3-1} above implies that the interaction $x_1 x_2 x_3$ and the constant term are aliased, in particular the vector obtained by evaluating $x_1 x_2 x_3$ at the design points is the unit vector, $(1, \ldots, 1)$. Theorem 2.1 generalises this observation.

Theorem 2.1 *Let the design D be known to satisfy the polynomial equation*

$$h(x) = 0$$

in $\mathbf{R}^d$ and let τ be a monomial ordering. Let $M(\neq \varnothing)$ be the set of monomials with non zero coefficients in h. Then

$$M \not\subseteq \text{Est}_\tau(D).$$

Proof. This is by contradiction. Suppose $M \subseteq \text{Est}_\tau(D)$. Then

$$h(x) = \sum_{\alpha \in M \subseteq L} \varphi_\alpha x^\alpha$$

where $\text{Est}_\tau(D) = \{x^\alpha : \alpha \in L\}$ and all $\varphi_\alpha \neq 0$ for α such that $x^\alpha \in M$. But this is false since all x^α are linearly independent over D and $h(x) = 0$ for all $x \in D$.

A proof relying on more classical arguments from matrix theory is as follows. Let $h(x) = \sum_{\alpha \in M} \varphi_\alpha x^\alpha$ and consider the matrix $X = \{x^\alpha\}_{x \in D, \alpha \in M}$ with columns $X(\alpha) = \{x^\alpha\}_{x \in D}$. The matrix X is singular because the condition $h(x) = 0$ implies that the linear combination of the columns of X, $\{\sum_{\alpha \in M} \varphi_\alpha X(\alpha)\}$ is the zero vector. Thus the monomials x^α, $\alpha \in M$ are linearly dependent over D and cannot all be included in a model identifiable by D. □

To repeat the result of the theorem: any M must have at least one "non zero" term not in $\text{Est}_\tau(D)$. Note also that if $x^\beta \notin \text{Est}_\tau(D)$ it also follows from Gröbner basis theory that $x^\gamma \notin \text{Est}_\tau(D)$ for all $\gamma > \beta$ component wise.

Corollary 2.1 *If $h(x) = x^\alpha - c$ for some index α and constant c then if $D \neq \varnothing$, x^α cannot be in $\text{Est}_\tau(D)$.*

Proof. Since D is not empty the constant must be in $\mathrm{Est}_\tau(D)$. This follows from Step 6 (iii). Thus by Theorem 2.1, $x^\alpha \notin \mathrm{Est}_\tau(D)$. □

Typically in construction we may know that $h_j(x) = 0$ on D for $j = 1, \ldots, r$. Then Theorem 2.1 applies to each corresponding M_j.

Sometimes we may construct designs as exactly all solutions of $h_j(x) = 0$, $j = 1, \ldots, r$

$$D = \{x : h_j(x) = 0, \qquad j = 1, \ldots, r\}.$$

However it is advisable to replace this by the Gröbner basis representation to obtain a more tractable description of aliasing.

Example 1 Factorial design. The corollary makes a strong connection to the fractional factorial construction mentioned above since those consist typically of solving sets of equations of the form

$$\{x^{\alpha(j)} - c_j : j = 1, \ldots, r\}.$$

3   Further examples

Example 2 Mixture. Here a basic equation is $\sum x_i - 1 = 0$. Theorem 2.1 simply says that not all of 1 and x_i, $i = 1, \ldots, d$ can be in $\mathrm{Est}_\tau(D)$ confirming the standard redundancy in this case. See Giglio *et al.* (2000).

Example 3 Other groups. Any design D invariant under a group G on $\mathbf{R}^d$ will preserve the maximal invariants, $\pi_j(x)$, under G. Thus candidates for $h_j(x)$ are

$$h_j(x) = \pi_j(x) - c_j.$$

As a very simple example consider designs on a circle in $\mathbf{R}^2$ satisfying

$$x_1^2 + x_2^2 = 1.$$

Then we can conclude that both x_1^2 and x_2^2 cannot be in $\mathrm{Est}_\tau(D)$. Since maximal invariants are constant on orbits, any design constructed as an orbit will be invariant

$$D = \{x = G(x_0) : \text{ for a point } x_0 \in D\}.$$

In the above example one can easily construct arbitrary large designs in this way and still not have x_1^2 and x_2^2 in $\mathrm{Est}_\tau(D)$. This can easily be extended to rotations in $\mathbf{R}^d$.

An important class of groups in design theory are reflection groups. Indeed the conditions above $x_1^2 = x_2^2 = x_3^2 = 1$ and $x_1 x_2 x_3 = 1$ are precisely a set of invariants for the subgroup generated by the reflections

$$(x_1, x_2, x_3) \longrightarrow \begin{cases} (-x_1, -x_2, x_3) \\ (-x_1, x_2, -x_3) \end{cases}$$

and the design $D = \{(1,1,1), (1,-1,-1), (-1,1,-1), (-1,-1,1)\}$ is an orbit.

Example 4 Lattices. One generator lattice designs are equally spaced designs on the integer grid defined as

$$D = \{gk \pmod n : k = 0, \ldots, n-1\}$$

where g is a vector of integers. If the components of g and n have greatest common divisor equal to one, then D has exactly n distinct points. For $g = (1,2)$ and $n = 5$, $D = \{(0,0), (1,2), (2,4), (3,1), (4,3)\}$. The Gröbner basis computed modulo 5 and with respect to any term-ordering for which x_2 is smaller than x_1, includes the polynomial $x_1 + 2x_2$. Thus every point in D has to satisfy the equation $x_1 + 2x_2 = 0 \pmod 5$. The full Gröbner basis is

$$x_1 + 2x_2,$$
$$x_2^5 - x_2$$

and $\text{Est}_\tau(D)$ is $\{1, x_2, x_2^2, x_2^3, x_2^4\}$. This shows algebraically that modulo 5 the design D is a one dimensional object.

Over the real numbers and with respect to a lexicographic term ordering (see Cox *et al.*, 1996) with again x_2 smaller than x_1 the Gröbner basis is

$$g_1(x) = x_2^5 - 10x_2^4 + 35x_2^3 - 50x_2^2 + 24x_2,$$
$$g_2(x) = x_1 + 5/6x_2^4 - 20/3x_2^3 + 50/3x_2^2 - 83/6x_2$$

with the same $\text{Est}_\tau(D)$. The condition $\pi(x) = x_1 + 2x_2$ is rewritten over $\text{Est}_\tau(D)$ as

$$-5/6x_2^4 + 20/3x_2^3 - 50/3x_2^2 + 95/6x_2 = \pi(x) - g_2(x).$$

With respect to an ordering that does not favour either x_1 or x_2 so strongly, namely tdeg (see Char *et al.*, 1991) the set $\text{Est}_\tau(D)$ is $\{1, x_1, x_2, x_1x_2, x_2^2\}$. This example shows that term orderings can be chosen to determine the structure of $\text{Est}_\tau(D)$ as far as the design allows.

4 Conclusion

The theory and examples in this paper are relatively simple but, we hope, show the power of the method. The challenge is to revisit many of the classical and some of the more recent constructions in design, such as lattices, to relate the special algebra used in each case to the wider Gröbner basis theory. The list should include notions such as blocking, dummying, trend resistance and cross-over which are of considerable practical importance, but where aliasing is not yet fully understood.

References

Char, B., Geddes, K., Gonnet, G., Leong, B., and Monogan, M. (1991). *MAPLE V Library Reference Manual.* Springer-Verlag, New York.

Cox, D., Little, J., and O'Shea, D. (1996). *Ideals, Varieties, and Algorithms*, 2nd edition. Springer, New York.

Giglio, B., Riccomagno E., and Wynn, H.P. (2000). Gröbner basis methods in mixture experiments and generalisations. In: Atkinson, A., Bogacka, B., and Zhigljavsky, A. (eds.), *Optimum Design 2000. Proceedings of "Optimum experimental designs: Prospects for the new millennium", Kluwer, Dordrecht.*

Pistone, G., Riccomagno E., and Wynn, H.P. (2000). *Algebraic Statistics.* Chapman and Hall, New York.

Average D-optimum Design for Randomly Varying Experimental Conditions

L. Pronzato

ABSTRACT: We consider a regression model $\eta(\vartheta, \xi)$ for which the experimental conditions ξ_k used at the experimentation stage may randomly fluctuate around the values ζ_k specified at the design stage. The problem consists in choosing N values $\zeta_1, \ldots, \zeta_N$ to estimate the parameters ϑ from observations y_k, $k = 1, \ldots, N$, with $y_k = \eta(\vartheta, \xi_k) + \varepsilon_k$ and $\{\varepsilon_k\}$ an i.i.d. sequence. We assume that the ξ_k's are observed, so that the accuracy of the estimation can be characterized by the usual information matrix. We design the experiment by maximizing the expected value of its determinant. We give the expression of this expected determinant in the case where the ξ_k's are stochastically independent. Deterministic exchange algorithms can then be used to determine the optimal choice of the ζ_k's. Two examples are presented.

KEYWORDS: D-Optimum design; expected determinant; random experiments; random matrix

1 Introduction

We consider experimental design for parameter estimation in a regression model, in the situation where the experimental conditions used at the observation stage may differ from those specified at the design stage. The observations y_k, $k = 1, \ldots, N$, are given by

$$y_k = \eta(\bar{\vartheta}, \xi_k) + \varepsilon_k \tag{1.1}$$

with $\eta(\vartheta, \xi)$ the model response for parameters ϑ and experimental conditions ξ, $\bar{\vartheta} \in \mathbb{R}^p$ the unknown true value of the model parameters and $\{\varepsilon_k\}$ a sequence of i.i.d. measurement errors with zero mean and variance σ^2. The parameters are estimated by least-squares.

mODa6, A.C.Atkinson, P.Hackl and W.G.Müller, eds., Physica, Heidelberg, 2001.

If the experimental conditions could be set without error, a classical approach for designing the experiment would be to maximise the determinant of the information matrix (D-optimal design)

$$\mathbf{M}_N = \mathbf{M}(\vartheta, \Xi_1^N) = \sum_{k=1}^N \mathbf{z}_k \mathbf{z}_k^\top,$$

where

$$\mathbf{z}_k = \mathbf{z}(\vartheta, \xi_k) = \frac{\partial \eta(\vartheta, \xi_k)}{\partial \vartheta}.$$

When $\eta(\vartheta, \xi)$ is nonlinear in ϑ, $\mathbf{M}_N$ depends on ϑ. Prior uncertainty on the value of ϑ can be taken into account e.g. by considering the expected value of $\det[\mathbf{M}(\vartheta, \Xi_1^N)]$, or $\log \det[\mathbf{M}(\vartheta, \Xi_1^N)]$, with respect to ϑ for a given prior distribution, see Fedorov (1980) and Pronzato and Walter (1985). However, only local design, based on the evaluation of $\mathbf{M}(\vartheta, \Xi_1^N)$ at a given nominal value of ϑ, will be considered in this paper.

We assume that, at the design stage, the ξ_k's are stochastically independent random variables, the distribution of ξ_k being dependent on a design variable ζ_k. For instance, ξ_k may be normally distributed $\mathcal{N}(\zeta_k, \rho^2(\zeta_k))$, or uniformly distributed in $[\zeta_k - u(\zeta_k), \zeta_k + u(\zeta_k)]$. We also assume that ξ_k is known when y_k is observed, so that the precision of the estimation can still be characterized by $\mathbf{M}_N$. A natural extension of D-optimal design for choosing $\zeta_1, \ldots, \zeta_N$ is then to maximise

$$ED(\zeta_1, \ldots, \zeta_N) = E_{\mathbf{z}_1^N}\{\det[\mathbf{M}_N]\},$$

with $\mathbf{z}_1^N = (\mathbf{z}_1, \ldots, \mathbf{z}_N)$.

Note that the situation would be quite different if the observations were taken with uncertain experimental conditions: randomness in the controllable variables ξ_k should then be taken into account at the estimation stage. For instance, Fedorov (1974) assumes that $E\{\xi_k\} = 0$, $\mathrm{var}\{\xi_k\} = \rho^2 \mathbf{D}(\zeta_k)$ and estimates ϑ by a reweighted least-squares procedure, based on the expansion of the mean and variance of y_k in powers of ρ; Keeler and Reilly (1992) use an error-in-variable approach, and estimate ϑ by maximum likelihood, with ξ_k assumed normally distributed.

The analytic expression of $ED(\zeta_1, \ldots, \zeta_N)$ is given in Section 2. It makes experimental design based on this criterion easier than with criteria such as $E_{\mathbf{z}_1^N}\{\log \det[\mathbf{M}_N]\}$ or $1/E_{\mathbf{z}_1^N}\{\det[\mathbf{M}_N^{-1}]\}$ (also note that the latter is undefined when the probability density function of the random variable $D = \det[\mathbf{M}_N]$ is bounded away from zero at $D = 0$, and both are undefined when the probability of the event $\{\det[\mathbf{M}_N] = 0\}$ is positive). In particular, design algorithms of the exchange type, see Fedorov (1972), Mitchell (1974), can be used for the maximisation of $ED(\zeta_1, \ldots, \zeta_N)$. Two examples are presented in Section 3. They illustrate the importance of taking the variability of the ξ_k's into account at the design stage when this variability is large.

2 Expected determinants

We first give an extension of a result given in Pronzato (1999). The proof is by induction on N and is omitted (it simply relies on repeated use of the so-called Bartlett-Sherman-Morrison-Woodbury formula).

Theorem 2.1 *For any regular matrix* $\mathbf{Q} \in \mathbb{R}^{p \times p}$,

$$\det[\mathbf{Q} + \sum_{l=1}^{N} \mathbf{z}_l \mathbf{z}_l^T] = \det[\mathbf{Q}] \times \sum_{l=0}^{\min(N,p)} \sum_{S_l} Q_l(S_l)$$

with $Q_0 = 1$ *and*

$$Q_l(S_l) = \sum_{n_1,\dots,n_l} (-1)^{l + \sum_{i=1}^{l} n_i} \sum_{\sigma_{1^{n_1}\dots l^{n_l}}(S_l)} \tau(1^{n_1}) \cdots \tau(l^{n_l}), \quad l \geq 1, \quad (2.1)$$

where: S_l *is one of the* $N!/[(N-l)!l!]$ *subsets of* $\{1, \dots, N\}$ *with* l *elements,* $\sigma_{1^{n_1}\dots l^{n_l}}(S_l)$ *denotes one of the* $l!/[(1^{n_1} \cdots l^{n_l})(n_1! \cdots n_l!)]$ *partitions of a given type of* S_l, *with* n_i *blocks of size* i *and* $(i-1)!$ *cyclic permutations of each block,* $\tau(i^{n_i}) = t(j_{1_1}, \dots, j_{1_i}) \times \cdots \times t(j_{n_{i1}}, \dots, j_{n_{ii}})$, *the* j_i*'s being elements of the blocks of size* i, *and* $t(i_1, \dots, i_n) = \text{trace}[\mathbf{Q}^{-1} \mathbf{z}_{i_1} \mathbf{z}_{i_1}^T \times \cdots \times \mathbf{Q}^{-1} \mathbf{z}_{i_n} \mathbf{z}_{i_n}^T]$. *The summation* $\sum_{n_1,\dots,n_l}$ *in (2.1) is over all non-negative integers* $n_1, \dots, n_l$ *such that* $\sum_{i=1}^{l} i n_i = l$, *that is, over all partitions of* l *which have* n_i *parts of size* i.

For instance, the expressions of Q_l for $l = 1, \dots, 4$ are as follows:

$$Q_0 = 1, \quad Q_1(\{1\}) = t(1), \quad Q_2(\{1,2\}) = t(1)t(2) - t(1,2),$$

$$Q_3(\{1,2,3\}) = t(1)t(2)t(3) + t(1,2,3) + t(2,1,3)$$
$$-t(1)t(2,3) - t(2)t(1,3) - t(3)t(1,2),$$

$$Q_4(\{1,2,3,4\}) = t(1)t(2)t(3)t(4) - t(1)t(2)t(3,4) - t(1)t(3)t(2,4)$$
$$-t(1)t(4)t(2,3) - t(2)t(3)t(1,4) - t(2)t(4)t(1,3) - t(3)t(4)t(1,2)$$
$$+t(1)t(2,3,4) + t(1)t(3,2,4) + t(2)t(1,3,4) + t(2)t(3,1,4) + t(3)t(1,2,4)$$
$$+t(3)t(2,1,4) + t(4)t(1,2,3) + t(4)t(2,1,3) + t(1,2)t(3,4) + t(1,3)t(2,4)$$
$$+t(1,4)t(2,3) - t(1,2,3,4) - t(2,1,3,4) - t(1,3,2,4)$$
$$-t(3,1,2,4) - t(1,4,2,3) - t(1,4,3,2).$$

By taking $\mathbf{Q} = \alpha\mathbf{I}$ and letting α tend to zero, we get the following corollary.

Corollary 2.1 *For any* $N \geq p = \dim(\mathbf{z}_l)$,

$$\det[\sum_{l=1}^{N} \mathbf{z}_l \mathbf{z}_l^T] = \sum_{S_p} Q_p(S_p),$$

where S_p and $Q_p(S_p)$ are defined as in Theorem 2.1, with

$$t(i_1,\ldots,i_n) = \mathrm{trace}[\mathbf{z}_{i_1}\mathbf{z}_{i_1}^\top \times \cdots \times \mathbf{z}_{i_n}\mathbf{z}_{i_n}^\top].$$

Note that Corollary 2.1 tells us more than the well-known Binet-Cauchy Lemma (see Gantmacher (1966), vol. 1, p. 9), which simply gives in this case

$$\det[\sum_{l=1}^{N} \mathbf{z}_l\mathbf{z}_l^\top] = \sum_{S_p}\{\det[\sum_{j \in S_p} \mathbf{z}_j\mathbf{z}_j^\top]\}^2.$$

From the result above one directly obtains the expression of $ED(\zeta_1,\ldots,\zeta_N)$ in the case where the regressors $\mathbf{z}_l$ are stochastically independent, which forms an extension of the result in Pronzato (1998).

Corollary 2.2 *Assume that the vectors $\mathbf{z}_l \in \mathcal{Z} \subset \mathbb{R}^p$, $l = 1,\ldots,N$, are stochastically independent and distributed with probability measures μ_l such that $\bar{\mathbf{M}}_l = \int \mathbf{z}_l\mathbf{z}_l^\top \mu_l(d\mathbf{z}_l)$ exists for $l = 1,\ldots,N$; then for $N \geq p$,*

$$ED(\zeta_1,\ldots,\zeta_N) = E_{\mathbf{z}_1^N}\{\det[\sum_{l=1}^{N}\mathbf{z}_l\mathbf{z}_l^\top]\} = \sum_{S_p} Q_p(S_p), \qquad (2.2)$$

where S_p and $Q_p(S_p)$ are defined as in Theorem 2.1, with

$$t(i_1,\ldots,i_n) = \mathrm{trace}[\bar{\mathbf{M}}_{i_1} \times \cdots \times \bar{\mathbf{M}}_{i_n}].$$

One may notice that a direct application of (2.2), or of the Binet-Cauchy Lemma, shows that the usual property of multiplication of determinants for replicated experiments does not hold for expected determinants. Indeed, consider the experiments Ξ, defined by $\mathbf{z}_1^N = (\mathbf{z}_1,\ldots,\mathbf{z}_N)$, and Ξ', defined by $\mathbf{z}'_{l_i}$, $l = 1,\ldots,N$, $i = 1,\ldots,q$ (all $\mathbf{z}_i$, $\mathbf{z}'_j$ are stochastically independent, $\mathbf{z}_l$ and $\mathbf{z}'_{l_i}$ are distributed with the same measure μ_l). Then,

$$E_{\Xi'}\{\det[\sum_{i=1}^{q}\sum_{l=1}^{N}\mathbf{z}'_{l_i}\mathbf{z}'_{l_i}^\top]\} \geq q^p E_{\Xi}\{\det[\sum_{l=1}^{N}\mathbf{z}_l\mathbf{z}_l^\top]\}.$$

The reason is that replications yield sets of indices S_p in (2.2) that correspond to vectors $\mathbf{z}'_{l_i}$, $\mathbf{z}'_{l_j}$ with the same distribution, which, in general, give a strictly positive contribution to the value of the expected determinant. Moreover, numerical examples show that duplicating one of the p observations in an optimal experiment with p points may be optimal or not depending which one is duplicated, whereas all possible duplications give the same result for D-optimality (Atkinson and Hunter, 1968). One can also prove (by induction on N), that, in the situation of Corollary 2.2, $ED(\zeta_1,\ldots,\zeta_N) \leq \det[E_{\mathbf{z}_1^N}\{\sum_{l=1}^{N}\mathbf{z}_l\mathbf{z}_l^\top\}] = \det[\sum_{l=1}^{N}\bar{\mathbf{M}}_l]$. This is illustrated in the following Example.

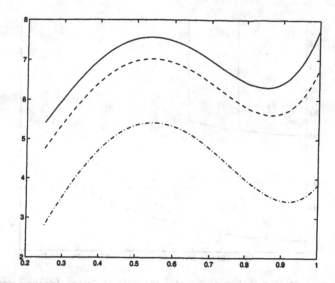

FIGURE 1. $\det[E\{M_{12}\}] = 4^3 \times \det[E\{M_3\}]$ (full line), $E\{\det[M_{12}]\}$ (dashed line) and $4^3 \times E\{\det[M_3]\}$ (dash-dotted line) as functions of ζ in Example 1.

3 Examples

Example 1. Consider a linear regression model, with quadratic effect:

$$y_i = z_i^\top \bar{\vartheta} + \varepsilon_i,$$

where $\{\varepsilon_i\}$ is an i.i.d. sequence of measurement errors, with zero mean and variance σ^2, and $z_i = [1, \zeta_i + \omega_i, (\zeta_i + \omega_i)^2]^\top$ with ω_i a disturbance corrupting the design variable ζ_i. The sequences $\{\varepsilon_i\}$ and $\{\omega_i\}$ are independent. The D-optimal experiment for $N = 12$ observations in $[0, 1]$ corresponds to four replications of the design points $0, 1/2$ and 1, which gives $\det[M_{12}^*] = 4$. Assume that ω_i is normally distributed $\mathcal{N}(0, \rho^2(\zeta_i))$, with $\rho^2(\zeta) = 0.03\zeta^2$. Figure 1 presents $\det[E\{M_{12}\}]$ (full line) and $E\{\det[M_{12}]\}$ (dashed line), as functions of ζ, when the three design points are $0, 1$ and ζ (the similarity to the usual plot of derivative of a design criterion is coincidental). The maximum of $\det[E\{M_{12}\}]$ is obtained for $\zeta = 1$, whereas the maximum of $E\{\det[M_{12}]\}$ is obtained at $\zeta \simeq 0.5400$. Figure 1 also gives $4^3 \times E\{\det[M_3]\}$, obtained without replications of the design points $0, 1$ and ζ (dash-dotted line). The maximum of this curve is obtained at $\zeta \simeq 0.5442$. Note that, as indicated in the previous section, $4^3 \times \det[E\{M_3\}] = \det[E\{M_{12}\}] > E\{\det[M_{12}]\} > 4^3 \times E\{\det[M_3]\}$.

Consider now the case where ω_i is uniformly distributed in $[-\alpha\zeta_i, \alpha\zeta_i]$, $0 \le \alpha \le 1$. Figure 2 presents $E\{\det[M_6]\}/10$ for the D-optimal design consisting of two replications of the design points $0, 1/2$ and 1 (dash-dotted line) and for the optimum design (dashed line). The support points

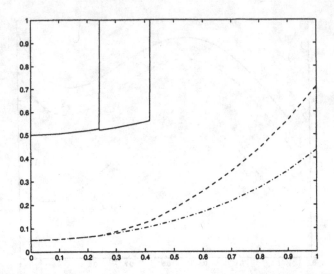

FIGURE 2. $E\{\det[M_6]\}$ ($\times 0.1$) for the D optimal design (dashed-dotted line) and the optimal design (dashed line) as functions of α in Example 1. The support points of the optimum design are in full line.

$\zeta_1, \ldots, \zeta_6$ of this optimum design are also plotted on Figure 2 as functions of α (full line): $\zeta_1 = \zeta_2 = 0$, $\zeta_5 = \zeta_6 = 1$ for any $\alpha \in [0,1]$, $\zeta_3 = \zeta_4$ for $\alpha \le 0.243$ and $\zeta_4 = 1$ for $\alpha > 0.243$, $\zeta_3 = 1$ for $\alpha > 0.417$.

Example 2. We consider the following compartmental model, taken from D'Argenio (1990), describing the plasma concentration of an orally administrated drug:

$$\eta(\vartheta, \xi) = \frac{DK_a}{V(K_a - K_e)}[\exp(-K_e\xi) - \exp(-K_a\xi)],$$

where D is the bolus dose of the drug, V represents the distribution volume of the sampled (plasma) compartment, K_a is the absorption rate from the input to the sampled compartment and K_e is the elimination rate from the sampled compartment. The design variable ξ corresponds to the sampling time for plasma samples. The dose D is known, and the parameters to be estimated are $\vartheta = (V, K_a, K_e)^\mathsf{T}$. The experiment is designed for the nominal value $V = 15(l)$, $K_a = 2.0(h^{-1})$ and $K_e = 0.25(h^{-1})$. The (local) D-optimal experiment with $N = n \times p$ then consists of n replications at each of the design points 0.43, 1.80 and 6.16.

We assume that plasma samples can be collected at time $\xi = \zeta$ when $\zeta \in [0, T)$ but the sampling time ξ is uniformly distributed in $[T, 10]$ if $\zeta \ge T$. Figure 3 presents $E\{\det[M_3]\}$ as a function of T for the D-optimal design (dashed line) and for the design that maximises $E\{\det[M_3]\}$ for each value of T (full line). The first two sampling times of this optimal design

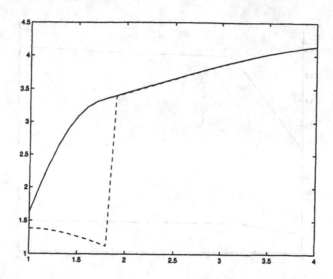

FIGURE 3. $E\{\det[M_3]\}$ ($\times 10^{12}$) for the D optimal design (dashed line) and the optimal design (full line) as functions of T in Example 2.

are presented in Figure 4 (the third one is random in $[T, 10]$).
The curve in dashed line in Figure 3 is discontinuous at $T = 1.8$, one of the sampling times of the D-optimal design, since for $T > 1.8$ the second sample is collected precisely at the D-optimal time. The initial decrease of the curve is due to the importance of taking the second sample early enough, and the probability of this event decreases when T increases. For the curve in full line, the second sampling time is always smaller than T, close to T for $1 \leq T \leq 1.8$, and then close to the value 1.8 of the second D-optimal sampling time, see Figure 4.

References

Atkinson, A.C. and Hunter, W.G. (1968). The design of experiments for parameter estimation. *Technometrics*, **10**, 271-289.

D'Argenio, D.Z. (1990). Incorporating prior parameter uncertainty in the design of sampling schedules for pharmacokinetic parameter estimation experiments. *Mathematical Biosciences*, **99**, 105-118.

Fedorov, V.V. (1972). *Theory of Optimal Experiments*. Academic Press, New York.

Fedorov, V.V. (1974). Regression problems with controllable variables subject to error. *Biometrika*, **61**(1), 49-56.

Fedorov, V.V. (1980). Convex design theory. *Math. Operationsforsch. & Statist.*, Ser. Statistics, **11**, 403-413.

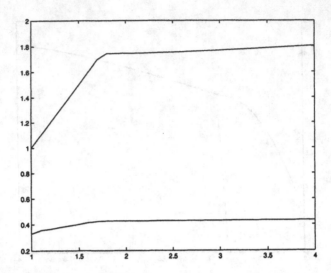

FIGURE 4. First two sampling times of the optimal design (full line) as functions of T in Example 2.

Gantmacher, F.R. (1966). *Théorie des Matrices*. Dunod, Paris.

Keeler, S. and Reilly, P. (1992). The design of experiments when there are errors in all the variables. *Can. J. Chem. Eng.*, **70**, 774-778.

Mitchell, T. (1974). An algorithm for the construction of "D-optimal" experimental designs. *Technometrics*, **16**, 203-210.

Pronzato, L. (1998). On a property of the expected value of a determinant. *Statistics and Probability Letters*, **39**, 161-165.

Pronzato, L. (1999). Sequential selection of observations in randomly generated experiments. *Tatra Mountains Math. Pub.*, **17**, 167-175.

Pronzato, L. and Walter, E. (1985). Robust experiment design via stochastic approximation. *Mathematical Biosciences*, **75**, 103-120.

Constrained Bayesian Optimal Designs for Phase I Clinical Trials: Continuous Dose Space

W.F. Rosenberger
L.M. Haines
I. Perevozskaya

ABSTRACT: We derive constrained Bayesian *D*-optimal designs for efficient estimation in phase I clinical trials with binary responses on a continuous dose space. The constraint is based on ethical considerations that patients cannot be assigned to highly toxic doses. We find that a naive restriction on the dose space, requiring patients to be assigned below the mean of a quantile, is about as effective as the more computationally intensive constrained designs.

KEYWORDS: Bayesian optimal design; constrained optimal design; ethics; sequential design

1 Motivation

Phase I clinical trials are typically very small, uncontrolled sequential studies of human subjects designed to determine the maximum tolerated dose (MTD) of an experimental drug. While performed in many areas of medicine, phase I clinical trials are particularly important in cancer, because of the severity of side effects of cytotoxic drugs to treat cancers. Considering the severity of adverse events associated with cytotoxic agents and the impact that such an episode may have on a volunteer's quality of life, ethical considerations would dictate that such trials be efficient in terms of obtaining maximum information with few patients, and every effort should be made to protect volunteers from exposure to highly toxic dose levels.

Schacter, *et al.* (1997) define the MTD as that dose which, if exceeded,

---

mODa6, A.C.Atkinson, P.Hackl and W.G.Müller, eds., Physica, Heidelberg, 2001.

would put patients at unacceptable risk for toxicity. Such a definition is vague from a statistician's point of view, because "unacceptable risk" is not defined quantitatively. If risk can be observed from patient data, then the MTD can simply be identified from the data and no further statistical considerations are warranted. On the other hand, if risk is defined as the probability of toxicity, and unacceptable risk as a specified probability, then the MTD is an unknown parameter corresponding to that specified probability and must be estimated. These vagaries have led to different philosophies in designing phase I clinical trials. Those who believe that the MTD can simply be identified from patient data favor a design approach that has been called *a conventional method* or *the standard method* (see Simon *et al.*, 1997). Those who believe that the MTD is a parameter to be estimated have treated it as a quantile of a monotonic dose-response curve. Phase I clinical trial designs to estimate a target quantile include the *continual reassessment method* (O'Quigley *et al.*, 1990), *escalation with overdose control* (Babb *et al.*, 1998), and *random walk rules* (Durham and Flournoy, 1994). Although these designs are used in practice, they are not based on formal optimality criteria. In this paper, we treat the MTD as a quantile of a parametric dose-response curve, and use Bayesian constrained optimal design theory to derive an efficient design for estimating the quantile. This is the first time, to our knowledge, that constrained Bayesian optimal design theory has been applied to this important biomedical problem.

2 Constrained Bayesian D-optimal designs

Consider a sample space of design points Ξ and let μ be a quantile of interest. Two problems can be considered: (1) the design points have not been specified, and they must be determined; i.e., Ξ is a compact subset of $\Re$ [e.g., log(dose)], and (2) the design points have been preselected; i.e., $\Xi = \{d_1, ..., d_K\}$ [e.g., a set of doses is predetermined based on animal studies or other pharmacological considerations, and we assume that μ is properly contained in Ξ]. In this paper, we deal with scenario (1). Consider a design measure ξ on Ξ, consisting of design points $x_1, ..., x_m$ and weights $w_1, ..., w_m, 0 \le w_i \le 1$, such that $\sum_{i=1}^{m} w_i = 1$. Since we have a finite number of patients, it may not be possible to obtain an exact design, and we will have to resort to integer approximation.

Define $Y_1, ..., Y_n$ to be binary indicators of toxicity ($Y_j = 1$ if toxicity; 0 if no toxicity) for patients $j = 1, ..., n$. The common assumed model is $p_i = \Pr\{Y_j = 1 | x_i\}, i = 1, ..., m$. A rich class of parametric models, the location-scale family, is often used to characterize p_i as follows:

$$p_i = F\left(\frac{x_i - \alpha}{\beta}\right),$$

where F is a distribution function [e.g., logistic] and α and β are unknown parameters, $\beta > 0$. Let $\Gamma \in [0, 1]$ be a specified probability of toxicity. Then the quantile μ is given by

$$\mu = \alpha + \beta F^{-1}(\Gamma).$$

Let $\vartheta = (\alpha, \beta)$ be contained in the space Θ, let f be the density of F, and $z_i = (x_i - \alpha)/\beta$. The standardized observed Fisher information matrix is given by

$$I(\xi, \vartheta) = \sum_{i=1}^{m} w_i I(z_i, \vartheta),$$

where

$$I(z_i, \vartheta) = \frac{1}{\beta^2} \frac{[f(z_i)]^2}{F(z_i)[1 - F(z_i)]} \begin{bmatrix} 1 & z_i \\ z_i & z_i^2 \end{bmatrix}.$$

Our optimality criterion will be to minimize logarithm of the determinant of the inverse of the information matrix. This approach will lead to a design that depends on the unknown parameter ϑ and can only be implemented at a best guess of ϑ. Such designs are called *locally optimal*. Alternatively, we can specify a prior density $g(\vartheta)$ for $\vartheta \in \Theta$. The Bayesian D-optimal design would be the design ξ_D^* that minimizes

$$\Psi_D(\xi) = \int_\Theta \log |I(\xi, \vartheta)^{-1}| g(\vartheta) d\vartheta.$$

This design will be independent of ϑ and generally have more support points than the locally optimal design, which can be thought of as a special case of the Bayesian optimal design with a point prior (Chaloner and Larntz, 1989).

We now introduce a constraint to effectively protect patients from being assigned highly toxic doses while allowing efficient estimation of the model parameters. In this setting, the design points are now dose levels of a drug, and μ is the MTD. For locally optimal designs in the binary case, Mats *et al.* (1998) found the D-optimal design over a restricted design space. Let μ_R be a quantile corresponding to a probability of toxicity Γ_R. The constraint they chose is that $x_i \le \mu_R, i = 1, ..., m$. This does not have a natural extension to Bayesian optimality, because μ_R is then no longer a constant, but has a prior probability distribution. A naive extension would be to establish sets such as $\Xi_R = \{\xi : x_i \le E(\mu_R)\}$, or other sets involving the mode, median, minimum, or maximum of the distribution of μ_R. Overall the corresponding Bayesian D-optimal design can be formulated as

$$\xi_{D(R)}^* = \arg \min_{\xi \in \Xi_R} \Psi_D(\xi),$$

where Ξ_R represents the class of designs defined on a chosen restricted design space.

A more sophisticated approach to the problem of accommodating reasonable dose levels in a clinical trial is to introduce the weighted sum of probabilities

$$\Psi(\xi) = \sum_{i=1}^{m} w_i \Pr\{\mu_R \leq x_i\}.$$

The individual probabilities are immediately calculated using the prior density $h(\mu_R)$ (derived from $g(\vartheta)$) as

$$\Pr\{\mu_R \leq x_i\} = \int_{-\infty}^{x_i} h(\mu_R)d\mu_R, i = 1, ..., m.$$

This constraint is similar to the concept of *overdose control*, introduced by Babb *et al.* (1998), but otherwise our methodology is quite different. An appropriate constrained Bayesian D-optimal design problem can then be formulated as

$$\text{minimize } \Psi_D(\xi) \text{ subject to } \Psi(\xi) \leq \varepsilon,$$

where ε represents a small probability that can be chosen according to the level of the investigator's concern in assigning high doses. Note that if $\varepsilon = 0$, we have the naive design with the upper bound $\min(\mu_R)$ over the prior distribution, if such a minimum exists. Standard optimization methods invoking Lagrangian theory (Pshenichnyi, 1971; Fedorov, 1992; Clyde and Chaloner, 1996) can then be applied. Global optimality can be verified by plotting directional derivatives and applying an equivalence theorem of Whittle (1973).

3 Numerical methods

The procedure was programmed in MATLAB using a quasi-Newton optimization algorithm with the Broyden-Fletcher-Goldfarb-Shanno Hessian update and finite-difference approximation to derivatives. For each function evaluation, we used an adaptive Newton-Cotes integration rule. In retrospect, a numerical integration algorithm using a fixed grid, such as the midpoint rule or Simpson's rule, would have been preferable, since adaptive rules add variability that could slow the convergence of the optimization algorithm.

For concentrated priors, the results for local and Bayesian optimal designs are similar, and therefore results from the local optimal design could be used as starting values for the Bayesian optimal design. However, this is not the case for disperse priors, where optimal designs require more points. Consequently, we added points sequentially where the directional derivative was maximized at each stage to obtain more appropriate starting values. In general, selecting appropriate starting values was the most intensive part for the constrained problem, in which the procedure could converge to local minima rather than global minima. Once appropriate starting values were found, convergence was obtained in a reasonable time frame.

4 Results

In this section, we present results for independent uniform priors for α and β. We examined three sets of priors which are increasingly disperse, with ranges (1) $[1.5, 2.5]$, $[0.9, 1.1]$ (2) $[1.7, 2.3]$, $[0.1, 0.25]$ and (3) $[1.0, 3.0]$, $[0.125, 0.167]$ for α and β, respectively. These correspond to $E(\mu_R)$ of (1) 2.85, (2) 2.15, and (3) 2.12. We examined the unconstrained design ($\varepsilon = 1$), the naive design, and three constrained designs $\varepsilon = 0, 0.05, 0.10$. For each example, we set $\Gamma_R = 0.70$.

Table 1 gives the optimal design, values of the criterion function, and values of the constraint for prior (3). The unconstrained design consists of seven points symmetric about 2. The most extreme constrained design, with $\varepsilon = 0$, is a two-point design with equal weights at 0.81 and 1.11. There is very little difference between designs for $\varepsilon = 0.05$ and 0.10, indicating that the design is not particularly sensitive to the choice of ε above 0.05. Each results in a five-point design with very small weights at the upper three points. Finally, the naive design seems to be a reasonable candidate, as it is a four-point design with fairly large weights for each, and a criterion value that is less than that for $\varepsilon = 0.05$ and only moderately larger than that for $\varepsilon = 0.10$. Values of the constraint function indicate approximately 1/3 chance of being assigned above $E(\mu_R)$ for the naive design. Ethical considerations may make this undesirable, in which case a constrained design with small ε would be more appropriate. Figure 1 plots the directional derivatives for $\varepsilon = 0.05$. One can see the dramatic effect the constraint has on the problem. Comparing the value of the criterion function for unconstrained and constrained designs in Table 1 indicates the loss of efficiency from imposing a constraint on the problem. Finally, it is clear from Figure 1 how one can verify global optimality by visual inspection, using equivalence theory.

We do not report in tabular form results from priors (1) and (2). For prior (1), the criterion value for the unconstrained design is 3.02. Criterion values for the constrained designs ranged from 3.24 for $\varepsilon = 0.10$ to 3.42 for $\varepsilon = 0$. The criterion value for the naive design was 3.13, indicating that it is more efficient that the constrained designs, but the constraint value of 0.25 indicates a 25 percent chance of assigning a dose above $E(\mu_R)$. The naive design is symmetric at design points $(0.12, 2.85)$. The unconstrained design is symmetric at $(0.45, 3.55)$. For prior (2), the criterion value for the unconstrained design is -3.73. Again, the naive design has a smaller criterion value (-3.56) than all the constrained designs, but with a 25 percent chance of assigning doses above $E(\mu_R)$. The naive design is symmetric at $(1.67, 2.15)$, and is only slightly less efficient than the unconstrained design, which is a three-point design at $(1.66, 2.00, 2.34)$ with weights $(0.38, 0.24, 0.38)$. One can compare priors (1) and (2) with the more disperse prior (3),which has more variability in efficiency for different values of ε and the number of design points varies more among choices of ε.

Design	Support points and weights	$\Psi_D(\xi^*)$	$\Psi(\xi^*)$
$\varepsilon = 1.00$	(1.05, 1.44, 1.74, 2.00, 2.26, 2.56, 2.95)	-3.15	0.44
	(0.13, 0.16, 0.15, 0.14, 0.15, 0.16, 0.12)		
$\varepsilon = 0.00$	(0.81, 1.11)	6.88	0.00
	(0.50, 0.50)		
$\varepsilon = 0.05$	(0.83, 1.12, 1.68, 2.11, 2.49, 2.91)	-0.80	0.05
	(0.21, 0.69, 0.05, 0.03, 0.02, 0.01)		
$\varepsilon = 0.10$	(0.85, 1.12, 1.62, 2.06, 2.47, 2.91)	-1.66	0.10
	(0.16, 0.61, 0.12, 0.06, 0.04, 0.02)		
Naive	(1.07, 1.40, 1.74, 2.12)	-1.57	0.31
	(0.12, 0.14, 0.39, 0.36)		

TABLE 1. D-optimal designs ξ^* for $(\alpha, \beta) \sim Unif[1,3] \times [0.125, 0.167]$, corresponding to $E\mu_R = 2.1237$.

5 Conclusions

We have described, for the first time, how constrained Bayesian optimal design theory may be applied to an important design problem in medicine, namely that of efficiently estimating the MTD in phase I clinical trials. We have also provided some examples for continuous dose spaces with uniform priors. Other prior distributions, including dependent priors, are not considered here due to space limitations. While the constrained optimal designs are more finely tuned to the specific ethical problem, naive designs perform quite well in terms of efficiency, and are easy to implement.

Because of space constraints, we were unable to describe similar c-optimal designs, which minimize the variance of the quantile estimator. We could also extend these results to ordinal response models, to characterize the severity of toxic response. One could use the proportional odds model, for instance, to describe such data. Details on c-optimality and constrained Bayesian optimal designs for the proportional odds model can be found in Perevozskaya (2000).

The optimal designs developed thus far do not accomodate the fact that patients are assigned sequentially in a phase I clinical trial, and hence data accrue sequentially. We could implement a two-stage sequential procedure,

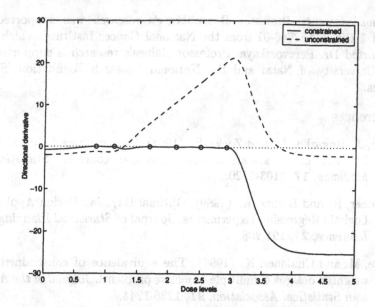

FIGURE 1. Directional derivatives for the Bayesian D-optimality criterion. The solid line represents the constrained criterion, and the dotted line the unconstrained criterion. Note that, while a six point design is optimal for the constrained problem, the directional derivative exceeds the zero bound for the unconstrained problem, and hence shifts the design to higher dose levels. Here we have priors $(\alpha, \beta) \sim Unif[1, 3] \times [0.125, 0.167]$ and $\varepsilon = 0.05$.

similar in principle to recent work of Haines (1998) and Sitter and Wu (1999). Specifically the first stage is to construct a Bayesian optimal design $\xi_{opt.}$ which incorporates the prior distribution on the parameters. In practice, an integer approximation to $\xi_{opt.}$ with $[n_0 w_i]$ patients from a total of n_0 allocated to the optimal doses $x_i, i = 1, ..., m$. One can think of the first n_0 patients, in a sense, as a pilot study, and, on completion, the data, denoted $\mathcal{D}_0$ can be used to update the prior $g(\vartheta)$ to a posterior distribution $g(\vartheta|\mathcal{D}_0)$. The second stage in the procedure involves the stepwise allocation of individuals or groups of patients to appropriate optimal dose levels. After n_0 patients have responded in the first stage, we find the value of x which maximizes

$$\int_{\Theta} \log |n_0 I(\xi_{opt.}, \vartheta) + I(x, \vartheta)| g(\vartheta|\mathcal{D}_0) d\vartheta$$

subject to the constraint that

$$\Pr\{\mu_R \leq x\} \leq \varepsilon,$$

where the probability is taken with respect to $h(\mu_R|\mathcal{D}_0)$. We can think of this as a simple one-point addition design. We leave this for future research.

Acknowledgments: Professor Rosenberger's research was supported by grant 1 R01 CA87746-01 from the National Cancer Institute, which also supported Dr. Perevozskaya. Professor Haines's research is supported by the University of Natal and the National Research Foundation, South Africa.

References

Babb, J., Rogatko, A., and Zacks, S. (1998). Cancer phase I clinical tials: Efficient dose escalation with overdose controls. *Statistics in Medicine*, **17**, 1103-1120.

Chaloner, K. and Larntz, K. (1989). Optimal Bayesian Design Applied to Logistic Regression Experiments. *Journal of Statistical Planning and Inference*, **21**, 191-208.

Clyde, M. and Chaloner, K. (1996). The equivalence of constrained and weighted designs in multiple objective problems. *Journal of the American Statistical Association*, **91**, 1236-1244.

Durham, S.D. Flournoy, N. (1994). Random walks for quantile estimation. In: Gupta, S.S. and Berger, J.O. (eds.), *Statistical Decision Theory and Related Topics V*. Springer, New York, 467-476.

Fedorov, V.V. (1992). *Optimal Design Construction With Constraints II.* School of Statistics, University of Minnesota (technical report 575).

Haines, L.M. (1998). Optimal design for neural networks. In: Flournoy, N., Rosenberger, W.F., and Wong, W.K. (eds.), *New Developments and Applications in Experimental Design*, Institute of Mathematical Statistics, Hayward, 152-162.

Mats, V.A., Rosenberger, W.F., and Flournoy, N. (1998). Restricted optimality for phase I clinical trials. In: Flournoy, N., Rosenberger, W.F., and Wong, W.K. (eds.), *New Developments and Applications in Experimental Design*, Institute of Mathematical Statistics, Hayward, 50-61.

O'Quigley, J., Pepe, M., and Fisher, L. (1990). Continual reassessment method: A practical design for phase I clinical trials in cancer. *Biometrics*, **46**, 33-48.

Perevozskaya, I. (1999). *Optimal Design for Quantile Estimation With Application to Phase I Clinical Trials.* University of Maryland Graduate School, Baltimore (doctoral thesis).

Pshenichnyi, B.N. (1971). *Necessary Conditions for an Extremum.* Marcel Dekker, New York.

Schacter, L., Birkhofer, M., Carter, S., Canetta, R., Hellmann, S., Onetto, N., Weil, C., Winograd, B., and Rozencweig, M. (1997). Anticancer drugs. In: O'Grady, J. and Joubert, P.H. (eds.), *Handbook of Phase I/II Clinical Trials*. CRC Press, Boca Raton, 523-534.

Simon, R., Freidlin, B., Rubinstein, L., Arbuck, S. G., Collins, J., and Christian, M.C. (1997). Accelerated titration designs for phase I clinical trials in oncology. *Journal of the National Cancer Institute*, **89**, 1138-1147.

Sitter, R.R. and Wu, C.F.J. (1999). Two-stage design of quantal response studies. *Biometrics*, **55**, 396-402.

Whittle, P. (1973). Some general points in the theory of optimal experimental design. *Journal of the Royal Statistical Society*, **B35**, 123-130.

Trend-Robust and Budget Constrained Optimum Designs

L. Tack
M. Vandebroek

ABSTRACT: Performing experiments in a time sequence may induce time dependence in the observed results. In practice, run orders that are optimally balanced for time trends are of little use because of economical reasons. This paper proposes an optimality criterion that strikes a balance between cost-efficiency and trend-resistance. The presented design algorithm serves as a proper tool for the construction of cost-efficient run orders with an optimal protection against time order dependence. The algorithm is intended to provide the experimenter with a general method for solving a wide range of practical problems. A real-life industrial case illustrates practical utility.

KEYWORDS: budget constraint; cost; trend-resistance; optimum design of experiments; run order

1 Introduction

When experiments are to be run in a time sequence, the experimenter may have reason to believe that the observed responses are influenced by unknown time trend effects. For instance, Joiner and Campbell (1976) mention an experiment to evaluate the amount of vitamin B_2 in turnips. A close examination of the data reveals that the vitamin B_2 values decay with the order in which they appear. Other examples that may lead to time dependence in the observed results include instrument drift, aging of material, warm-up effects, analyst fatigue, buildup of deposits in a test engine, poisoning of a catalyst, etc.

Section 2 deals with the construction of designs that are optimally balanced for time trends. Section 3 concentrates on cost-efficient design of experiments. Section 4 discusses trend-robust run orders under budget constraints and a practical example is given in Section 5.

mODa6, A.C.Atkinson, P.Hackl and W.G.Müller, eds., Physica, Heidelberg, 2001.

2 Time trends in design of experiments

Let the model for the response of interest y be of the following form:

$$y = \mathbf{f}'(\mathbf{x})\alpha + \mathbf{g}'(t)\beta + \varepsilon, \tag{2.1}$$

with $\mathbf{f}(\mathbf{x})$ the $p \times 1$ vector representing the polynomial expansion of design point $\mathbf{x}$ for the response model, $\mathbf{g}(t)$ the $q \times 1$ vector of the polynomial expansion for the time trend, expressed as a function of time t, α the $p \times 1$ vector of important parameters and β the $q \times 1$ vector of parameters of the polynomial time trend. In this paper, no interaction effects between $\mathbf{x}$ and t are considered, an assumption which holds in many practical situations. The independent error terms ε are assumed to have expectation zero and constant variance σ^2. It is convenient to rewrite (2.1) as

$$\mathbf{y} = \mathbf{F}\alpha + \mathbf{G}\beta + \varepsilon,$$

with $\mathbf{y}$ a column vector of n responses and $\mathbf{F}$ and $\mathbf{G}$ the $n \times p$ and the $n \times q$ extended design matrices respectively.

Bradley and Yeh (1980) call a design trend-free if the treatment effects are orthogonal to the polynomial trend components. In the context of regression designs, this condition comes down to $\mathbf{F}'\mathbf{G} = 0$.

Apart from Atkinson and Donev (1996), the literature on the existence and the construction of trend-robust run orders is mainly limited to two and three level factorial designs, equally spaced time points and regular design spaces. Atkinson and Donev (1996) present an algorithm for the construction of exact optimum designs that maximize the information on the important parameters α, whereas the q parameters modeling the time dependence are treated as nuisance parameters. The corresponding D_t-optimal design δ_{D_t} is found by maximizing

$$D_t = |\mathbf{F}'\mathbf{F} - \mathbf{F}'\mathbf{G}(\mathbf{G}'\mathbf{G})^{-1}\mathbf{G}'\mathbf{F}|^{\frac{1}{p}}. \tag{2.2}$$

In the absence of time trend effects, the D_t-optimal design equals the D-optimal design δ_D that maximizes $D = |\mathbf{F}'\mathbf{F}|^{\frac{1}{p}}$. Both designs are compared through the trend factor

$$\frac{D_t(\delta_{D_t})}{D(\delta_D)}. \tag{2.3}$$

It can easily be seen that trend-free designs have the maximum value $D_t(\delta_{D_t})/D(\delta_D) = 1$. In situations where it is impossible to obtain completely trend-free run orders, the value of (2.3) will be less than 1. Henceforth, (2.3) will be referred to as the degree of trend-resistance of the D_t-optimal design. Tack and Vandebroek (2000b) extend the approach of Atkinson and Donev (1996) to designs with either fixed or random block effects.

3 Cost-efficient design of experiments

Although D_t-optimal designs have good statistical proporties, practice has shown that the larger the protection against unknown trend effects, the larger the total cost of the run order. As a result, many trend-robust run orders may not be fit for use because of economical reasons. Until recently, cost considerations have not been dealt with in optimum design. The cost associated with factor level combination x_i will be referred to as the measurement cost $c^m(x_i)$. Measurement costs include equipment costs, the cost of material and personnel, etc. The total measurement cost C^m of a design then equals

$$C^m = \sum_{i=1}^{d} n_i c^m(x_i),\qquad (3.1)$$

with d the number of different design points and n_i the number of observations to be taken at design point x_i. Next, the cost for changing the factor levels of design point x_i in the previous run to the factor levels of design point x_j in the next run will be called the transition cost $c^t(x_i, x_j)$. Examples of transition costs include the cost for changing temperature in a blast-furnace, the cost for changing line set-up, the time needed for a system to return to steady state after a change in the parameter settings, etc. The total transition cost C^t of a run order equals

$$C^t = \sum_{i=1, j=1}^{d} n_{i,j} c^t(x_i, x_j),\qquad (3.2)$$

with $n_{i,j}$ the number of transitions from design point x_i to design point x_j in the considered run order. In many practical situations, the costs $c^m(x_i)$ and $c^t(x_i, x_j)$ are known exactly. For instance, changing the temperature of a blast-furnace takes almost one day. Consequently, the transition cost associated with the temperature change equals the loss in profit resulting from the fact that the blast-furnace will be suspended during one day. Note that contrary to the total measurement cost C^m, the total transition cost C^t depends on the order in which the experiments are performed. The total cost C of a run order simply equals the sum of the total measurement cost (3.1) and the total transition cost (3.2). A run order is said to be cost-optimal if it has the lowest total cost C among all possible run orders.

4 Trend-resistant design of experiments under budget constraints

In practice, the difficulty is to strike a balance between cost-efficiency and trend-resistance. The trade-off can be made by stating the design problem as follows: for a fixed number of observations $n = \sum_{i=1}^{d} n_i$, determine the number of replicates n_i at the d different design points $\mathbf{x}_i$ and the sequence in which the observations have to be performed in order to maximize the amount of information per unit cost. The corresponding (D_t, C)-optimality criterion equals

$$(D_t, C) = |\mathbf{F}'\mathbf{F} - \mathbf{F}'\mathbf{G}(\mathbf{G}'\mathbf{G})^{-1}\mathbf{G}'\mathbf{F}|^{\frac{1}{p}}/C. \tag{4.1}$$

The power $1/p$ ensures that the amount of information obtained has the dimension of variance. The resulting run order $\delta_{(D_t,C)}$ is called the (D_t, C)-optimal run order. A major drawback of the (D_t, C)-optimality criterion is that budget constraints are not allowed for. In practice, the experimenter usually has the disposal of a budget B to perform the experiments. Now, the aim is the construction of run orders that maximize the protection against time order dependence subject to the constraint that the total cost C of the experiments has to be lower than the experimenter's budget at hand. The fact that this budget constrained problem is of a discrete form in the numbers $\{n_i\}$ makes it impossible to apply the classical Lagrangian method for continuous functions. However, a good approximation can be obtained by computing run orders that maximize

$$k|\mathbf{F}'\mathbf{F} - \mathbf{F}'\mathbf{G}(\mathbf{G}'\mathbf{G})^{-1}\mathbf{G}'\mathbf{F}|^{\frac{1}{p}} - C \tag{4.2}$$

for different values of the strictly positive weighting coefficient k. This weighting coefficient k acts as a tuning constant between cost-efficiency and trend-resistance. Large values of k put more emphasis on trend-resistance whereas low k-values put more weight on cost-efficiency. For instance, it can easily be seen that if $k \to \infty$, criterion (4.2) is nothing else than the D_t-optimality criterion (2.2). If $k \to 0$, the resulting design is the cost-optimal design. It can easily be shown that the (D_t, C)-optimal run order equals the run order that maximizes (4.2) for

$$k = \frac{C(\delta_{(D_t,C)})}{D_t(\delta_{(D_t,C)})}.$$

It can be proven that the total cost C and the D_t-value of the optimal run orders under criterion (4.2) are non-decreasing functions of k. This means that an increased k-value leads to an optimal run order with the same or a higher degree of trend-resistance. However, this non-decrease in protection against time order dependence goes at the expense of the total cost of the optimal run order. It follows that the relation between the total cost C and the D_t-value of the optimal run orders under criterion (4.2) is a strictly increasing function. Based on this property, it can also be proven that our generic algorithm that iteratively computes optimal run orders for different k-values, converges to a run order with an optimal protection against time order dependence and a total cost C very close to the budget at hand.

Roughly speaking, the algorithm proceeds as follows. Firstly, an optimal run order is computed for an initial user-specified k-value. If the total cost C of the resulting optimal run order exceeds the available budget B, the k-value must be decreased in the next iteration. On the other hand, if the total cost C is lower than the budget B, one takes advantage of increasing k in the next iteration. In each iteration, the optimal run order is constructed from a randomly selected starting run order by sequentially exchanging design points and time points in such a way as to maximize (4.2). For a detailed description of the exchange procedures we refer the interested reader to Tack and Vandebroek (2000c).

5 The cryogenic flow meter experiment

The aim of this section is to demonstrate the wide range of practical design problems that can be solved with our design algorithm. An experiment with $n = 20$ observations has to be designed to evaluate the accuracy of flow meters for use with cryogenic fluids such as liquid oxygen or liquid nitrogen. The experimenter's budget B equals 800 and the accuracy of the flow meters is supposed to be sensitive to the temperature x_1, the pressure x_2, the flow rate of the liquid x_3 and the total weight x_4 of the liquid pumped during a test. Management decides to restrain the factor levels to those shown in Table 1.

factor	coded factor levels
x_1	-1, 1
x_2	-1, 1
x_3	-1, 0, 1
x_4	-1, 0, 1

TABLE 1. Coded factor levels in cryogenic flow experiment

Run orders will be computed for the following response models:

$$(F_1) \quad \mathbf{f}'(\mathbf{x}) \quad = \quad (1 \; x_1 \; x_2 \; x_3 \; x_4),$$
$$(F_2) \quad \mathbf{f}'(\mathbf{x}) \quad = \quad (1 \; x_1 \; x_2 \; x_3 \; x_4 \; x_1x_2 \; x_1x_3 \; x_1x_4 \; x_2x_3 \; x_2x_4 \; x_3x_4),$$
$$(F_3) \quad \mathbf{f}'(\mathbf{x}) \quad = \quad (1 \; x_1 \; x_2 \; x_3 \; x_4 \; x_3^2 \; x_4^2),$$
$$(F_4) \quad \mathbf{f}'(\mathbf{x}) \quad = \quad (1 \; x_1 \; x_2 \; x_3 \; x_4 \; x_1x_2 \; x_1x_3 \; x_1x_4 \; x_2x_3 \; x_2x_4 \; x_3x_4 \; x_3^2 \; x_4^2).$$

Besides, the flow meters are known to deteriorate with time and the postulated time trend is $\mathbf{g}(t) = t$. The measurement costs are described by $c^m(\mathbf{x}) = 20 + 5x_1 + 5x_2 - 5x_3 + 5x_4$, since raised temperatures x_1 and/or higher pressures x_2 load the flow meter more heavily and so give an increased measurement cost. Raising the total weight x_4 of the liquid pumped prolongs the total execution time of the experiment and increases the measurement cost. The opposite holds for the flow rate x_3. Besides, changing the temperature x_1 from the high level to the low level or vice versa takes a half day and corresponds to a loss in profit of about 100. Changing the pressure level only takes some hours and amounts to a cost of 50. Changes in flow rate x_3 and weight x_4 could be made almost instantaneously. As a consequence, the transition costs for the latter two factors x_3 and x_4 are set equal to zero.

The performance of the D_t-, the (D_t, C)-, and the budget constrained optimal run orders is shown in Table 2.

		F_1	F_2	F_3	F_4
D_t-optimality	cost	1,900	1,480	2,360	2,150
	degree of trend-resistance	100	99.99	99.99	99.99
(D_t, C)-optimality	cost	280	600	310	480
	degree of trend-resistance	49.17	89.97	59.25	73.11
$B = 800$	cost	800	780	745	800
	degree of trend-resistance	100	99.91	99.99	99.70

TABLE 2. Performance of the optimal run orders

By definition, the D_t-optimal run orders have the best protection against time trend effects, whereas the (D_t, C)-optimal run orders are very cheap. The budget constrained run orders are somewhere in between. Remark that as compared to the D_t-optimal run orders, the decrease in the total cost of the budget constrained run orders only leads to a small loss in the degree of trend-robustness. It is shown in Tack and Vandebroek (2000a) that the results are insensitive to small changes in the cost functions. The same holds for small changes in the available budget.

time point	x_1	x_2	x_3	x_4	time point	x_1	x_2	x_3	x_4
1	-1	-1	-1	0	11	-1	-1	1	1
2	-1	1	0	1	12	-1	1	-1	1
3	1	-1	1	0	13	1	1	1	1
4	-1	1	1	-1	14	-1	1	-1	-1
5	1	1	1	-1	15	-1	-1	0	-1
6	1	1	-1	1	16	1	-1	-1	1
7	1	-1	0	1	17	-1	-1	-1	1
8	1	-1	-1	-1	18	1	1	0	0
9	-1	-1	1	-1	19	1	-1	1	-1
10	1	1	-1	-1	20	-1	1	1	0

TABLE 3. The D_t-optimal run order

As an illustration, the optimal run orders for response model F_4 are shown in Table 3 to Table 5. It can be seen that the large cost of the D_t-optimal run order follows from the large total transition cost: the number of level changes for factors x_1 and x_2 equals 23, whereas for the two other optimal run orders, these numbers only equal 3 and 6. Besides, the low measurement cost of the (D_t, C)-optimal order is among other things a result of the fact that it replicates the cheap design points $(-1, -1, 0, -1)$ and $(-1, -1, 1, -1)$ respectively two and three times, whereas the D_t-optimal run order and the budget constrained run order do not replicate these design points.

time point	x_1	x_2	x_3	x_4	time point	x_1	x_2	x_3	x_4
1	-1	1	1	-1	11	-1	-1	1	1
2	-1	1	1	1	12	-1	-1	1	-1
3	-1	1	-1	0	13	-1	-1	1	-1
4	-1	-1	0	-1	14	-1	-1	0	-1
5	-1	-1	1	0	15	1	-1	1	-1
6	-1	-1	1	-1	16	1	-1	1	1
7	-1	-1	-1	-1	17	1	-1	-1	0
8	-1	-1	-1	1	18	1	1	-1	1
9	-1	-1	0	0	19	1	1	-1	-1
10	-1	-1	-1	-1	20	1	1	1	0

TABLE 4. The (D_t, C)-optimal run order

time point	x_1	x_2	x_3	x_4	time point	x_1	x_2	x_3	x_4
1	1	1	-1	1	11	-1	1	-1	0
2	1	1	1	-1	12	-1	1	0	-1
3	1	-1	0	-1	13	-1	1	1	1
4	1	-1	-1	0	14	-1	-1	-1	1
5	1	-1	1	1	15	-1	-1	1	-1
6	-1	-1	1	1	16	1	-1	1	-1
7	-1	-1	-1	-1	17	1	-1	-1	1
8	-1	-1	0	0	18	1	1	1	0
9	-1	1	1	-1	19	1	1	0	1
10	-1	1	-1	1	20	1	1	-1	-1

TABLE 5. The budget constrained run order

It is also worth mentioning that the performance of the computed run orders depends on the postulated time trend. The presented results are useful as long as one has certainty about the linearity of the unknown time trend. In order to circumvent this difficulty, Tack and Vandebroek (2000d) use a semiparametric approach for the construction of trend-robust run orders.

6 Conclusion

This paper has presented a generic design algorithm for the construction of cost-efficient or budget constrained run orders with an optimal protection against time trend effects. It enables the practitioner to tackle a wide range of practical design problems. The cryogenic flow meter experiment has shown that the algorithm serves as a proper tool for the construction of trend-resistant run orders under budget constraints.

References

Atkinson, A.C. and Donev, A.N. (1996). Experimental Designs Optimally Balanced for Trend. *Technometrics*, **38**, 333-341.

Bradley, R.A. and Yeh, C.-M. (1980). Trend-Free Block Designs: Theory. *The Annals of Statistics*, **8**, 883-893.

Joiner, B.L. and Campbell, C. (1976). Designing Experiments When Run Order is Important. *Technometrics*, **18**, 249-259.

Tack, L. and Vandebroek, M. (2000a). (D_t, C)-Optimal Run Orders. To appear in *Journal of Statistical Planning and Inference*.

Tack, L. and Vandebroek, M. (2000b). Trend-Resistant and Cost-Efficient Fixed or Random Block Designs. Research Report 0021, K.U.Leuven, Belgium.

Tack, L. and Vandebroek, M. (2000c). Trend-Resistant Design of Experiments under Budget Constraints. Research Report 0035, K.U.Leuven, Belgium.

Tack, L. and Vandebroek, M. (2000d). Semiparametric Exact Optimum Designs. To appear as Research Report, K.U.Leuven, Belgium.

Minimax Designs for Logistic Regression in a Compact Interval

B. Torsney
J. López–Fidalgo

ABSTRACT: Minimax designs for logistic regression with one design variable are considered where the design space is a general interval $[a, b]$ on the straight real line. This corresponds to a weighted linear regression model for a specific weight function. Torsney and López–Fidalgo (1995) computed maximum variance (MV–) optimal designs for simple linear regression for a general interval. López–Fidalgo *et al.* (1998) gave MV–optimal designs for symmetric weight functions and symmetric intervals, $[-b, b]$. Computations for general intervals for logistic regression are much more complex. Using similar approaches to these two papers and the equivalence theorem, explicit minimax optimal designs are given for different regions of points (a, b) in the semi–plane $b > a$ for the logistic model. This complements Dette and Sahm (1998) who give a method for computing MV–optimal designs without any restriction on the design space. From this, Standardized Maximum Variance (SMV–) optimal designs immediately follow. The same approach could be used for probit, double exponential, double reciprocal and some other popular models in the biomedical sciences.

KEYWORDS: *c*–optimality; equivalence theorem; minimax criteria; MV–optimality; optimal design.

1 Introduction

We consider the problem of finding minimax designs for a weighted linear regression model. In particular we consider the model

$$E(y) = \gamma\sqrt{w(x)} + \delta x\sqrt{w(x)}, \quad \sigma^2(x) = 1, \quad x \in [a, b] \qquad (1.1)$$

where y is a response variable, x is a design variable restricted to the interval $[a, b]$, γ, δ are unknown parameters and $w(x) = f^2(x)/\{F(x)[1 - F(x)]\}$, with $F(x) = e^x/(1 + e^x)$, which is the standardized logistic distribution

---

mODa6, A.C.Atkinson, P.Hackl and W.G.Müller, eds., Physica, Heidelberg, 2001.

function and $f(x)$ is the corresponding density function. We wish to find for all $[a, b]$ the probability measure on $[a, b]$ which solves

$$\min_{p(\cdot)} \max \{ \mathrm{var}_{p(\cdot)}(\widehat{\gamma}), \mathrm{var}_{p(\cdot)}(\widehat{\delta}) \}.$$

The method of obtaining the estimates $\widehat{\gamma}$, $\widehat{\delta}$ will be discussed below. The model is a transformation of a logistic regression model. Ford et $al.$ (1992) exploit this. Suppose that, conditional on a value z for a design variable, a binary response U has model: $U \sim Bi(1, F(\alpha + \beta z))$, $c \le z \le d$. Then under a design $\xi(\cdot)$ on $[c, d]$, the asymptotic Fisher information matrix of the maximum likelihood estimates of α and β is,

$$E_{\xi(\cdot)} \left[w(\alpha + \beta z) \begin{pmatrix} 1 \\ z \end{pmatrix} (1, z) \right].$$

Under the parameter dependent linear transformation $x = \alpha + \beta z$ local D–optimal and c–optimal design problems transform to equivalent design problems for the above linear model.

We have already considered the case of a symmetric $w(x)$ with $a = -b$; see López–Fidalgo et $al.$ (1998), while Torsney and López–Fidalgo (1995) consider the constant variance case, $w(x) = 1$, for all $[a, b]$.

Pioneering and comprehensive work in the area of minimax designs includes Elfving (1959). He uses minimax designs, that is, E–optimal designs, aiming to minimize the maximum of the variances of any unitary linear combination of the parameters. Dette et $al.$ (1995), generalize the Minimax criterion to any norm over Euclidean space,

$$\Phi_{|\cdot|}[M(\xi)] = \max \{ c'M^{-1}(\xi)c : c \in R^m, |c| = 1 \}.$$

In particular, for the ℓ_1-norm, $|c|_1 = \sum_{i=1}^{m} |c_i|$, we have the MV–optimality criterion,

$$\Phi_{|\cdot|_1}[M(\xi)] = \max_i e_i'M^{-1}(\xi)e_i = \max_i \mathrm{var}_\xi(\alpha_i)$$

where ξ is a design over a compact set X and $\alpha_i, i = 1, 2, \ldots, m$, are the parameters of the model.

Dette (1997) further consider a minimax criterion of standardized variances, called SMV–optimality. López–Fidalgo and Wong (2000) consider the efficiencies of the MV- and SMV-optimal designs with respect to each other's criterion and study the sensitivity of both designs with respect to the choice of the initial values of the parameters in binary response models.

Ford (1976) found the directional derivative at M in the direction of N,

$$\partial \Phi_{MV}(M, N) = \{ M^{-1} - M^{-1}NM^{-1} \}_{ss}$$

where $\{M^{-1}\}_{ss}$ is the biggest diagonal element of M^{-1}. If there are r coincident biggest diagonal elements $ss_1, \ldots, ss_r$, say, then we choose s such that, $\{M^{-1}NM^{-1}\}_{ss} = \min_{ss_i}\{M^{-1}NM^{-1}\}_{ss_i}$. We will say that a matrix N has the ss_j property if,

$$\partial\Phi_{MV}(M, N) = \{M^{-1} - M^{-1}NM^{-1}\}_{ss_j}.$$

Ford (1976), denotes by $\mathcal{M}'$ the convex sub-set of $\frac{1}{2}m(m+1)$ dimensional space whose points have as components the elements of the upper triangular parts of the matrices of which $\mathcal{M}$ is composed. To check optimality of a matrix M, where Φ_{MV} is not differentiable at M, we only need to calculate $\partial\Phi_{MV}[M, N]$, for those $N \in \mathcal{M}$ which are the generators of the convex sets $\mathcal{M}'_{ss_i}, j = 1, \ldots, m$, and which are feasible optimal solutions. The directional derivative must be non-negative for optimal designs.

2 Some basic results

If $[a, b] \subseteq [-1, 1]$ then MV–optimality is the same as c_2–optimality. If $b \leq -1$ or $a \geq 1$ then MV–optimality is the same as c_1–optimality. In other cases equal variance designs must be considered.

Note that for many weight functions including the above there need be no restrictions on a, b. In the case of $w(x) = f^2(x)/\{F(x)[1-F(x)]\}$ the widest possible choice of $[a, b]$, say X_w, will be the sample space of the random variable with density function $f(x)$. Thus, $X_w = \Re$ in the case of the logistic distribution. Both of the functions $\sqrt{w(x)}$ and $x\sqrt{w(x)}$ converge to zero as $x \to \pm\infty$.

For choices of $[a, b]$ other than the above a minimax design is either the c_1–optimal design or the c_2–optimal design or an optimal equal variance design. It will be the latter if the common value of the two c–criteria at this design is smaller than their values at each other's optimal design; i.e. than the value of the c_1–criterion at the c_2–optimal design and vice–versa. We need to be clear about the structure of c–optimal designs and optimal equal variance designs on each interval $[a, b]$.

2.1 c–Optimal designs

For c–optimal designs we cite Ford *et al.* (1992). In summary they establish for many weight functions that any c–optimal design on an interval $[a, b]$ has either one or two support points, although we could also appeal to theorem 3.1.4 of Fellman (1974) on this point.

Now if a c–optimal design is to be an MV–optimal design it must ensure estimation of both parameters and so have at least two support points. Ford *et al.* (1992) show that these two support points are a fixed pair of points, the same for any c. Their structure is as follows

$$\text{Supp}\{p^*\} = \begin{cases} \{a^*, b^*\} & a \le a^*, \ b \ge b^* \\ \{a, \min\{b, b^*(a)\}\} & a \ge a^*, \ b \ge b^* \\ \{\max\{a, a^*(b)\}, b\} & a \le a^*, \ b \le b^* \\ \{a, b\} & a \ge a^*, \ b \le b^* \end{cases}$$

where $\text{Supp}(p)$ denotes the set of support points of the design p; and, assuming $X_w = [A, B]$, where A and B are chosen as big as necessary in absolute value, a^*, b^* are the support points on X_w; $b^*(a)$ is the upper support point on $[a, b]$; $a^*(b)$ is the lower support point on $[A, b]$. In the case of the logistic $a^* = -2.4$, $b^* = 2.4$.

2.2 Optimal equal variance designs

Now we wish to find a design, if it exists, on $[a, b]$ which minimizes the two variances subject to them being equal. We argue that the structure of these designs must be similar. There must be at least two support points. There may be three. For the moment we focus on optimal two–point equal variance designs denoted by $\hat{p}_{EV}$. We argue that,

$$\text{Supp}\{p^{**}\} = \begin{cases} \{a^{**}, b^{**}\} & a \le a^{**}, \ b \ge b^{**} \\ \{a, \min\{b, b^{**}(a)\}\} & a \ge a^{**}, \ b \ge b^{**} \\ \{\max\{a, a^{**}(b)\}, b\} & a \le a^{**}, \ b \le b^{**} \\ \{a, b\} & a \ge a^{**}, \ b \le b^{**} \end{cases}$$

where a^{**}, b^{**} are the support points on $X_w = [A, B]$ (so that A is the smallest possible value of a and B is the largest possible value of b); $b^{**}(a)$ is the upper support point on $[a, B]$; $a^{**}(b)$ is the lower support point on $[A, b]$. In the case of the logistic model $a^{**} = -1$, $b^{**} = 1$. See López–Fidalgo et al. (1998) .

3   MV–optimal designs

We wish to find MV–optimal designs for all design intervals $[a, b]$. It helps to recall the solution for $w(x) = 1$ in Torsney and López–Fidalgo (1995). The relationship of (a,b) to two loci of points determines which design is MV-optimal. These are the circle of radius 2 and the hyperbolas $ab = 1$ and $ab = -1$.

In the case of weighted regression corresponding boundaries will again be crucial in (potentially) determining when the MV–optimal design changes from a c–optimal to an optimal equal variance design. In particular, c_1–optimal designs produce equal variances on the line $h_1(a)/|a| = -h_1(b)/|b|$ and c_2–optimal designs produce equal variances on the line $h_1(a) = -h_1(b)$, where $h_1(x) = (x^2 - 1)\sqrt{w(x)}$.

To these, however we must add the functions $a^*(b)$, $b^*(a)$, $a^{**}(b)$, $b^{**}(a)$. These render parts of the equal variance boundaries irrelevant. They cut away areas of the plane $\{(a, b) : a \leq b\}$ since designs have fixed support points in these areas.

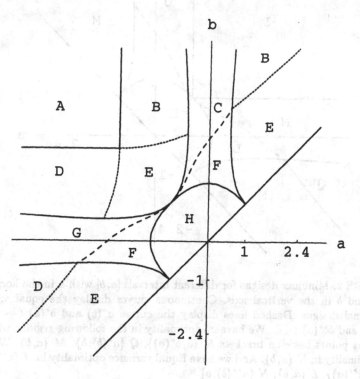

FIGURE 1. Minimax designs for different values of a (horizontal axis), b (vertical axis). Continuous curves display the equal variance c-optimal designs. Dashed lines display the curves $a^*(b)$ and $b^*(a)$ (- - -) and $a^{**}(b)$ and $b^{**}(a)$ (· · ·). We have c_1-optimality in the following regions with the support points between brackets $A\{a^*, b^*\}$, $B\{a, b^*(a)\}$, $D\{a^*(b), b\}$, $E\{a, b\}$. We have c_2-optimality in $H\{a, b\}$. And we have Equal variance optimality in: $C\{a, b^{**}(a)\}$, $F\{a, b\}$, $G\{a^{**}(b), b\}$.

We illustrate the final solution in the case of the logistic model in Figures 1 and 2. Figure 1 partitions the plane into regions in which the supports of either c_1-optimal or c_2-optimal or optimal two point equal variance designs are identified. In Figure 2 the nature and supports of MV-optimal designs are identified.

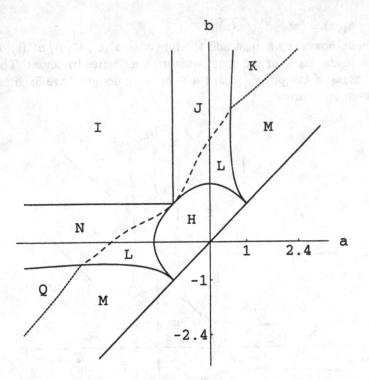

FIGURE 2. Minimax designs for different intervals $[a, b]$ with a in the horizontal axis and b in the vertical axis. Continuous curves display the equal variance c-optimal designs. Dashed lines display the curves $a^*(b)$ and $b^*(a)$ (- - -) and $a^{**}(b)$ and $b^{**}(a)$ ($\cdots$). We have c_1-optimality in the following regions with the support points between brackets K $\{a, b^*(a)\}$, Q $\{a^*(b), b\}$, M $\{a, b\}$. We have c_2-optimality in H $\{a, b\}$. And we have Equal variance optimality in: I $\{a^{**}, b^{**}\}$, J $\{a, b^{**}(a)\}$, L $\{a, b\}$, N $\{a^{**}(b), b\}$

4   Conditions of optimality for MV-criterion

We have already noted that if the MV-optimal design is a non-equal variance one it is a c-optimal design and Ford *et al.* (1992) have solved this problem. Therefore we need only consider equal variance designs. In contrast to a non-equal variance design on two support points the MV-criterion can, at equal variance designs, be non-differentiable. To check for optimality we need to examine the directional derivative of information matrices of these designs in the direction of the generators of $\mathcal{M}'_{ss_1}$ and $\mathcal{M}'_{ss_2}$. Thus, let the following matrix be,

$$M = \begin{pmatrix} e & f \\ f & e \end{pmatrix}.$$

It is necessary to check the directional derivative, $\partial\Phi(M,N)$, for two types of generators of $\mathcal{M}'_{ss_1}$ and $\mathcal{M}'_{ss_2}$,

1. $N = vv'$, $v' = (\sqrt{w(x)}, x\sqrt{w(x)})$, 1 point design,

$$\partial\Phi(M,N) = \begin{cases} \frac{e}{e^2-f^2} - \frac{w(x)(e-fx)^2}{e^2-f^2} & \text{if } x^2 > 1 \\ \frac{e}{e^2-f^2} - \frac{w(x)(ex-f)^2}{e^2-f^2} & \text{if } x^2 < 1 \end{cases}$$

This must be greater than or equal to zero for all $x \in [a,b]$.

2. N corresponding to a 2 point equal variance design,

$$\partial\Phi(M,N) = \frac{(e,f)\begin{pmatrix} e-g & h \\ h & -g-e \end{pmatrix}\begin{pmatrix} e \\ f \end{pmatrix}}{(e^2-f^2)^2}, \qquad N = \begin{pmatrix} gh \\ hg \end{pmatrix}$$

This must be greater than or equal to zero for $|x_1| \le 1, |x_2| \ge 1$.

For more details see López–Fidalgo *et al.* (1998).

5 Asymmetric design intervals

This general case is non-trivial. There needs to be reliance on numerical analysis both in determining the functions $a^*(b)$, $a^{**}(b)$, $b^*(a)$, $b^{**}(a)$ and in calculating the directional derivatives of Section 4. We report some basic facts.

Recall that we are restricting attention to two point equal variance designs. The equal variance constraint then uniquely determines the weights. If the two support points are a, b the design is

$$\xi = \left\{\begin{array}{cc} a & b \\ w(b)(b^2-1)/D & w(a)(1-a^2)/D \end{array}\right\}$$

where $D = w(a)(1-a^2) + w(b)(b^2-1)$.

Note that the weights are only nonnegative either if $|a| < 1$ and $b > 1$ in which case $b^{**}(a)$ minimizes the common variance over $b > 1$ for given lower support point a; or if $a < -1$ and $|b| \le 1$ in which case $a^{**}(b)$ minimizes the common variance over $a < -1$ for a given upper support point b.

We have checked for MV-optimality of the two point equal variance designs for $a < -1$, $b > 1$ and we found MV-optimality. We have checked also in some choices of the interval $[a,b] = [-0.6,3]$; $[-0.6,4]$; $[0.2,3]$; $[-0.6,2.3]$; $[0.6,2.7]$; $[0.9,1.1]$. These cover different possibilities for two-point equal variance designs, i.e., different regions shown in Figure 2. To check the

Equivalence theorem we have used the formulae given in Section 4. In all these cases we have found optimality. Of course, given the limitation of these calculations, it is possible that a three point equal variance design is needed in some regions.

Remark. We will report more general results elsewhere encompassing other weight functions and the issue of three point equal variance designs. We note also that the above approach can be used for any two c-criteria, and in particular for standardized c-criteria, such as those considered by Dette (1997).

References

Dette, H. (1997). Designing experiments with respect to 'standardized' optimality criteria. *Journal of the Royal Statististical Society*, **B59**, 97-110.

Dette, H., Heiligers, B., and Studden, W.J. (1995). Minimax designs in linear regression models. *The Annals of Statistics*, **23**, 30-40.

Dette, H. and Sahm, M. (1998). Minimax optimal designs in nonlinear regression models. *Statistica Sinica*, **8(4)**, 4249-4264.

Elfving, G. (1959). Design of linear experiments. In: Grenander, U. (ed.), *Probability and Statistics. The Harald Cramér Volume*. Almagrist and Wiksell, Stockholm, 58-74.

Fellman, J. (1974). On the allocation of linear observations. *Soc. Sci. Fenn. Comment. Phys. Math.*, **44**, 27-77.

Ford, I. (1976). *Ph.D. Thesis*. University of Glasgow.

Ford, I., Torsney, B., and Wu, C.F.J. (1992). The use of a canonical form in the construction of locally optimal designs for non-linear problems. *Journal of the Royal Statistical Society*, **B54**, 569-583.

López–Fidalgo, J., Torsney, B., and Ardanuy, R. (1998).   MV–optimization in weighted linear regression. In: Atkinson, A.C., Pronzato, L., and Wynn, H.P. (eds.), *MODA 5 – Advances in Model-Oriented Data Analysis and Experimental Design*. Springer-Verlag, New York, 39-50.

López–Fidalgo, J. and Wong, W.K. (2000). A comparative study of MV- and SMV-optimal designs for binary response models. *Advances in Stochastic Simulation Methods*, **386**, 135-152.

Torsney, B. and López–Fidalgo J. (1995). MV–optimization in simple linear regression. In: Kitsos, C.P. and Müller, W.H. (eds.), *MODA 5 – Advances in Model-Oriented Data Analysis*. Springer-Verlag, New York, 57-69.

Sensor Motion Planning with Design Criteria in Output Space

D. Uciński

ABSTRACT: Motion planning of pointwise scanning sensors while estimating unknown parameters in models described by partial differential equations is addressed. In contrast to the common approach based on parameter-space criteria, close attention is paid to criteria in output space, which are of interest if the purpose of parameter estimation is to accurately predict system outputs. Some performance indices are proposed to quantify the prediction accuracy. Then the approach is to convert the problem to a state-constrained optimal-control one in which both the control forces of the sensors and the initial sensor positions are optimized. A method of successive linearizations is then employed for numerical solution.

KEYWORDS: distributed-parameter systems; parameter estimation; sensor location; spatial statistics

1 Introduction

The states of many real systems, usually called distributed-parameter systems (DPS's), vary spatially as well as temporally, leading to mathematical models in the form of partial differential equations (PDE's). In order to estimate the unknown parameters in these models, the system's behaviour or response is observed with the aid of some suitable collection of sensors. The inability to take distributed measurements leads, however, to the question where to locate the sensors so that the information content of the resulting signals be as high as possible.

The sensor location problem was attacked from various angles, but the results communicated by most authors are limited to the selection of stationary sensor positions (for reviews, see Kubrusly and Malebranche, 1985; Uciński, 1999; 2000b). A generalization which imposes itself is to apply sensors which are capable of tracking points providing at a given moment the best information about the parameters. However, communications in this field are rather limited. Rafajlowicz (1986) considers the determinant of

mODa6, A.C.Atkinson, P.Hackl and W.G.Müller, eds., Physica, Heidelberg, 2001.

the Fisher Information Matrix (FIM) associated with the parameters to be estimated as a measure of the identification accuracy and looks for an optimal time-dependent measure, rather than for the trajectories themselves. On the other hand, Uciński (1999; 2000a; 2000b), apart from generalizations of Rafajlowicz's results, develops some computational algorithms based on the FIM. He reduces the problem to a state-constrained optimal-control one for which solutions are obtained via gradient techniques capable of handling various constraints imposed on sensor motions.

Another deficiency of the existing approaches is that they determine sensor locations in experiments performed for the most accurate determination of parameter values which may have some physical significance. In some applications, however, the reliability of model predictions is sometimes more important than the accuracy of model parameters, because the ultimate objective in modelling is the prediction or forecast of the system states (Müller, 1998). The topic was discussed to some extent in Sun (1994, p.201), but without connection to constructive solution methods. This failing constitutes the main motivation for the present study undertaken in order to extend sensor motion planning techniques set forth in Uciński (1999; 2000a; 2000b). For that purpose, three output criteria are proposed as measures of the prediction accuracy. The main difficulty in solving the resulting optimal-control problem is that two criteria are not Fréchet differentiable. A method of circumventing this inconvenience is proposed and then an effective method of successive linearizations is employed to find numerical solutions.

2 Sensor location problem

Consider the scalar (possibly non-linear) distributed system

$$\frac{\partial y}{\partial t} = \mathcal{F}\left(x, t, y, \frac{\partial y}{\partial x_1}, \frac{\partial y}{\partial x_2}, \frac{\partial^2 y}{\partial x_1^2}, \frac{\partial^2 y}{\partial x_2^2}, \vartheta\right), \quad x \in \Omega, \quad t \in T \qquad (2.1)$$

with initial and boundary conditions of the general form

$$y(x, 0) = y_0(x), \quad x \in \Omega \qquad (2.2)$$

$$\mathcal{E}(x, t, y, \vartheta) = 0, \quad x \in \partial\Omega, \quad t \in T \qquad (2.3)$$

where $\Omega \subset \mathbb{R}^2$ is a fixed, bounded, open set with sufficiently smooth boundary $\partial\Omega$, the points of which will be denoted by $x = (x_1, x_2)$, $\mathcal{F}$ and $\mathcal{E}$ are some known functions, y_0 is a given initial state, t stands for time, $T = (0, t_f)$, and $y = y(x, t)$ signifies the state variable with values in $\mathbb{R}$. We assume that y depends on the vector $\vartheta \in \mathbb{R}^m$ of unknown parameters to be estimated from measurements made by N moving pointwise sensors. Let $x^j : T \longrightarrow \bar{\Omega} = \Omega \cup \partial\Omega$ be the trajectory of the j-th sensor. Our basic assumption is that the observations are of the form

$$z^j(t) = y(x^j(t), t) + \varepsilon(x^j(t), t), \quad t \in T, \quad j = 1, \ldots, N \qquad (2.4)$$

where $\varepsilon = \varepsilon(x,t)$ is a Gaussian white noise process whose statistics are

$$\mathrm{E}\{\varepsilon(x,t)\} = 0, \quad \mathrm{E}\{\varepsilon(x,t)\varepsilon(x',t')\} = \sigma^2 \delta(x-x')\delta(t-t') \qquad (2.5)$$

$\sigma > 0$ being a constant and δ the Dirac delta function.

We assume that the parameter estimate $\widehat{\vartheta}$, defined as the solution to the usual output least-squares formulation of the parameter estimation problem, is to provide a basis for prediction of certain variables depending on spatial location and/or time. Since in general the conditions applied for prediction may differ from the conditions of the experiment, the prediction equations need not be the same as (2.1)–(2.3), nor need the variables to be predicted coincide with the state y. Let the solution to the prediction problem in context be a scalar quantity $q = q(x,t;\vartheta)$. We are interested in selecting the sensors' trajectories in such a way as to maximize the accuracy of q in a given compact spatio-temporal domain $Q = \mathcal{X} \times \mathcal{T}$. Clearly, in order to compare different trajectories, a quantitative measure of the 'goodness' of particular trajectories is required. A logical approach is to choose a measure related to the expected accuracy of prediction.

For a given $(x,t) \in Q$, the variance of q obtained by a first-order expansion around a preliminary estimate ϑ^0 of ϑ (Mehra, 1974) has the form

$$\mathrm{var}(q(x,t;\widehat{\vartheta})) = \mathrm{E}\left((q(x,t;\vartheta) - q(x,t;\widehat{\vartheta}))^2\right)$$

$$\approx (\nabla_\vartheta q(x,t;\vartheta^0))^{\mathrm{T}} \mathrm{cov}(\widehat{\vartheta}) \nabla_\vartheta q(x,t;\vartheta^0) \qquad (2.6)$$

where we write $\nabla_\vartheta q$ for the gradient of q with respect to ϑ. It is customary to choose ϑ^0 as a nominal value of ϑ or a result of a preliminary experiment. As for $\mathrm{cov}(\widehat{\vartheta})$, under some assumptions (Uciński, 1999; 2000b) it can be approximated by the inverse of the *Fisher Information Matrix* (FIM) whose normalized version can be written down as (Rafajlowicz, 1986)

$$M(s) = \frac{1}{Nt_f} \sum_{j=1}^{N} \int_0^{t_f} g(x^j(t),t) g^{\mathrm{T}}(x^j(t),t)\,dt \qquad (2.7)$$

where $s(t) = (x^1(t), x^2(t), \ldots, x^N(t))$ and $g(x,t) = \nabla_\vartheta y(x,t;\vartheta^0)$ (we require g and $\nabla_x g$ to be continuous). Consequently, we get

$$\mathrm{var}(q(x,t;\widehat{\vartheta})) \sim (\nabla_\vartheta q(x,t;\vartheta^0))^{\mathrm{T}} M^{-1}(s) \nabla_\vartheta q(x,t;\vartheta^0). \qquad (2.8)$$

Some criteria may now be set up such that the 'optimal' trajectories $x^1, \ldots, x^N$ minimize $\mathrm{var}(q(x,t;\widehat{\vartheta}))$ over Q. Based on the suggestions of Fedorov and Hackl (1997, p.25), in the sequel the following choices are considered:

$$\Psi_1[M(s)] = \max_{(x,t)\in Q} \mathrm{var}(q(x,t;\widehat{\vartheta})) = \max_{(x,t)\in Q} \mathrm{tr}\left\{A(x,t)M^{-1}(s)\right\} \qquad (2.9)$$

$$\Psi_2[M(s)] = \max_{x\in\mathcal{X}} \int_{\mathcal{T}} \mathrm{var}(q(x,t;\widehat{\vartheta}))\,dt = \max_{x\in\mathcal{X}} \mathrm{tr}\left\{B(x)M^{-1}(s)\right\} \qquad (2.10)$$

$$\Psi_3[M(s)] = \iint_Q \mathrm{var}(q(x,t;\widehat{\vartheta}))\,dx\,dt = \mathrm{tr}\left\{CM^{-1}(s)\right\} \qquad (2.11)$$

where

$$A(x,t) = \left(\nabla_\vartheta q(x,t; \vartheta^0)\right) \left(\nabla_\vartheta q(x,t; \vartheta^0)\right)^{\mathrm{T}}$$

$$B(x) = \int_T \left(\nabla_\vartheta q(x,t; \vartheta^0)\right) \left(\nabla_\vartheta q(x,t; \vartheta^0)\right)^{\mathrm{T}} dt$$

$$C = \iint_Q \left(\nabla_\vartheta q(x,t; \vartheta^0)\right) \left(\nabla_\vartheta q(x,t; \vartheta^0)\right)^{\mathrm{T}} dx \, dt.$$

3 Optimal-control formulation

We assume that the dynamics of sensor motions is given by

$$\dot{s}(t) = f(s(t), u(t)) \quad \text{a.e. on } T, \quad s(0) = s_0 \tag{3.1}$$

where $f : I\!R^{n+r} \to I\!R^n$ ($n = \dim s(t) = 2N$) is required to be continuously differentiable, $s_0 \in I\!R^n$, $u : T \to I\!R^r$ is a measurable control which satisfies

$$u_l \le u(t) \le u_u \quad \text{a.e. on } T \tag{3.2}$$

for fixed $u_l, u_u \in I\!R^r$. Given any s_0 and u, there is a unique absolutely continuous trajectory $s : T \to I\!R^n$ which satisfies (3.1) a.e. on T. Additionally, several pathwise state inequality constraints have to be included, e.g. the requirement of keeping sensors away from one another (for independence of measurements) while staying within a region where measurements are allowed. Hence, for the general case, introduce

$$\gamma_\ell(s(t)) \le 0, \quad \forall t \in T \tag{3.3}$$

for $\ell = 1, \ldots, \nu$, where γ_ℓ's are continuously differentiable. In the sequel, we set $\bar{\nu} = \{1, \ldots, \nu\}$.

Consequently, we wish to solve the following problem:

$$\min_{(s_0, u) \in \mathcal{P}} J(s_0, u) \tag{3.4}$$

subject to the inequality constraint

$$h(s_0, u) \le 0 \tag{3.5}$$

where $J(s_0, u) = \Psi_i[M(s)]$, $h(s_0, u) = \max_{(\ell, t) \in \bar{\nu} \times T} \{\gamma_\ell(s(t))\}$ and $\mathcal{P} = \{(s_0, u) \mid s_0 \in \bar{\Omega}^N, u : T \to I\!R^r \text{ is measurable}, u_l \le u(t) \le u_u \text{ a.e. on } T\}$. A severe difficulty is that Ψ_1 and Ψ_2 are not Fréchet differentiable. In order to reduce this case to the framework handled by the algorithm outlined in what follows (which necessitates such differentability of the performance index), an additional control parameter w is introduced for $i = 1$ or 2 and the equivalent problem of minimizing

$$\mathcal{J}(s_0, u, w) = w \tag{3.6}$$

is considered subject to (3.5) and the additional inequality state constraint

$$e(s_0, u, w) = \Psi_i[M(s)] - w \leq 0 \qquad (3.7)$$

(Due to the differentiability of Ψ_3, its minimization can be performed in much the same way as in Uciński (1999; 2000a; 2000b).)

4 Minimization algorithm

Given $(s_0^0, u^0) \in \mathcal{P}$ which determines a trajectory s^0, consider the linearized problem: Find $\delta s_0^1 \in \mathbb{R}^n$, $\delta u^1 \in L^\infty(T; \mathbb{R}^r)$ and $\delta w^1 \in \mathbb{R}$ to minimize

$$J(s_0^0 + \delta s_0^1, u^0 + \delta u^1, w^0 + \delta w^1) = w^0 + \delta w^1 \qquad (4.1)$$

subject to the constraints

$$\begin{cases} h(s_0^0, u^0, w^0) + \delta h(s_0^0, u^0, w^0; \delta s_0^1, \delta u^1, \delta w^1) \leq 0 \\ e(s_0^0, u^0, w^0) + \delta e(s_0^0, u^0, w^0; \delta s_0^1, \delta u^1, \delta w^1) \leq 0 \\ u_l \leq u^0 + \delta u^1 \leq u_u \\ \|\delta u^1\| \leq \eta_u, \quad \|\delta s_0^1\| \leq \eta_s, \quad |\delta w^1| \leq \eta_w \end{cases} \qquad (4.2)$$

where $w^0 = J(s_0^0, u^0)$, and η_u, η_s and η_w are sufficiently small positive numbers. Here δh and δe denote the Gâteaux differentials of h and e, respectively. Applying the Lagrangian approach to calculation of reduced gradients (Pytlak, 1999; Uciński, 1999; 2000a; 2000b), we conclude that

$$\delta h(s_0^0, u^0, w^0; \delta s_0, \delta u, \delta w)$$
$$= \max_{(\ell, t) \in S} \left\{ \langle \zeta_h^\ell(0; t), \delta s_0 \rangle + \int_0^{t_f} \langle f_u^T(\tau) \zeta_h^\ell(\tau; t), \delta u(\tau) \rangle \, d\tau \right\} \qquad (4.3)$$

where $S = \{ (\ell, t) \in \bar{\nu} \times T : \gamma_\ell(s(t)) = h(s_0, u) \}$, $\zeta_h^\ell(\cdot\,; t)$ is the solution to

$$\frac{d\zeta_h^\ell(\tau; t)}{d\tau} + f_s^T(\tau) \zeta_h^\ell(\tau; t) = -\delta(\tau - t) \nabla_s \gamma_\ell(s^0(\tau)), \quad \zeta_h^\ell(t_f; t) = 0 \qquad (4.4)$$

with $f_s(t) = \partial f(s^0(t), u^0(t))/\partial s$, $f_u(t) = \partial f(s^0(t), u^0(t))/\partial u$. In much the same way, for the criterion Ψ_1 it can be shown that

$$\delta e(s_0^0, u^0, w^0; \delta s_0, \delta u, \delta w) = -\delta w + \max_{(x, t) \in S_1} \{ \langle \zeta_e(0; x, t), \delta s_0 \rangle$$
$$+ \int_0^{t_f} \langle f_u^T(\tau) \zeta_e(\tau; x, t), \delta u(\tau) \rangle \, d\tau \} \qquad (4.5)$$

where $S_1 = \{ (x, t) \in Q : \text{tr}\left[A(x, t) M^{-1}(s^0) \right] = e(s_0^0, u^0, w^0) \}$, $\zeta_e(\cdot\,; x, t)$ denotes the solution to the Cauchy problem

$$\frac{d\zeta_e(\tau; x, t)}{d\tau} + f_s^T(\tau) \zeta_e(\tau; x, t) = -\sum_{i=1}^m \sum_{j=1}^m d_{ij} \nabla_s M_{ij}(s^0(\tau), \tau) \qquad (4.6)$$

with $\zeta_e(t_f; x, t) = 0$, $D_1(s, x, t) = [d_{ij}] = -M^{-1}(s)A(x,t)M^{-1}(s)$,

$$M_{ij}(s(t), t) = \frac{1}{Nt_f} \sum_{\ell=1}^{N} g_i(x^\ell(t), t)g_j(x^\ell(t), t).$$

In the case of Ψ_2, it suffices to replace D_1 by $D_2 = -M^{-1}(s)B(x)M^{-1}(s)$ and S_1 by $S_2 = \{x \in \mathcal{X} : \operatorname{tr}[B(x)M^{-1}(s^0)] = e(s_0^0, u^0, w^0)\}$.

The non-optimality of s_0^0 and u^0 implies the existence of $s^1 = s_0^0 + \delta s_0^1$ and $u^1 = u^0 + \delta u^1$ such that $J(s_0^1, u^1) < J(s_0^0, u^0)$. The process of minimization then resumes with s_0^1 and u^1 instead of s_0^0 and u^0, respectively. By repeating this procedure over and over, a sequence $\{(s_0^k, u^k)\}$ is thus obtained. Proceeding on the same lines as in Galicki and Uciński (2000), we can prove that the sequence of the corresponding trajectories is convergent.

Through a finite-dimensional approximation (e.g. with controls being piecewise linear polynomials), the algorithm reduces to solving a sequence of finite-dimensional linear-programming problems.

5 Numerical example

Consider the two-dimensional diffusion equation

$$\frac{\partial y(x,t)}{\partial t} = \frac{\partial}{\partial x_1}\left(\varkappa(x)\frac{\partial y(x,t)}{\partial x_1}\right) + \frac{\partial}{\partial x_2}\left(\varkappa(x)\frac{\partial y(x,t)}{\partial x_2}\right)$$
$$+ 20\exp\left(-50(x_1 - t)^2\right) \quad \text{on } \Omega \times T = (0,1)^3 \tag{5.1}$$

subject to homogeneous initial and boundary conditions, where

$$\varkappa(x) = \vartheta_1 + \vartheta_2 x_1 + \vartheta_3 x_2, \quad \vartheta_1^0 = 0.1, \quad \vartheta_2^0 = -0.05, \quad \vartheta_3^0 = 0.2$$

ϑ_1^0, ϑ_2^0 and ϑ_3^0 being treated as nominal and known to the experimenter prior to the identification itself. The forcing term in (5.1) imitates a line source whose support is constantly oriented along the x_2-axis and moves with constant speed to the right. Our purpose is to estimate y inside the circle $\mathcal{X}$ with centre $(0.5, 0.5)$ and radius 0.2 during the time horizon $\mathcal{T} = T$, but in the course of the experiment, the sensors must not enter $\mathcal{X}$.

Assuming that the sensor dynamics are not of primary concern, we set

$$\dot{s}(t) = u(t), \quad s(0) = s_0.$$

Moreover, we impose the following constraints on u:

$$|u_i(t)| \leq 0.7, \quad \forall t \in T, \quad i = 1, \dots, 6.$$

The controls were considered in the class of piecewise linear polynomials (with a uniform partition of T into 40 subintervals). The parameters η_u,

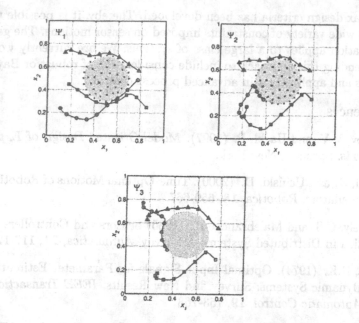

FIGURE 1. Optimal sensor trajectories for different criteria. The lightly shaded circle constitutes the forbidden region $\mathcal{X}$ where spatial prediction is required.

η_s and η_w were gradually decreased from 0.05 to 0.01. For simplicity, the constraints on the minimum allowable distance between the sensors were not considered and only the constraints forcing the sensors to remain in $\bar{\Omega}$ were imposed.

Figure 1 shows the calculated approximations to the optimal motions of the sensors. Fifty crosses in $\mathcal{X}$ denote the discretization points introduced to solve numerically the minimax problems; additionally, ten points forming a uniform partition on $\mathcal{T}$ were considered for criterion Ψ_1 to discretize the time horizon. Open circles, squares and triangles indicate the consecutive sensor positions at time steps corresponding to multiplicities of the period of 0.125 (they are marked in order to reflect the sensor speeds). Furthermore, the same symbols corresponding to the sensors' positions at $t = 0$ are filled with black colour. In principle, the shapes of the trajectories obtained for different criteria are similar (the sensors follow the moving source).

6   Conclusion

This paper formulates the optimal sensor location problem in an experiment performed for prediction of certain variables depending on spatial location and/or time. An iterative algorithm capable of handling the adopted minimax design criteria has been developed. Thereby, it is possible to enforce a wide variety of constraints imposed on sensor motions. The general formulation applies to a large class of systems. We are currently working on extending this approach to include consideration of robust or Bayesian designs and applications in advanced process control.

References

Fedorov, V.V. and Hackl, P. (1997). *Model-Oriented Design of Experiments*. Springer, New York.

Galicki, D. and Uciński, D. (2000). Time-Optimal Motions of Robotic Manipulators. *Robotica*, **18**, 659-667.

Kubrusly, C.S. and Malebranche H. (1985) Sensors and Controllers Location in Distributed Systems: A Survey. *Automatica*, **21**, 117-128.

Mehra, R.K. (1974). Optimal Input Signals for Parameter Estimation in Dynamic Systems: Survey and New Results. *IEEE Transactions on Automatic Control*, **19**, 753-768.

Müller W.G. (1998). *Collecting Spatial Data. Optimum Design of Experiments for Random Fields*. Physica-Verlag, Heidelberg.

Pytlak R. (1999). *Numerical Methods for Optimal Control Problems with State Constraints*. Springer-Verlag, Berlin.

Rafajlowicz, E. (1986). Optimum Choice of Moving Sensor Trajectories for Distributed Parameter System Identification. *International Journal of Control*, **43**, 1441-1451.

Sun, N.-Z. (1994). *Inverse Problems in Groundwater Modeling*. Kluwer Academic Publishers, Dordrecht.

Uciński, D. (1999). *Measurement Optimization for Parameter Estimation in Distributed Systems*. Technical University Press, Zielona Góra.

Uciński, D. (2000a). Optimization of Sensors' Allocation Strategies for Parameter Estimation in Distributed Systems. *Systems Analysis–Modelling–Simulation*, **37**, 243-260.

Uciński, D. (2000b) Optimal Sensor Location for Parameter Estimation of Distributed Processes. *International Journal of Control*, **73**, 1235-1248.

Quality Improvement of Signal-Dependent Systems

I.N. Vuchkov
L.N. Boyadjieva

ABSTRACT: This paper presents a model-based approach to quality improvement of signal-dependent systems. Mean and variance models for such systems are given, including the case of performance characteristic's variance depending on time. It is shown that the additional information obtained from measurable external variables, called signals, can improve process quality considerably.

KEYWORDS: quality improvement; signal-dependent systems; design of experiments

1 Introduction

Consider a class of systems with performance characteristics, which depend on factors measurable during experimentation, mass production or usage. These factors are called *signal factors*. Sometimes the signal factors can also be controllable in mass production. Some typical examples are given below.

i) Taguchi (1986) considers a class of products or processes with performance characteristics changing due to user's intention, which is transmitted to the system through a signal factor. Taguchi refers to such products as "dynamic systems". An example given by Taguchi is a car braking system, which has to reduce the car's speed in a manner desired by the driver. The Driver's intention is transmitted to the car by pressing the brake pedal. The braking system must be designed to provide stable behaviour of the car while stopping. Factors such as road conditions, tyre type, front tyre air pressure, etc. may act as perturbations (external noises) to the braking system. Its parameters must be chosen to minimize the noise effect. In this example the signal factor is also controllable.

Note that Taguchi uses the term "dynamic systems" in a different way than its use in control science, where a dynamic characteristic is always a

mODa6, A.C.Atkinson, P.Hackl and W.G.Müller, eds., Physica, Heidelberg, 2001.

function of time or frequency and describes the transient behaviour of a system. While keeping the signal factor according to the design, Taguchi studies the steady state and not the transient process of the system.

ii) Consider a process with performance characteristics that depend on environmental factors, some of them being observable but uncontrollable, others being neither observable nor controllable. All environmental factors can be considered as noise factors and the system can be made robust to them. Box and Jones (1992) proposed models for the mean and variance of the performance characteristics of such systems. The variance can be minimized while keeping the mean value on target.

Another approach is to explicitly include the observable but uncontrollable factors in the mean and variance models. These factors can be considered as signal factors and the additional information that they bring to the system enables us to reduce performance characteristic's variation (Pledger, 1996).

iii) In many cases product or process performance characteristics are considered as functions of time, but they also depend on some product parameters. The means and variances of the performance characteristics are also functions of time. In this case, time can be considered as a signal. We refer to these systems as signal-dependent dynamic systems. An example is the tyre production process. The starting moment, the time and rate of curing, as well as the ability of the rubber composition for further processing, depend on the rheological properties of the composition. They are functions of time, but also depend on the proportions of mixture components in the rubber.

There are many examples of signal-dependent systems in control engineering. Modelling of signal-dependent systems is considered by Velev *et al.* (1992). Variation of their characteristics in time is still not studied. Several authors consider quality improvement of signal-dependent systems. Miller and Wu (1996) call them signal-response systems.

In this paper a model-based approach to quality improvement of signal-dependent systems is proposed.

2 Mean and variance models of the performance characteristics of signal-dependent systems

One way of modelling signal-dependent systems is to express the model coefficients as functions of signals. If the signals can be measured during the production process the model can be adjusted and used to control the system appropriately (Vuchkov *et al.* (1992). Similar ideas emerge in Miller and Wu (1996), where a two-stage design of experiments procedure is used for quality improvement of Taguchi's "dynamic" systems. In this paper we propose a one-stage approach for quality improvement of signal-dependent systems. Consider a product or system with product parameters

$\mathbf{p} = (p_1 p_2 ... p_m)^T$. They may be disturbed by errors in mass production and let the vector of errors be $\mathbf{e} = (e_1 e_2 ... e_m)^T$. Assume also that there are some external noise factors $\mathbf{n} = (n_1 n_2 ... n_q)^T$ as well. Let $\mathbf{s} = (s_1 s_2 ... s_l)^T$ be the vector of signal factors.

Various models can be used to describe the performance characteristic. They are used to derive mean or variance models for the performance characteristic during mass production or usage (Vuchkov and Boyadjieva, 2000). They are sensitive to the structure of the original models, which should be carefully tested for lack of fit. In order to be more specific, and in relation to the example in Section 4, we consider a second order polynomial as follows:

$$\hat{y} = y(\mathbf{p}, \mathbf{n}, \mathbf{s}) = \beta_0' + \beta_p^T \mathbf{p} + \mathbf{p}^T \mathbf{B} \mathbf{p} + \alpha_n^T \mathbf{n} + \mathbf{n}^T \mathbf{A} \mathbf{n} + \mathbf{p}^T \mathbf{G} \mathbf{n} + \beta_s^T \mathbf{s} +$$

$$+ \mathbf{s}^T \mathbf{B}_{ss} \mathbf{s} + \mathbf{s}^T \mathbf{B}_{sp} \mathbf{p} + \mathbf{s}^T \mathbf{A}_{sn} \mathbf{n} + \varepsilon \quad (2.1)$$

where ε is an independent random disturbance, $\beta_p = (\beta_1 \beta_2 ... \beta_m)^T$, $\alpha_n = (\alpha_1 \alpha_2 ... \alpha_q)^T$, $\beta_s = \left(\beta_1^{(s)} \beta_2^{(s)} ... \beta_l^{(s)} \right)^T$, $\mathbf{B}$ is an $(m \times m)$ matrix, $\mathbf{A}$ is a $(q \times q)$ matrix, $\mathbf{A}_{sn}$ is an $(l \times q)$ matrix, $\mathbf{B}_{ss}$ is an $(l \times l)$ matrix, $\mathbf{B}_{sp}$ is an $(l \times m)$ matrix, and $\mathbf{G}$ is an $(m \times q)$ matrix.

A combined array can be used and least-squares estimates of the model coefficients can then be calculated. The performance characteristic's model(2.1) can be rewritten as follows:

$$y(\mathbf{p}, \mathbf{n}, \mathbf{s}) = \beta_0' + \beta_p^T \mathbf{p} + \mathbf{p}^T \mathbf{B} \mathbf{p} + \alpha_n^T \mathbf{n} + \mathbf{n}^T \mathbf{A} \mathbf{n} + \mathbf{p}^T \mathbf{G} \mathbf{n} + \varepsilon \quad (2.2)$$

where $\beta_0 = \beta_0' + \beta_s^T \mathbf{s} + \mathbf{s}^T \mathbf{B}_{ss} \mathbf{s}$, $\beta^T = \beta_p^T + \mathbf{s}^T \mathbf{B}_{sp}$, and $\alpha^T = \alpha_n^T + \mathbf{s}^T \mathbf{A}_{sn}$. Suppose that the noise variances are $\Sigma_e = diag(\sigma_1^2 \sigma_2^2 ... \sigma_m^2)$ and $\Sigma_n = diag(\sigma_{n1}^2 \sigma_{n2}^2 ... \sigma_{nq}^2)$.

Vuchkov and Boyadjieva (1992) proposed the following models of performance characteristic's mean and variance:

$$\tilde{y}(\mathbf{p}, \mathbf{s}) = \beta_0 + \beta^T \mathbf{p} + \mathbf{p}^T \mathbf{B} \mathbf{p} + tr(\mathbf{B}\Sigma_e) + tr(\mathbf{A}\Sigma_n), \quad (2.3)$$

$$\sigma^2(\mathbf{p}, \mathbf{s}) = (\beta + 2\mathbf{B}\mathbf{p})^T \Sigma_e (\beta + 2\mathbf{B}\mathbf{p}) + (\alpha + \mathbf{G}\mathbf{p})^T \Sigma_n (\alpha + \mathbf{G}\mathbf{p}) + HOT + \sigma_\varepsilon^2 \quad (2.4)$$

where

$$HOT = 2 \sum_{i=1}^{m} \beta_{ii}^2 \sigma_i^4 + \sum_{i=1}^{m-1} \sum_{j=1+1}^{m} \beta_{ij}^2 \sigma_i^2 \sigma_j^2 + 2 \sum_{i=1}^{q} \alpha_{ii}^2 \sigma_{ni}^4 +$$

$$+ \sum_{i=1}^{q-1} \sum_{j=1+1}^{q} \alpha_{ij}^2 \sigma_{ni}^2 \sigma_{nj}^2 + \sum_{i=1}^{m} \sum_{j=1+1}^{q} \gamma_{ij}^2 \sigma_i^2 \sigma_{nj}^2. \quad (2.5)$$

In this equation β_{ii} and $\beta_{ij}/2$ are diagonal and off-diagonal elements of $\mathbf{B}$ respectively, α_{ii} and $\alpha_{ij}/2$ are the corresponding elements of $\mathbf{A}$, while the γ_{ij} are elements of $\mathbf{G}$.

One can see that the coefficients in (2.2), (2.3) and (2.4) are functions of the signals. As the signals are measurable in production or usage, these models can be used for adjustment of mean and variance models according to the observed signal values. With the adjusted models we reduce the quality improvement problem to the problem of variance minimization while keeping the mean value on a target. In Taguchi's "dynamic" systems the signal is controllable and can be chosen so as to minimize the variance and to adjust the mean to the desired target value.

If the system parameters can be adapted according to the changes of the signal factors, we call it an *adaptive* system in contrast to *robust* systems that consider the external disturbances as noise factors.

3 Variance models for performance characteristics depending on time

Consider a system with performance characteristic depending on several parameters $\mathbf{p} = (p_1 p_2 ... p_m)^T$ and time t, which is considered as a signal factor. Other signal factors $\mathbf{s}$ may also exist. Generally a state-space model is able to describe the process:

$$x(t+1) = \mathbf{A}(\mathbf{p}, \mathbf{s}) x(t) + \mathbf{B}(\mathbf{p}, \mathbf{s}) u(t) + \varepsilon(t),$$

$$y(t) = \mathbf{C}(\mathbf{p}, \mathbf{s}) x(t) + \nu(t).$$

In this model $y(t)$ is the system output, $u(t)$ the input, $x(t)$ a state variable, $\varepsilon(t)$ the output noise, $\nu(t)$ is measurement noise, and $\mathbf{A}(\mathbf{p}, \mathbf{s})$, $\mathbf{B}(\mathbf{p}, \mathbf{s})$ and $\mathbf{C}(\mathbf{p}, \mathbf{s})$ are matrices whose elements depend on $\mathbf{p}$ and $\mathbf{s}$. Suppose that system parameters $\mathbf{p}$ can be tuned to minimize the output variance, provided that $\mathbf{s}$ and t are known. In tyre production (see the example in Section 1) mixture proportions are the product parameters, while the time is a signal factor. One may wish to obtain minimum output variance at a given instant of time.

There are many discrete-time models that can be considered as special cases of the state-space model (Ljung 1987). The most popular are: Auto-Regression (AR), Moving Average (MA), Auto-Regressive Moving Average with Exogenous Input (ARMAX) and Finite Impulse Response (FIR). Systems without input signals (such as the tyre production example) are best described using AR. For the sake of simplicity we shall consider an AR model, but the same approach is also applicable to the rest of the models mentioned above.

Let y_r be a performance characteristic of the system at time $t_r = r\Delta t$, where Δt is a constant time interval. An auto-regressive model for a parameter-dependent system can be written as follows:

$$y_r = -\alpha_1(\mathbf{p}) y_{r-1} - ... - \alpha_n(\mathbf{p}) y_{r-n} + \varepsilon_r.$$

Vuchkov *et al.* (1992) proposed methods for estimation of parametrically dependent models. Suppose $\widehat{a}_\nu(\mathbf{p})$, $\nu = 1, 2, ...n$ are the estimates of the model coefficients and that they can be expressed as polynomial functions of the parameters:

$$\widehat{a}_\nu(\mathbf{p}) = \widehat{\delta}_{\nu o} + \sum_{i=1}^{m-1} \widehat{\delta}_{\nu i} p_i + \sum_{i=1}^{m-1} \sum_{j=1+1}^{m} \widehat{\delta}_{\nu ij} p_i p_j + \sum_{i=1}^{m} \widehat{\delta}_{\nu ii} p_i^2.$$

Using these estimates, the AR-model becomes:

$$\widehat{y}_r = -\widehat{a}_1(\mathbf{p}) \widehat{y}_{r-1} - ... - \widehat{a}_n(\mathbf{p}) \widehat{y}_{r-n} =$$

$$= c_{ro} + \sum_{i=1}^{m} c_{ri} p_i + \sum_{i=1}^{m-1} \sum_{j=1+1}^{m} c_{rij} p_i p_j + \sum_{i=1}^{m} c_{rii} p_i^2, \qquad (3.1)$$

where

$$c_{ro} = -\sum_{\nu=1}^{n} \widehat{\delta}_{\nu 0} \widehat{y}_{r-\nu} \; , c_{ri} = -\sum_{i=1}^{m} \sum_{\nu=1}^{n} \widehat{\delta}_{\nu i} \widehat{y}_{r-\nu}, \; c_{rij} = -\sum_{i=1}^{m-1} \sum_{j=1+1}^{m} \sum_{\nu=1}^{n}$$

$$\widehat{\delta}_{\nu ij} \widehat{y}_{r-\nu}, \; c_{rii} = -\sum_{i=1}^{m} \sum_{\nu=1}^{n} \widehat{\delta}_{\nu ii} \widehat{y}_{r-\nu} \; .$$

Suppose the process parameters are subject to errors $e_i, i = 1, 2, ...m$. One can replace p_i by $p_i + e_i$ and p_j by $p_j + e_j$ in (3.1) and calculate a response variance at moment $t_r = r\Delta t$. Let the errors in the process parameters be uncorrelated, with normal distribution, zero mean and variance σ_i^2. Using a formula given in Vuchkov and Boyadjieva (1988) one can calculate the variance caused by the errors in the process parameters as follows:

$$\sigma_r^2 = \sum c_{ri} \sigma_i^2 \left(c_{ri} + 2 c_{rii} p_r + \sum_{i=1, j \neq i}^{m} c_{rij} p_i \right) + HOT + \sigma_\varepsilon^2, \qquad (3.2)$$

where σ_ε^2 is the response error variance and

$$HOT = 2 \sum_{i=1}^{m} c_{rii}^2 \sigma_i^4 + \sum_{i=1}^{m-1} \sum_{j=1+1}^{m} c_{rij}^2 \sigma_i^2 \sigma_j^2. \qquad (3.3)$$

Formulae (3.2) and (3.3) can be used to calculate the variance at a given instant of time or to calculate an integral characteristic of variance over a given period of time $T = h\Delta t$:

$$I = \frac{1}{h} \sum_{r=1}^{k} \sigma_r^2. \qquad (3.4)$$

The quality improvement problem comprises selection of process parameters that minimize (3.4).

4 Example: quality improvement of a process with an observable and uncontrollable factor

If a noise factor is observable, but uncontrollable in the production process, its observed value can be used in the mean and variance models and this may improve the performance characteristics. Following the approach of Section 2 the observable uncontrollable factors can be considered as signals and equations (2.3) and (2.4) are used to calculate performance characteristic's mean and variance.

Consider the following example. Suppose that a medicine is produced by a chemical reaction, which depends on stock and environmental temperature. Target value for the performance characteristic y is 120. The environmental temperature is an observable but uncontrollable factor and varies between $10°C$ and $26°C$. If it is normally distributed its standard deviation is $2.67°C$. During the production process the stock is perturbed by a normally distributed random error with standard deviation $2kg$.

An experiment is carried out in which the stock is varied in the interval $160kg$ to $200kg$ and the environmental temperature in the interval $10°C$ to $26°C$. The design of experiments in coded factors is given in Table 1.

No	p	n	y	No	p	n	y
1	-1	-1	111.361	6	1	0	124.968
2	1	-1	133.435	7	-1	0	122.098
3	-1	1	128.483	8	0	1	121.262
4	1	1	121.768	9	0	-1	113.875
5	0	0	116.487				

TABLE 1. Experimental design for the example.

The following regression model was obtained:

$$\hat{y} = 116.866 + 3.038p + 2.140n - 7.197pn + 6.478p^2 + 0.513n^2. \quad (4.1)$$

The model is characterized by the following estimated values: multiple correlation coefficient 0.98; $F = 14.71$, residual variance $\sigma_e^2 = 5.09$. The coded variance of the stock is $\sigma_e^2 = 0.01$ and the coded variance of the environmental temperature is $\sigma_n^2 = \frac{1}{9}$.

The environmental temperature can be considered as external noise (n) or as signal factor (s). We examine both options.

i) If the environmental temperature is considered as an external noise factor (n) one can find a model of the mean value in mass production using formula (2.3):

$$\tilde{y} = 116.988 + 3.038p + 6.478p^2$$

where the intercept is calculated as follows:

$$\hat{\beta}_0 = \hat{\beta}_0' + \hat{\beta}_{pp}\sigma_e^2 + \hat{\alpha}_{nn}\sigma_n^2 = 116.866 + 6.478 \times 0.01 + 0.513/9 = 116.988.$$

The variance calculated by formulae (2.4) and (2.5) is:

$$\sigma^2 = 5.763 - 2.635p + 7.434p^2. \tag{4.2}$$

Since the target value is 120, one can obtain the optimal parameter value from the equation $6.478p^2 + 3.038p - 3.012 = 0$. There are two solutions: $p_{1*}^n = 0.478$ and $p_{2*}^n = -0.956$. By substituting these values in (4.2) we obtain $\sigma^2(p_{1*}^n) = 6.243$ and $\sigma^2(p_{2*}^n) = 15.076$. The optimal solution is $p_{1*}^n = 0.478$ because it provides smaller variance than does p_{2*}^n.

ii) Consider the option where the environmental temperature is a signal factor (s). Suppose that a temperature of 24.4 $^\circ C$ was measured, and its coded value is $s = 0.8$. With $n = s = 0.8$ the regression (4.1) is now

$$\hat{y} = 118.906 - 2.720p + 6.478p^2$$

where the model coefficients were calculated as follows:

$$\hat{\beta}_0 = \hat{\beta}_0' + \hat{\beta}_s s + b_{ss}s^2 = 116.866 + 2.14 \times 0.8 + 0.513 \times 0.8^2 = 118.906,$$

$$\hat{\beta}_1 = \hat{\beta}_p + \hat{\beta}_{ps}s = 3.038 - 7.197 \times 0.8 = -2.720.$$

The mean value in the production process can be calculated by (2.3):

$$\tilde{y} = 118.971 - 2.720p + 6.478p^2.$$

Formulae (2.4) and (2.5) are used to find the variance model

$$\sigma^2 = 5.172 - 0.705p + 1.679p^2. \tag{4.3}$$

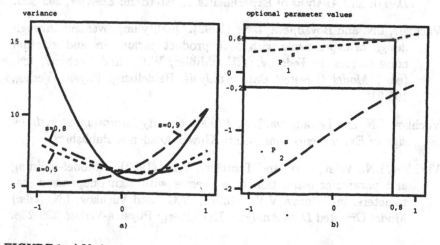

FIGURE 1. a) Variance versus p: robust system-solid line, adaptive system-dotted lines b) Optimal parameter values for target 120 as a function of the signal s.

Two parameter values provide performance characteristics equal to the target: $p_{1*}^s = 0.66$ and $p_{2*}^s = -0.240$. The corresponding values of the variance are: $\sigma^2(p_{1*}^s) = \sigma^2(p_{2*}^s) = 5.439$.

Plots of variances calculated from equations (4.2) and (4.3) are given in Figure 1a). Figure 1b) shows the optimal parameter values against the signal. One can see that an adaptive system that utilizes the information provided by the signal considerably reduces the variance compared to a robust system that treats the environmental temperature as a noise factor.

References

Box, G.E.P. and Jones, S. (1992). Designing products that are robust to environment. *Total Quality Management*, **3**, 265 - 282.

Ljung, L. (1987). *System Identification*. Englewood Cliffs, N.J.: Prentice Hall.

Miller, A. and Wu, C.F.J. (1996). Parameter design for signal-response systems: a different look at Taguchi's dynamic parameter design. *Statistical Science*, **11**, 122-136.

Pledger, M. (1996). Observable uncontrollable factors in parameter design. *Journal of Quality Technology*, **28**, 153-162.

Taguchi, G. (1986). *Introduction to Quality Engineering*. Tokyo: Asian Productivity Organization.

Vuchkov, I.N. and Boyadjieva, L.N. (1988). The robustness against tolerances of performance characteristics described by second order polynomials. In Dodge, Y., Fedorov, V.V., and Wynn, H.P. (eds.) *Optimal Design and Analysis of Experiments*. Amsterdam: Elsevier, 293-309.

Vuchkov, I.N. and Boyadjieva, L.N. (1992). Quality improvement through design of experiments with both product parameters and external noise factors. In Fedorov, V.V., Müller, W.G., and Vuchkov, I.N. (eds.) *Model Oriented Data Analysis*. Heidelberg: Physica-Verlag, 195-212.

Vuchkov, I.N. and Boyadjieva, L.N. (2000). *Quality Improvement with design of Experiments*. Dordrecht: Kluwer Academic Publishers.

Vuchkov, I.N., Velev, K.D., and Tzotchev, V.K. (1992). Model-building and parameter estimation of processes with signal-dependent parameters. In Fedorov, V.V., Müller, W.G., and Vuchkov, I.N. (eds.) *Model Oriented Data Analysis*. Heidelberg: Physica-Verlag, 229-238.

Recursive Algorithm for Digital Diffusion Networks and Applications to Image Processing

G. Yin
P.A. Kelly
M.H. Dowell

ABSTRACT: This work aims to develop a class of recursive algorithms for a digital diffusion network, which is an approximation of an analog diffusion machine. Our main effort is to prove the convergence of a continuous-time interpolation of the discrete-time algorithm to that of the analog diffusion network via weak convergence methods. The parallel processing feature of the network makes it attractive for solving large-scale optimization problems, and applicable not only to image estimation problems but also to optimal design, optimal control, and related fields. As an application, we consider image estimation problems and present simulation results.

KEYWORDS: convergence; digital diffusion network; image estimation; recursive algorithm

1 Introduction

This paper considers a class of stochastic optimization algorithms. Although the primary motivation of our study stems from applications in image estimation, the techniques used and the numerical methods developed may be useful for applications arising from optimal design and other related fields.

Recently, Wong (1992) suggested an analog diffusion machine based on modifications of the Langevin algorithm and the Hopfield network. Let $\mathcal{E} : [0,1]^r \mapsto \mathbb{R}$ be an "energy" function defined on the hypercube $[0,1]^r = [0,1] \times \cdots \times [0,1]$. Find the global minimizer of $\mathcal{E}$ using a neural network. Suppose that for all $t \geq 0$, each $\alpha = 1, \ldots, r$, $v_\alpha(t) \in [0,1]$ is the state at node α at time t and $v = (v_1, \ldots, v_r)^\tau \in [0,1]^r$ is an r-dimensional col-

mODa6, A.C.Atkinson, P.Hackl and W.G.Müller, eds., Physica, Heidelberg, 2001.

umn vector (z^τ denotes the transpose of z). By injecting noise into a Hopfield network, the dynamics of the αth node are given by $v_\alpha(t) = g(u_\alpha(t))$, $du_\alpha(t) = -(\partial/\partial v_\alpha)\mathcal{E}(v(t))dt + \tilde{a}_\alpha(u(t))dw_\alpha(t)$, where for $\alpha \leq r$, w_α are independent (standard and real-valued) Brownian motions, and $\tilde{a}_\alpha$ and g are appropriate functions. Choosing $\tilde{a}_\alpha$ to be $\tilde{a}_\alpha(u(t)) = [(2T)/g'(u_\alpha(t))]^{1/2}$ (where g' denotes the derivative of g), v is a stationary Markov process with stationary density $p_\infty(v) = (1/Z)\exp(-(1/T)\mathcal{E}(v))$, where Z is an appropriate normalizing factor so that $\int p_\infty(v)dv = 1$. Furthermore, by selecting $f(x) = g'(g^{-1}(x))$ (for each $x \in \mathbf{R}$), for each $\alpha \leq r$, $dv_\alpha(t) = -f(v_\alpha(t))(\partial/\partial v_\alpha)\mathcal{E}(v(t))dt + Tf'(v_\alpha(t))dt + \sqrt{2Tf(v_\alpha(t))}dw_\alpha(t)$, where T goes to zero sufficiently slowly. By appropriate choice of g, v is stationary, one need not worry about the reflected boundary. It is interesting to note that $\sqrt{2Tf(v_\alpha(t))}$ depends only on the αth node, and the noise of the system is "decoupled" among different processors, which allows us to efficiently use parallel processors and to simplify the computational tasks. Subsequent work on diffusion networks was carried out in Kesidis (1995). He noted that an analog implementation of the network appears to be impractical for large-scale problems. To take advantages of Wong's diffusion network and to overcome the difficulties of the analog implementation, Cai et al. (1995) proposed a digital diffusion network (a discrete algorithm). Based on such an idea, we propose a recursive algorithm in this paper and show that an appropriately scaled sequence of the discrete iterates converges to Wong's diffusion machine.

The rest of the paper is arranged as follows. Section 2 presents the recursive algorithm. It also gives the asymptotic results concerning the digital diffusion network. It is shown that a continuous-time interpolation of the discrete sequence converges to a diffusion, and the machinery used to establish the desired result is the method of weak convergence; see Kushner and Yin (1997). Section 3 then proceeds with applications of image estimation. Finally, we close the paper with some further remarks in Section 4.

2   Properties of the digital diffusion networks

First let $\mathcal{E} : \mathbf{R}^r \mapsto \mathbf{R}$. Suppose $f : \mathbf{R} \mapsto \mathbf{R}$, and $g : \mathbf{R} \mapsto \mathbf{R}$. Throughout the paper, unless otherwise noted, a Greek letter α or β denotes a component (or corresponding to a processor) and i, k and n denote the indices of the iterations. Inspired by the ideas in stochastic approximation (see, for example, the up-to-date treatment of Kushner and Yin, 1997) and the simulated annealing algorithm discussed in Gelfand and Mitter (1991), we consider a recursive algorithm with a periodic restarting device. The idea is to partially reset the step size sequence once in a while. For each $\iota \geq 0$, and each $\alpha \leq r$, construct the estimates

$$v_{\alpha,\iota n+k+1} = v_{\alpha,\iota n+k} - a_{\iota n+k}f(v_{\alpha,\iota n+k})\frac{\partial}{\partial v_{\alpha}}\mathcal{E}(v_{\iota n+k})$$
$$+c_{\iota n+k}f'(v_{\alpha,\iota n+k}) + b_{\iota n+k}\sqrt{f(v_{\alpha,\iota n+k})}W_{\alpha,\iota n+k}, \tag{2.1}$$

where $\varepsilon > 0$ is a small parameter, $A_0 > 1$, and

$$a_{\iota n+k} = \varepsilon, \quad b_{\iota n+k} = \frac{\sqrt{2\varepsilon}}{\sqrt{\ln[\varepsilon k + A_0]}}, \quad c_{\iota n+k} = \frac{\varepsilon}{\ln[\varepsilon k + A_0]}. \tag{2.2}$$

(A) f, g are continuously differentiable and $\mathcal{E}_{vv}$ (the second partials of the function $\mathcal{E}$) is continuous and bounded such that $\mathcal{E}(v) \geq 0$ for all $v \in \mathbf{R}^r$ and $\min_v \mathcal{E}(v) = 0$; $f(x) \geq 0$ for all $x \in \mathbf{R}$, and f is bounded with bounded derivative, the inverse g^{-1} exists and is continuously differentiable; $f(x) = g'(g^{-1}(x))$. For each $\alpha \leq r$, $\{W_{\alpha,k}\}$ is a sequence of independent and identically distributed random variables such that $EW_{\alpha,k} = 0$; $EW_{\alpha,k}^2 = 1$; For $\alpha \neq \beta$, $W_{\alpha,k}$ and $W_{\beta,k}$ are independent.

The smoothness of f and g is verified for the segmentation problems treated in the next section as well as in many other applications. Owing to the conditions, $\mathcal{E}$ and f, verify a Lipschitz condition in the domain of interest. As a consequence, (2.3) has a unique solution for each initial condition. For each $\alpha \leq r$, define $w_\alpha^\varepsilon(t) = \sqrt{\varepsilon}\sum_{i=0}^{t/\varepsilon-1} W_{\alpha,n+i}$, and denote $w^\varepsilon(t) = (w_1^\varepsilon(t), \ldots, w_r^\varepsilon(t))^\tau$.

Our analysis is for a fixed ι, without loss of generality, set $\iota = 1$ henceforth. Define $v_\alpha^\varepsilon(t)$ as $v_\alpha^\varepsilon(0) = v_{\alpha,n}$, and for $t \in [\varepsilon k, \varepsilon(k+1))$, $v_\alpha^\varepsilon(t) = v_{\alpha,n+k}$, and denote $v^\varepsilon(t) = (v_1^\varepsilon(t), \ldots, v_r^\varepsilon(t))^\tau$. Note that $v^\varepsilon(\cdot) \in D^r[0,\infty)$, the space of functions that are right continuous, and have left limits endowed with the Skorokhod topology. Under (A), we are able to show that $\{v^\varepsilon(\cdot)\}$ is tight in $D^r[0,\infty)$, and any weakly convergent subsequence has a limit $v(\cdot)$, whose components are the solutions of

$$dv_\alpha(t) = -f(v_\alpha(t))\frac{\partial}{\partial v_\alpha}\mathcal{E}(v(t))dt + T(t)f'(v_\alpha(t))dt + \sqrt{2T(t)f(v_\alpha(t))}dw_\alpha(t),$$
$$\tag{2.3}$$

for $\alpha = 1, \ldots, r$, with $T(t) = 1/\ln(t + A_0)$

Next, we aim to obtain the tightness of the sequence $\{v_{n+k}\}$. Denote

$$F(v) = \text{diag}(f(v_1), \ldots, f(v_r)), \quad D(v) = (f'(v_1), \ldots, f'(v_r))^\tau,$$
$$\Sigma(v) = \text{diag}(\sqrt{f(v_1)}, \ldots, \sqrt{f(v_r)}), \quad \mathcal{E}_v(v) = (\frac{\partial}{\partial v_1}\mathcal{E}(v), \ldots, \frac{\partial}{\partial v_r}\mathcal{E}(v))^\tau,$$

where $\text{diag}(d_1, \ldots, d_r)$ denotes a diagonal matrix with diagonal entries d_1 through d_r. In view of the notation above, (2.3) can be written as

$$v_{n+k+1} = v_{n+k} - \varepsilon F(v_{n+k})\mathcal{E}_v(v_{n+k}) + c_{n+k}D(v_{n+k}) + b_{n+k}\Sigma(v_{n+k})W_{n+k}. \tag{2.4}$$

Assume (A), $EV(v_0) < \infty$, and suppose that there is a twice continuously differentiable Liapunov function $V : \mathbf{R}^r \mapsto \mathbf{R}$ for (2.3) such that $V(v) \geq 0$ for all v, $V(v) \to \infty$ as $|v| \to \infty$; $|V_v(v)| \leq K(1 + V^{1/2}(v))$, $|V_{vv}(\cdot)| \leq K$, and $|\mathcal{E}_v(v)| \leq K(1 + V^{1/2}(v))$; $V_v^\tau(v)F(v)\mathcal{E}_v(v) \geq \lambda_0 V(v)$ for some $\lambda_0 > 0$ and $v \notin S = \{v \in \mathbf{R}^r; \ \mathcal{E}_v(v) = 0\}$; Then $EV(v_{n+k}) = O(1)$.

Our weak convergence gives a result for large but still bounded t. We proceed further by treating the large-time behavior of the algorithm when the energy function is confined to the domain $[0,1]^r$, which enables us to use the results of Wong. Note that in this case, the iterates obtained from our algorithm can be shown to be bounded for sufficiently small ε.

As noted by Wong, if $f(0) = f(1) = 0$ then $v(\cdot)$ is stationary. The modification for $\mathcal{E}$ defined on $[0,1]^r$ can be done as in Yin et al. (2000). In fact, suppose (A) is satisfied with $\mathbf{R}^r$ replaced by $[0,1]^r$, $\mathcal{E}$ is defined on $[0,1]^r$, f satisfies $f(0) = f(1)$, $\{W_k\}$ is a sequence of bounded random variables, $\{v_{n+k}\} \in [0,1]^r$, and there is a unique minimizer $v^* = \mathrm{argmin}\mathcal{E}(v)$. Then $v^\varepsilon(\cdot + \widehat{q}_\varepsilon) \overset{\varepsilon}{\longrightarrow} v^*$ in probability for any $\widehat{q}_\varepsilon \to \infty$ as $\varepsilon \to 0$.

3   Application to image estimation

Using the algorithm developed, we can carry out the computational tasks for image estimation problems. One of the important features of the diffusion network is the possibility of fast, parallel optimization of systems having many variables. The main ingredient is to recast the desired optimization problem as minimization of an energy function over $[0,1]^r$. Image estimation belongs to such a class of optimization algorithms. The desired outcome is a Bayesian estimate in the form of a segmentation–a partition of the image into a small number of classes, or a restoration–recovery of image data from corrupted observations. For images modeled by Markov Random Fields, the estimation is accomplished by minimizing a Gibbs distribution energy over an appropriate space; see Hokland and Kelly (1996), Derin et al. (1990) and Manjunath et al. (1990), among others. It is well known that the usual sequential implementations of simulated annealing are too slow for many practical applications, while faster deterministic approximations often fail to achieve good estimates. It has been shown by Manjunath et al. (1990) that for a number of segmentation problems, the space over which the Gibbs distribution energy must be minimized can be taken as the discrete set $\{0,1\}^r$ (where r is an integer that is generally larger than the number of pixels in the image). In Manjunath et al. (1990), parallel computations of approximations to simulated annealing are performed over this space by means of a modified Hopfield network or a stochastic network similar to a Boltzman machine. In this section, we use a similar approach to consider some examples of the application of diffusion networks (operating over $[0,1]^r$) for image segmentation and restoration.

Let Ω denote the $K \times L$ lattice of pixels on which images are defined, and let the pixels be indexed by $\ell = 1, \ldots, KL$. Suppose that the desired image consists of M constant-intensity regions. Let $\{\mu_m : m = 1, \ldots, M\}$ denote the set of region intensities. We define $X = \{X(\ell) : \ell = 1, \ldots, KL\}$ to be the field of *region labels*. That is, each $X(\ell)$ takes a value in the set $\{1, \ldots, M\}$; and if $X(\ell) = x(\ell)$, then the mean intensity at pixel ℓ is $\mu_{x(\ell)}$. To impose on the model the spatial continuity inherent in image regions, we assume that X has a Gibbs distribution in the form $P(X = x) \propto \exp(\sum_{c \in C} \{V_c(x(m) : m \in c)\})$ where c is a subset of $\{1, \ldots, KL\}$ called a *clique*; C is the collection of all cliques; and V_c is some function of x restricted to c. Let $\eta_\ell = \{m : m \text{ is in a clique with } \ell\}$ (η_ℓ is called the *neighborhood* of ℓ). As is shown in Geman and Geman (1984), X has the Markov property $P(X(\ell) = x(\ell) \mid X(m) = x(m), \ m \neq \ell) = P(X(\ell) = x(\ell) \mid X(m) = x(m), \ m \in \eta_\ell)$. An example of a Gibbs distribution, commonly used in image processing, is $P(X = x) \propto \exp\{-2\beta \sum_{\ell=1}^{KL} \sum_{m \in \eta_\ell} [1 - \delta(x(\ell) - x(m))]\}$ where β is a positive constant, η_ℓ is the set of four or eight pixels nearest to ℓ, and δ is the Kronecker delta function.

In a recent paper (Yin *et al.*, 2000), we treated several examples of image segmentation with Gaussian white noise. In this paper we will consider segmentation problems with speckle-like noise. Such problems arise in, for example, radar imaging, in which the observed image has the form of exponentially-distributed speckle with different means in different regions; see Derin *et al.* (1990). That is, $P(Y = y \mid X = x) = \prod_{\ell=1}^{KL} \sigma_{x(\ell)} e^{-\sigma_{x(\ell)} y(\ell)}$, $0 \leq y(\ell) < \infty$, where $\mu_j = \sigma_j^{-1}$ is the mean of the jth region. Our objective is to obtain a *maximum a posteriori* (MAP) segmentation estimate. That is, we want to find x that maximizes $P(Y = y \mid X = x)P(X = x)$. It then follows that the energy function to be minimized for MAP estimation is

$$U(x) = \sum_{\ell=1}^{KL} \{-\ln(\sigma_{x(\ell)}) + \sigma_{x(\ell)} y(\ell) + 2\beta \sum_{m \in \eta_\ell} [1 - \delta(x(\ell) - x(m))]\}. \quad (3.1)$$

To find the MAP estimate using the diffusion network, define the M-vector $v(\ell) = [v_1(\ell), \ldots, v_M(\ell)]^\tau$ for each pixel, where $v(\ell) = e_m$ (the mth unit vector in $\mathbf{R}^M$) if $x(\ell) = m$. Let $v = \{v(\ell); \ \ell = 1, \ldots, KL\}$, let $\sigma = [\sigma_1, \ldots, \sigma_M]^\tau$, and define $\sigma(\ell) = y(\ell)\sigma$ and $\gamma = [\ln(\sigma_1), \ldots, \ln(\sigma_M)]^\tau$. Let λ be some large constant, and let u be the M-vector of all 1's. Then we can set the energy function to be minimized by the diffusion network to be

$$\mathcal{E}(v) = \sum_{\ell=1}^{KL} \{\langle \sigma(\ell) - \gamma, v(\ell) \rangle + \beta \sum_{m \in \eta_\ell} |v(\ell) - v(m)|^2 \\ + \lambda[\langle v(\ell), u - v(\ell) \rangle + (1 - \langle v(\ell), u \rangle)^2]\}. \quad (3.2)$$

The term in (3.2) multiplied by λ is to ensure that minimization of (3.2) over the hypercube occurs at a corner point; the remaining terms ensure that the minimizing corner point corresponds to the x that minimizes (3.1). Parallel operation of the constant step-size network defined by (2.1) was simulated on a sequential computer, and applied to several test segmentation problems for speckled images using the energy function (3.2). In the tests we set $f(v) = \frac{1}{\pi}\cos^2\pi(v - 0.5)$. The $\{W_{\alpha,k}\}$ sequences in (2.1) were taken to be independent Bernoulli random variables. We set $c_{\iota n+k} = c_k = \varepsilon T_k$ and $b_{\iota n+k} = b_k = \sqrt{2c_k}$, where $T_k = \frac{200}{199+k}$ and $\varepsilon = 0.01$. For the Gibbs distribution terms we set $\beta = 0.3$ and let η_ℓ consist of the eight nearest neighbors to pixel ℓ. Although in principle, λ in (3.2) should be fixed at some "large enough" value, for certain constrained optimization problems, it was found to be more effective to have the constraint term multiplier start at a small value and grow slowly with the iteration number; hence, we replaced λ with $\lambda_k = \log(k)$. The network was run for 1000 iterations, at the end of which each $v(\ell)$ was set to the nearest element of $\{e_1, \ldots, e_M\}$, and the estimated label at pixel ℓ was set to j if $v(\ell) = e_j$. For the first test we set $K = L = 64$, $M = 2$, and $\sigma = \{1, 6\}$. The results are displayed in Figure 1. From left to right, the figure shows the true region labels, the speckled observation, and the segmentation generated by the network. The pixel error rate in this case is 3.5% (assuming equal-sized regions, the error rate with a maximum likelihood (ML) estimator would be 20.9%). Figure 2 shows results for a test with $M = 4$, $\sigma = \{0.125, 1, 8, 64\}$. Again, from left to right the figure shows the true labels, the speckled image, and the segmentation estimate. The pixel error rate for the segmentation is 8.8% (the ML error rate assuming equal-sized regions would be 26.3%).

FIGURE 1: Two-region segmentation.

FIGURE 2: Four-region segmentation.

Similar formulations can be used for applying the diffusion network to other image estimation problems – again, the key is to cast the problem as optimization over $[0, 1]^r$. For example, in applications including radar

(Munson and Visentin, 1989; Wu and Kelly, 1991) and medical ultrasound (Hokland and Kelly, 1996) imaging, the observed image is often corrupted by both noise and blur. The true underlying image G can often be modeled as a field of independent Gaussian random variables having different means and/or variances in different regions, with the observed image having the form $Y = h * G + N$, where h is a blurring filter due, for example, to the imaging system's point-spread function; the "$*$" denotes convolution; and N is a noise term. In such cases, we often wish to perform both restoration (that is, estimation of G) and segmentation (labeling of regions). Let X be a region field and $\{\mu_1, \ldots, \mu_M\}$ be a set of region intensities as before. When the region field realization is x, we let the true underlying image be given by $G(\ell) = \mu_{x(\ell)} + N_g(\ell)$ at each pixel ℓ, where N_g is a white Gaussian noise field, independent of X and having mean zero and variance σ_g^2 at each pixel (the resulting image statistics are similar to those of some "ideal" intensity images in multi-look synthetic aperture radar - see Derin *et al.* (1990)). As noted above, we assume that the observed image is obtained by convolving the true image with a blurring filter and by adding noise N. We assume that N is also a white Gaussian noise with mean zero and variance σ_N^2 at each pixel, and that N is independent of X and N_g. From this model, it follows that to find the joint MAP estimates of G and X (i.e., joint restoration and segmentation) from an observation y, we need to minimize the energy function over $\{0,1\}^{KL} \times \mathbf{R}^{KL}$,

$$
U(x,g) = \sum_{\ell=1}^{KL} \{ \frac{1}{2\sigma_N^2} |y(\ell) - \sum_m h(\ell - m)g(m)|^2 + \frac{1}{2\sigma_g^2} |g(\ell) - \mu_{x(\ell)}|^2 \\
+ 2\beta \sum_{m \in \eta_\ell} [1 - \delta(x(\ell) - x(m))] \}.
$$
(3.3)

As is shown in Yin *et al.* (2000), if $M = 2$ the minimization of (3.3) can be accomplished by a diffusion network having the energy function

$$
\mathcal{E}(v) = \sum_{\ell=1}^{KL} \{ \frac{1}{\sigma_N^2} |z(\ell) - b_1 \sum_m h(\ell - m)v_1(m)|^2 + \frac{b_1^2}{\sigma_g^2} |v_1(\ell) - a_1 v_2(\ell) - a_2|^2 \\
+ 4\beta \sum_{m \in \eta_\ell} (v_2(\ell) - v_2(m))^2 + \lambda v_2(\ell)[1 - v_2(\ell))] \},
$$
(3.4)

where $v(\ell) = [v_1(\ell), v_2(\ell)]^\tau$; $v_1(\ell) = \frac{g(\ell) - b_2}{b_1}$, with $b_1 = \mu_2 - \mu_1 + 5\sigma_g$ and $b_2 = \mu_1 - 2.5\sigma_g$; $v_2(\ell) = x(\ell) - 1$; $z(\ell) = y(\ell) - b_2 \sum_m h(m)$; $a_1 = \frac{\mu_2 - \mu_1}{b_1}$; and $a_2 = \frac{2.5\sigma_g}{b_1}$. For results of tests using (3.4) for joint image restoration and segmentation, we refer the reader to Yin *et al.* (2000).

To see the potential gains in processing speed that could be obtained by using a parallel diffusion network instead of sequential simulated annealing, the diffusion network was programmed on a fixed-point parallel-processing PC card. This card, the MM32k manufactured by Current Technologies,

Inc., contains 32,768 parallel processing elements and acts as a vector co-processor for a host computer. In our tests, the card was connected to a 25MHz, 486 host machine. To compare processing speeds, a standard sequential implementation of simulated annealing was programmed on the same machine. The parallel network and the simulated annealing algorithm were applied to binary segmentation of $K \times K$ images for different sizes K, and the number of iterations/second (where one iteration = one pass through the entire image) was measured. As expected, the number of iterations/second stayed constant with increasing K with the parallel processor, while it decreased as $1/K^2$ for the sequential algorithm. Even for images of moderate size, this leads to a substantial speed advantage for the parallel algorithm (e.g., approximately 40 times more iterations/second when $K = 128$). Since the parallel processor speed is largely determined by the speed of the host computer, similar advantages for the parallel processor implementation would be expected on faster host machines. For more details we refer the reader to Yin et al. (2000).

4   Further remarks

In this paper, we have developed an algorithm, which takes advantage of the structure of Wong's diffusion network and uses parallel processing methods. The parallel processing nature of the algorithm makes it especially suitable for application to large-scale optimization problems, including image estimation and optimal design problems.

In lieu of using the step-size (2.2), we may develop algorithms with decreasing step sizes. Consider (2.1) with decreasing stepsizes

$$a_{\iota n+k} = \frac{1}{(\iota n + k)^\gamma}, \quad b_{\iota n+k} = \frac{\sqrt{2a_{\iota n+k}}}{\sqrt{\widetilde{c}_{\iota n+k}}}, \quad c_{\iota n+k} = \frac{a_{\iota n+k}}{\widetilde{c}_{\iota n+k}},$$

where $\widetilde{c}_{\iota n+k} = \log[(\iota n + k)^{1-\gamma} - (\iota n)^{1-\gamma} + A_0]$, with $A_0 > 1$ and $1/2 < \gamma < 1$. Moreover, additional noisy effects such as observation noise can be considered. One of our current efforts is to obtain the rates of convergence of the digital diffusion network algorithms. For some related work on rates of convergence of global optimization algorithms, see Yin (1999).

References

Cai, X., Kelly, P., and Gong, W.B. (1995). Digital diffusion network for image segmentation. *Proc. IEEE Internat. Conf. Image Processing*, Vol. III, 73-76.

Derin, H., Kelly, P., Vezina, G., and Labitt, S. (1990). Modeling and segmentation of speckled images using complex data. *IEEE Trans. Geosci. Remote Sensing*, **28**, 76-87.

Geman, D. and Geman, S. (1984). Stochastic relaxation, Gibbs distributions, and the Bayesian restoration of images. *IEEE Trans. Pattern Anal. Machine Intelligence*, **6**, 721-741.

Hokland, J. and Kelly, P. (1996). Markov models of specular and diffuse scattering in restoration of medical ultrasound images. *IEEE Trans. Ultrasonics, Ferroelectrics and Frequency Control*, **43**, 660-669.

Kesidis, G. (1996). Analog optimization with Wong's stochastic neural network. *IEEE Trans. Neural Networks*, **6**, 258-260.

Kushner, H.J. (1987). Asymptotic global behavior for stochastic approximation and diffusions with slowly decreasing noise effects: Global minimization via Monte Carlo. *SIAM J. Applied Mathematics*, **47**, 169-185.

Kushner, H.J. and Yin, G. (1997). *Stochastic Approximation Algorithms and Applications*. Springer-Verlag, New York.

Manjunath, B., Simchony, T., and Chellappa, R. (1990). Stochastic and deterministic networks for texture segmentation. *IEEE Trans. Acoustics, Speech, Signal Processing*, **38**, 1039-1049.

Munson, D. and Visentin, R. 1989. A signal processing view of strip mapping synthetic aperture radar. *IEEE Trans. Acoustics, Speech, Signal Processing*, **37**, 2131-2147.

Wong, E. (1991). Stochastic neural networks. *Algorithmica*, **6**, 466-478.

Wu, B. and Kelly, P. (1991). Restoration and segmentation of complex-valued SAR images. *Proc. 25th Conference on Information Sci. Systems*, The Johns Hopkins University, Baltimore, 377-381.

Yin, G. (1999). Rates of convergence for a class of global stochastic optimization algorithms. *SIAM J. Optimization*, **10**, 99-120.

Yin, G., Kelly, P., and Dowell, M.H. (2000). Approximation of an analog diffusion network with applications to image estimation. *J. Optim. Theory Appl.* (submitted).

List of Authors

- Kasra Afsarinejad
  Biostatistics, AstraZeneca R& D Molndal, 43183 Molndal, Sweden
  kasra.afsarinejad@astrazeneca.com

- Stefanie Biedermann
  Ruhr-Universität Bochum, Fakultät für Mathematik, 44780 Bochum,
  Germany
  Stefanie.Biedermann@ruhr-uni-bochum.de

- Eric P.J. Boer
  University of Wageningen, sectie Wiskunde, Bode 3, Postbus 9101,
  6700 HB Wageningen, The Netherlands
  eric.boer@wts.wk.wau.nl

- Lidia N. Boyadjieva
  Higher Insitute of Chemical Technology, Bul. Kliment Ochridski 8,
  11256 Sofia, Bulgaria
  lnb@adm1.uctm.acad.bg

- Bronislaw Ceranka
  Department of Mathematical and Statistical Methods, Agricultural
  University of Poznan, Wojska Polskiego 28, 60–637 Poznan, Poland
  bronicer@owl.au.poznan.pl

- Kathryn Chaloner
  University of Minnesota, School of Statistics, 381 Ford Hall, 224
  Church Street SE, Minneapolis, MN 55455, USA
  kathryn@stat.umn.edu

- G. Peter Y. Clarke
  Agriculture Western Australia, Room B125a, 3 Baron Hay Court,
  South Perth WA 6151, Australia
  pclarke@agric.wa.gov.au

- Arjeh M. Cohen
  Eindhoven University of Technology, Department of Mathematics,
  P.O. Box 513, 5600 MB Eindhoven, The Netherlands
  amc@win.tue.nl

- Holger Dette
  Ruhr-Universität Bochum, Fakultät für Mathematik, 44780 Bochum,
  Germany
  Holger.Dette@ruhr-uni-bochum.de

- Alessandro Di Bucchianico
  EURANDOM and Eindhoven University of Technology, Faculty of
  Technology Management, Section Quality of Products and Processes,
  P.O. Box 513, 5600 MB Eindhoven, The Netherlands
  A.d.Bucchianico@tm.tue.nl

- Darryl Downing
  GlaxoSmithKline, Mailstop 280C, 709 Swedeland Road, P.O. Box
  1539, King of Prussia, PA 19406, USA
  darryl_j_downing@sbphrd.com

- Vladimir Dragalin
  GlaxoSmithKline, Mailstop 280C, 709 Swedeland Road, P.O. Box
  1539, King of Prussia, PA 19406, USA
  Vladimir_2_Dragalin@sbphrd.com

- Arkadii G. D'yachkov
  Moscow State University, Faculty of Mechanics & Mathematics, Department of Probability Theory, Moscow, 119899, Russia
  dyachkov@artist.math.msu.su

- Shenghua K. Fan
  Statistics and Applied Probability, National University of Singapore,
  3 Science Drive 2, Singapore 117543, Republic of Singapore
  shfan@stat.nus.edu.sg

- Valerii V. Fedorov
  GlaxoSmithKline, Mailstop 280C, 709 Swedeland Road, P.O. Box
  1539, King of Prussia, PA 19406, USA
  Valeri_V_Fedorov@sbphrd.com

- Klaus Felsenstein
  Department of Statistics, University of Technology Vienna, Wiedner
  Haupstraße 8-10, 1040 Vienna, Austria
  fels@statistik.tuwien.ac.at

- Leonid I. Galtchouk
  Department of Mathematics, Strasbourg University, 7, rue Rene Des-
  cartes, 67084, Strasbourg, France
  galtchou@math.u-strasbg.fr

- Peter Goos
  Katholieke Universiteit Leuven, Departement Toegepaste economis-
  che wetenschappen, Naamsestraat 69, B-3000 Leuven, Belgium
  Peter.Goos@econ.kuleuven.ac.be

- Eleanor Gouws
  Biostatistics, MRC, P.O.Box 17120, Congella 4013, South Africa
  eleanor.gouws@mrc.ac.za

- Ulrike Graßhoff
  Free University Berlin, Institute for Mathematics I, Arnimallee 2–6,
  D-14195 Berlin, Germany
  `grasshoff@math.fu-berlin.de`

- Heiko Großmann
  University of Münster, Institute for Psychology IV, Fliednerstr. 21,
  D-48149 Münster, Germany
  `grossman@psy.uni-muenster.de`

- Linda M. Haines
  University of Natal Pietermaritzburg, Department of Statistics and
  Biometry, Durban 4041, South Africa
  `haines@stat.unp.ac.za`

- Janis Hardwick
  University of Michigan, 3509 Creekside Drive, Ann Arbor, Michigan,
  48105 USA
  `jphard@umich.edu`

- Eligius M.T. Hendrix
  University of Wageningen, sectie Wiskunde, Bode 3, Postbus 9101,
  6700 HB Wageningen, The Netherlands
  `eligius.hendrix@alg.orl.wau.nl`

- Ralf-Dieter Hilgers
  Institut für Biometrie, Universitätsklinikum der RWTH Aachen, Pau-
  welsstraße 30, 52074 Aachen, Germany
  `ralf.dieter.hilgers@mbio.rwth-aachen.de`

- Alan Hoffman
  Department of Mathematical Sciences, Thomas J. Watson Research
  Center, IBM, P.O. Box 218, Yorktown Heights, New York 10598, USA
  `hoffa@watson.ibm.com`

- Krystyna Katulska
  Faculty of Mathematics and Computer Science, Adam Mickiewicz
  University, Jana Matejki 48/49, 60-769 Poznań, Poland
  `krakat@amu.edu.pl`

- Patrick A. Kelly
  Department of Electrical and Computer Engineering, University of
  Massachusetts, 215B Marcus Hall, Amherst, MA 10013, USA
  `kelly@ecs.umass.edu`

- Henning Läuter
  Institut für Mathematik, Universität Potsdam, PF 60 15 53, 144 15
  Potsdam, Germany
  `laeuter@rz.uni-potsdam.de`

- Jon Lee
  Department of Mathematical Sciences, Thomas J. Watson Research Center, IBM, P.O. Box 218, Yorktown Heights, New York 10598, USA
  jonlee@us.ibm.com

- Sergei Leonov
  GlaxoSmithKline, Mailstop 280C, 709 Swedeland Road, P.O. Box 1539, King of Prussia, PA 19406, USA
  sergei_2_leonov&sbphrd.com

- Jesús F. Lopez-Fidalgo
  University of Salamanca, Departamento de Estadistica, Edificio de la Merced, Plza. de la Merced, 1-4, Fac. Ciencias, Salamanca Spain
  fidalgo@gugu.usal.es

- Anthony J. Macula
  Department of Mathematics, State University of New York at Geneseo, South 325 D, 1 College Circle, Geneseo, New York 14454, USA
  macula@geneseo.edu

- Mikhail B. Malyutov
  Northeastern University, Department of Mathematics, 360 Huntington Avenue, Boston, MA 02115, USA
  mmaliout@lynx.dac.neu.edu

- Ignacio J. Martinez López
  University of Almeria, Edificio CITE-III, Despacho 2-32, Spain
  ijmartin@ual.es

- Viatcheslav B. Melas
  St. Petersburg State University, Department of Mathematics and Mechanics, Bibliotechnaya sq. 2, St. Petersburg, Petrodvoretz 198904, Russia
  melas@niimm.spb.su

- Christine H. Müller
  Fachbereich Mathematik, Carl von Ossietzky Universität Oldenburg, 26111 Oldenburg, BRD
  mueller@mathematik.uni-oldenburg.de

- Isabel M. Ortiz Rodriguez
  University of Almeria, Edificio CITE-III, Despacho 2-32, Almeria, Spain
  iortiz@ual.es

- Inna Perevozskaya
  Department of Mathematics and Statistics, University of Maryland, Baltimore County, 1000 Hilltop Circle, Baltimore, MD 21250, USA
  inna@math.umbc.edu

- Fortunato Pesarin
  Universitá degli Studi di Padova, Dipartimento di Scienze Statistiche,
  Via San Francesco, 33, 35121 Padova, Italia
  pesarin@hal.stat.unipd.it

- Giovanni Pistone
  Politecnico di Torino, Dipartimento di Matematica, Corso Duca degli
  Abruzzi, 24, 10129 Torino, Italy
  pistone@calvino.polito.it

- Luc Pronzato
  Laboratoire I3S, UNSA-CNRS, 2000, route des lucioles, Les Algo-
  rithmes - bât. Euclide B, BP.121-06903 Sophia Antipolis, France
  pronzato@i3s.unice.fr

- Dieter A.M.K. Rasch
  University of Wageningen, Wiskunde, Postbus 9101, 6700 HB Wa-
  geningen, The Netherlands
  RASCH@RCL.WAU.NL

- Eva Riccomagno
  EURANDOM and The University of Warwick, Department of Statis-
  tics, Gibbet Hill Road, Coventry CV4 7AL, United Kingdom
  E.Riccomagno@tue.nl

- Carmelo Rodriguez Torreblanca
  University of Almeria, Edificio CITE-III, Despacho 2-49, Almeria,
  Spain
  crt@ual.es

- William F. Rosenberger
  University of Maryland, Baltimore County, 1000 Hilltop Circle, Baltimore, MD 21250, USA
  billr@math.umbc.edu

- Luigi Salmaso
  Universitá degli Studi di Padova, Dipartimento di Scienze Statistiche,
  Via San Francesco, 33, 35121 Padova, Italia
  lsalmaso@stat.unipd.it

- Rainer Schwabe
  Universität Tübingen, Institut für Medizinische Biometrie, West-
  bahnhofstraße 55, D-72070 Tübingen, Germany
  rainer.schwabe@uni-tuebingen.de

- Lieven Tack
  Katholieke Universiteit Leuven, Departement Toegepaste economi-
  sche wetenschappen, Naamsestraat 69, B-3000 Leuven, Belgium
  Lieven.Tack@econ.kuleuven.ac.be

- Ben Torsney
  University of Glasgow, Department of Statistics, Mathematics Building, University Gardens, Glasgow G12 8QW, Great Britain
  bent@stats.gla.ac.uk

- David C. Torney
  Los Alamos National Laboratories, New Mexico 87545, USA
  dct@lanl.gov

- Ivan I. Tsitovich
  Institute for Information Transmission Problems, B. Karetny 19, Moscow, 101447, Russia
  cito@iitp.ru

- Dariusz Ucinski
  Department of Control and Computation Engineering, Technical University of Zielona Gora, ul. Podgorna 50, 65-246 Zielona Gora, Poland
  D.Ucinski@irio.pz.zgora.pl

- Martina Vandebroek
  Katholieke Universiteit Leuven, Departement Toegepaste economische wetenschappen, Naamsestraat 69, B-3000 Leuven, Belgium
  martina.vandebroek@econ.kuleuven.ac.be

- Pavel A. Vilenkin
  Moscow State University, Faculty of Mechanics & Mathematics, Department of Probability Theory, Moscow, 119899, Russia
  paul@vilenkin.dnttm.ru

- Ivan N. Vuchkov
  Department of Automation, University of Chemical Technology and Metallurgy, 1756 Sofia, Bulgaria
  vuchkov@adm1.uctm.acad.bg

- Joy Williams
  Level 3 Communications, 1025 Eldorado Blvd., Broomfield, Colorado, 80021, USA
  Joy.Williams@Level3.com

- Henry P. Wynn
  The University of Warwick, Department of Statistics, Gibbet Hill Road, Coventry CV4 7AL, United Kingdom
  hpw@stats.warwick.ac.uk

- George Yin
  Department of Mathematics, Wayne State University, 1217 Faculty/Administration Bldg, Detroit, MI 48202, USA
  gyin@math.wayne.edu

List of Referees

- Anthony C. Atkinson, London School of Economics

- Peter Bauer, University of Vienna

- Alessandro Di Bucchianico, EURANDOM and Eindhoven University of Technology

- Valerii V. Fedorov, GlaxoSmithKline, Philadelphia

- Klaus Felsenstein, Vienna University of Technology

- Peter Hackl, Vienna University of Economics and Business Administration

- Christos P. Kitsos, University of Thessally

- Henning Läuter, University of Potsdam

- Werner G. Müller, Vienna University of Economics and Business Administration

- Andrej Pázman, Comenius University Bratislava

- Jürgen Pilz, University of Klagenfurt

- Luc Pronzato, Université de Nice Sophia Antipolis

- Rainer Schwabe, University of Tübingen

- Ben Torsney, University of Glasgow

- Ivan N. Vuchkov, Sofia University of Chemical Technology and Metallurgy

- Henry P. Wynn, University of Warwick

- Anatoly A. Zhigljavsky, University of Cardiff

Printing: Weihert-Druck GmbH, Darmstadt
Binding: Buchbinderei Schäffer, Grünstadt